Laser Techniques for Extreme Ultraviolet Spectroscopy

(Boulder, 1982)

AIP Conference Proceedings
Series Editor: Hugh C. Wolfe
Number 90
Subseries on Optical Science and Engineering
Number 2

Laser Techniques for Extreme Ultraviolet Spectroscopy
(Boulder, 1982)

Editors

T. J. McIlrath
University of Maryland
and
R. R. Freeman
Bell Laboratories

American Institute of Physics
New York   1982

L.C. Catalog Card No. 82-073205
ISBN 0-88318-189-4
DOE CONF- 820336

TOPICAL MEETING ON

LASER TECHNIQUES FOR EXTREME

ULTRAVIOLET SPECTROSCOPY

March 8 - 10, 1982, Boulder, Colorado.

Sponsored by:
**Air Force Office of
Scientific Research
and
Optical Society of America**

TECHNICAL PROGRAM COMMITTEE

R. R. Freeman, *Conference Cochairman*
Bell Laboratories, Holmdel, NJ

T. J. McIlrath, *Conference Cochairman*
University of Maryland, College Park, MD

J. A. Armstrong
IBM Research Center, Yorktown Heights, NY

J. Bokor
Bell Laboratories, Holmdel, NJ

R. Compton
Oak Ridge National Laboratory, Oak Ridge, TN

S. E. Harris
Stanford University, Stanford, CA

K. Manes
Lawrence Livermore National Laboratory,
Livermore, CA

J. A. Paisner
Lawrence Livermore National Laboratory,
Livermore, CA

C. K. Rhodes
University of Illinois at Chicago Circle, Chicago, IL

M. A. Stroscio
Air Force Office of Scientific Research,
Bolling Air Force Base, Washington, D.C.

Table of Contents

Chapter 2 Direct Generation

Chapter 3 Plasmas

Chapter 5 Non-Linear Mixing

LASER TECHNIQUES FOR
EXTREME ULTRAVIOLET SPECTROSCOPY

by

Thomas J. McIlrath
University of Maryland, College Park, MD   20742

Richard R. Freeman
Bell Laboratories, Murray Hill, NJ   07974

In March of this year a meeting was held on the subject of Laser Techniques for Extreme Ultraviolet Spectroscopy. The meeting was topical in nature and international in attendance and was sponsored by the Optical Society of America and the Air Force Office of Scientific Research. It brought together researchers from a broad spectrum of backgrounds who have an interest in expanding our knowledge of atomic and molecular properties through high resolution studies of high lying energy levels; in this context, high energy means greater than 6eV. The meeting was specifically directed towards the use of laser techniques, broadly interpreted, for spectroscopic studies. The conference was concerned with the use and development of lasers in this energy region both directly and through up-conversion techniques, the study of high energy levels with lower frequency incoherent radiation by laser scattering and laser plasma generation, and optical techniques for working with high frequency radiation. Both examples of spectroscopy using laser systems and work on solving the problems involved in developing such systems were given, reflecting a characteristic feature of this subject that the study of spectroscopy is not only the subject to which the high frequency lasers are applied but it is also an integral part of the physics involved in designing and building the laser themselves.

Since the focus of the meeting was techniques for studying atomic and molecular energy levels, the first topic was the presentation of what has been accomplished in this area. It was clear from the work presented at the meeting that classical and modern laser techniques are best viewed as complementary rather than competing approaches.[1] The traditional techniques give very broad spectral converge with a resolution which is often adequate for high energy transitions involving predissociating and autoionizing states. A rapid survey of photographic or photoelectric data obtained with long focal length grating instruments allows identification of the specific areas where higher resolution is needed for

ISSN:0094-243X/82900001-08$3.00 Copyright 1982 American Institute of Physics extracting the important physics from the problem. Grating techniques, however, are near the limit of their resolution due to technical and cost factors, and interferometry is not feasible for the high energy regime, so laser techniques are the prime tool for very high resolution studies. Beyond providing higher resolution, the laser will bring its armament of nonlinear techniques, time-resolved spectroscopy and few atom detection techniques to bear to complement traditional photoabsorption and photoemission studies. Laser techniques are now to the point where one can say that in general they don't exist for the VUV but in particular they do.[2] These particular applications provide both guidance and stimulation for future work.

Essentially, all the work reported to date shows a classical heritage in that older spectra were used to identify the interesting problems and the problems that exist in attainable spectral regions. The workhorse of the vacuum ultraviolet is clearly the NO molecule.[2-4] It has transitions in the LiF VUV (>105 nm) with strong fluorescence below its ionization limit (134 nm) and heavily structured ionization spectrum above the limit. It possesses a large nonlinear susceptibility allowing efficient multiphoton excitation and its spectrum has been extensively analyzed from photographic data. Regions where perturbations had previously been identified have now been studied with fluorescence lifetimes measured using intense, high resolution, pulsed VUV sources generated by nonlinear mixing techniques in gases.[2] The perturbation of lifetimes of single rovibronic could thus be compared with energy level perturbations. A second state selective study of NO was performed using optical-optical double resonances where a first VUV source excites specific rovibronic levels while a second ionizes the excited state.[3] Detection is achieved by collecting the ions and intermediate states of known quantum number can be studied for perturbations and for studies of dynamic processes in excited states. The sensitivity of ionization studies outlined one of the major advantages of laser spectroscopy. Photoionization studies using incoherent sources achieve substantially lower resolving power than photographic studies because of the compromises necessary to attain adequate light intensity for strong signals. The availability of high brightness coherent sources allows both double resonance studies and direct photoionization and photodissociation to be carried out quickly on small samples with the same resolution achievable in absorption studies.

In addition to the fluorescence and ionization studies in NO, high resolution intensity measurements of photoabsorption were reported using tunable, coherent VUV to

give line strength spectra which complement the line
position data obtained photographically.

Although NO is the most extensively studied molecule,
work has now been reported on the CO excitation spectrum
using single photon excitation longward of 105 nm.[4]
Studies of higher energy levels in N_2 have been made
using multiphoton excitation combining a VUV beam and a
visible laser beam. Direct studies of very high resolution spectra at wavelengths below the LiF cutoff have
been obtained in H_2, again using radiation produced by
up-conversion in gases but in this case using a windowless apparatus.[5] The spectra obtained Doppler limited
line shape information for regions which had been studied
with lower resolution with a 10 m grating instrument. In
addition to higher resolution, the laser source and photoelectric detection provided accurate intensity information. Intensity information can only be obtained as a
crude approximation from photographic studies and with
generally a significant loss in resolutions using
incoherent sources and photoelectric methods, although,
at least one state-of-the-art spectrograph with a scanning photomultiplier has approached photographic resolution on a 6.7 m instrument.

Very high resolution studies of atomic systems using
laser sources are also being reported with atomic hydrogen being the most extensively studied.[6] The use of
optical-optical double resonance with a 121.5 nm beam
combined with a tunable visible laser to probe an atomic
beam of hydrogen has allowed detailed studies of high
Rydberg states in this system. Studies of excitations in
the rare gases also require high energies and these studies are primarily being pursued using multiphoton ionization.[7] Metal atoms continue to be studied using visible lasers and multiphoton excitation to probe high Rydberg and autoionizing resonances.[8,9] Studies of metal
atoms at still higher energies reveal inner shell transitions. Hybrid techniques using tunable up-conversion
through laser scattering off metastable He atoms has produced a tunable, high resolution, high brightness source
for studies near 55 nm. Spectra of K vapor have produced
intensity measurements on the broad features which have
previously been observed classically as well as new inner
shell transitions which were not observed in the older,
lower resolution, spectra.[10]

The study of atomic and molecular energy levels is being
pursued by both traditional, single photon absorption or
excitation and by nonlinear, multiphoton techniques.
These two techniques complement each other nicely. The
multiphoton techniques are in general more selective and
often more efficiently identify the levels being

studied[11] while the single photon techniques are much
more sensitive and are usable at lower powers. As VUV
power levels increase the multiphoton processes will be
applicable to higher energy levels and weaker transitions
and should play an increasing role in VUV spectroscopy.[12]

High brightness laser sources are now being applied to
diagnostic studies as well. The greatest effort has been
in observing atomic hydrogen by induced fluorescence and
the greatest stimulus has been the desire to probe hydrogen concentrations in Tokamak and other thermonuclear
plasmas.[13] The greatest benefit comes when the resolution is sufficient to study the Doppler distribution and
the first measurements on hydrogen in the preheat phase
of a Tokamak have now been reported. Another diagnostic
application reported here involves Doppler resolved studies of the products of laser induced photodissociation
in simple molecules.[6] The final state of the products
can be analyzed as a function of laser polarization by
studies of the velocity distribution of the dissociated
hydrogen. Other diagnostic studies currently being pursued involve detection of short-lived nuclear species and
of very small atomic contaminations in large samples.

While the usefulness of coherent VUV radiation is being
demonstrated, the systems for generating it are still
cumbersome, of limited range and the object of vigorous
research.[14,15] Current research in generating VUV was a
major topic of discussion at the conference. This
research involves direct production of VUV, still an
unrealized goal for must of the VUV, and generation by
up-conversion, often with intermediate amplification by
UV lasers. The work on direct generation generally
involves efforts at obtaining substantial inversions in
transient plasmas,[16] including recombination stages,
and work involving selective photoionization of gaseous
systems to produce inverted ion systems.

The research on transient plasmas is now systematic and
quantitative.[17] Inversions involving VUV transitions in
laser produced plasmas have been seen at densities too
low for useful gain. The conditions in these plasmas are
accurately predicted by computer simulations and these
models are the staging ground for designing higher gain
systems. Higher density recombination plasmas produced
by pulsed electrical discharges have produced laser in
the UV, visible and IR spectral region in several members
of isoelectronic sequences.[18] Efforts are underway to
scale these experiments to production of higher sequence
members to produce lasing in the VUV.

Research on optically pumped systems centers around photoionization of neutral vapors through strong autoionizing resonances which leave the final ion in an excited state.[18] Such pumping has been shown to produce lasing in the visible. A variation on this technique includes producing populations in metastable states embedded in the continuum which can be optically switched to radiating states with visible laser to produce high gain systems.[10] These approaches involve identification of strong optical transition ionic levels with appropriate radiative and autoionizing properties. This has led to a greatly increased interest in experimental and theoretical studies of the locations and properties of atomic states lying above the first ionization limit.[19]

A second result of the work on optically pumped VUV lasers is an interest in intense incoherent light sources, i.e. flashlamps. The most direct approach to flashlamp development involves laser generated plasmas.[20] A judicious choice of target material has been shown to produce an intense continuum over a limited spectral range. Matching the spectral output with broad autoionizing resonances could lead to efficient pumping of desired transitions with little pumping of interfering transitions. A large amount of both theoretical and experimental effort is being invested in studies of proper target materials, proper laser pulse shapes, laser timing, driving laser wavelengths and other factors which affect the dynamics of the plasma and the subsequent flashlamp output. The physics of these plasmas can be exotic as shown by experiments, now well explained theoretically, in which the plasma itself acts as a nonlinear medium and results in efficient generations of very high (more than 50) harmonics of the driving laser.[21] All of these flashlamp studies are closely related to other research work on laser fusion studies and other plasma interests.

More subtle means for producing VUV pump radiation comes from the technique of producing metastable atoms in gas discharges and releasing the energy in a very short pulse by anti-Stokes Raman scattering using long wavelength pulsed lasers.[22] Although this work began in the search for VUV flashlamps, it has subsequently led to useful applications as a pulsed, monochromatic, tunable VUV light source for high resolution spectroscopy and to research on the modification of gaseous discharges to produce extensive metastable populations.

Finally, the interest in direct generation and optical pumping of VUV lasers has stimulated research on VUV and x-ray lasers. This work motivated by several major applications, has made major advances in mirrors, lenses,

beam splitter, filters and transmission gratings in the
1-10 nm region and the field is still rapidly advancing.[23]

The actual generation of high resolution VUV radiation to
date has been achieved by frequency up-conversion in
gases and vapors using long wavelength laser drives. All
of the region above the 105 nm LiF cutoff and significant
portions of the region between 50 and 105 nm has been
covered by these techniques. The nonlinear materials
being used include metal vapors,[8] rare gases[14] and
molecular gases[24] and both resonant and nonresonant
systems are being developed. It is clear from the work
presented to date that different systems have different
tradeoffs between convenient resonant wavelengths (metal
vapors) and convenient materials (rare gases). The rare
gases and mercury vapors are popular choices for many
current workers, although, metal vapors were developed
first for resonant systems and are a developed item in
some established laboratories.

Up-conversion techniques provide resolutions equal to
those achievable with visible pulsed lasers with
wavelengths determined from visible measurements. These
techniques have already been applied to high resolution
spectroscopic studies of H_2, H, NO and CO among others.
Much of the current work is aimed at understanding the
physics and limitations of the basic processes and at
developing systems usable as routine spectroscopic
instruments.

Extreme ultraviolet spectroscopy using laser techniques
is now an established research area and vigorous work is
going on in parallel on the development and sophistication of systems and on the application of systems to
atomic and molecular structure studies. Now that the
field is fully born, the gaze has moved once again over
the horizon, past the extreme ultraviolet, to new schemes
aimed at the still virgin fields of coherent x-ray and
Γ-ray radiation.[25]

REFERENCES

[1] T. J. McIlrath, "High Resolution Spectroscopy"........9

[2] B. P. Stoicheff, "Tunable, Coherent Sources for
     High Resolution VUV and XUV Spectroscopy"............19

[3] W. Seaver, et al, "Optical-Optical Double Resonance
     Multiphoton Ionization Spectroscopy of NO"............45

[4] C. R. Vidal, "Spectroscopy of CO and NO Using
     Coherently Generated VUV"................................431

[5] C. K. Rhodes, "Direct Generation of XUV Radiation".. 112

[6] K. H. Welge, "Non-Linear Mixing"...................... 402

[7] M. S. Pindzola, "Two-Photon Excitation of Argon".....384

[8] F. S. Tomkins and R. Mahon, "Generation of Narrow-
     band Tunable Radiation in the 1200Å Spectral
     Region"...352

[9] R. R. Freeman and J. Bokor, "Generation of Coherent
     Radiation Below 100 nm in Hg Vapor"..................422

[10] S. E. Harris, et al, "Anti-Stokes Scattering as an
     XUV Radiation Source".....................................137

[11] R. N. Compton and J. C. Miller, "Multiphoton Ioni-
     zation of Atoms and Molecules"..........................319

[12] R. Hilbig and R. Wallenstein, "Generation of Nar-
     rowband Tunable VUV Radiation".........................442

[13] R. W. Dreyfuss, "Atomic H and D Concentrations and
     Velocities Measured"...................................... 57

[14] Y. M. Yiu, et al, "Resonant Four-Wave Mixing in
     Xenon"...478

[15] H. Egger, et al, "Generation of Tunable, Coherent
     79nm Radiation by Frequency Mixing".....................445

[16] W. T. Silfvast and O. R. Wood, II, "Recombination
     Lasers in the Vacuum Ultraviolet"........................128

[17] R. H. Dixon, et al, "Population Density and VUV
     Gain Measurement in Laser Produced Plasmas".........277

[18] J. Bokor and R. R. Freeman, "Photo-Autoionization
     Pumped Ba Ion Laser".......................................153

HIGH RESOLUTION SPECTROSCOPY

Thomas J. McIlrath
Institute for Physical Science and Technology,
University of Maryland, College Park, MD 20742
and National Bureau of Standards, Washington, D.C. 20234

ABSTRACT

The recent extension of laser sources to the VUV has provided unparalleled resolution in this region. However, classical instrumentation still provides sufficient resolution for many problems and has special features which complement laser techniques. Properties of grating instruments are reviewed and particular spectroscopic studies presented to show this complementarity and to illustrate the value of varying techniques with problems.

I.   INTRODUCTION

The advent of sources for generation of coherent radiation in the vacuum ultraviolet (VUV) has brought the promise of unprecedented resolution in this spectral region. However, conventional spectroscopy based on diffraction gratings is well developed and sophisticated and should continue to play a major role in atomic and molecular physics. This paper briefly reviews the state of conventional, high resolution, VUV spectroscopy in the context of three considerations: 1. What is the useable resolution of VUV grating instruments? 2. What resolution is needed for atomic and molecular physics studies? 3. How do grating instruments and laser source complement each other?

II. FUNDAMENTAL LIMITS TO RESOLUTION

The theoretical resolving power, $R = \lambda/\Delta\lambda = \nu/\Delta\nu$, of a plane diffraction grating is $R = mN$ with m the order number of the grating and N the total number of grooves.[1] If θ_i and θ_r are the angles of incidence and refraction respectively and d is the groove spacing, then the grating equation gives $m\lambda = d(\sin \theta_i + \sin \theta_r)$. Using $\Delta\nu = \nu/R = c/\lambda mN$ we obtain

$$\delta\nu = \frac{c}{W(\sin \theta_i + \sin \theta_r)} = \frac{1}{\tau} > \frac{c}{2W}$$

where W=Nd is the full width of the grating and τ is the time difference between rays diffracting from opposite ends of the grating. This reflects the fundamental fact that resolution is achieved in all instruments by comparing a wave with a time delayed portion of itself (an autocorrelation measurement). The maximum time delay achievable with a diffraction grating is the round trip time acrossed it width so that the resolution of a grating is fundamentally limited by its width.

An interferometer of finesse F and spacing L has a resolution $\delta\nu = c/2LF$[1] which again reflects the time delay of interfering waves stored in the interferometer. A laser is an interferometer with an active medium in the cavity. Non-linearities make laser linewidths ever narrower by significantly reducing them below the linewidth of the passive interferometer which forms the laser cavity. Again the fundamental limit on resolution is set by the physical size of the interferometer except that the multiple pass feature introduces an additional multiplicative factor, i.e. the finesse. Because the size of most laser cavities is greater than even the largest diffraction grating, it is easy to see that a single mode laser has an inherent linewidth which is much narrower than that of the best diffraction grating. There is no realistic option for making the resolution of grating instruments comparable to that of lasers. However, the more important question which we shall be considering is when does the higher resolution of lasers produce more physics and when is it simply an impressive tour de force?

The resolution limit discussed above refers to plane diffraction gratings using focusing mirrors. At short wavelengths the low reflectivity of all surfaces dictates the use of concave gratings in a Rowland circle mount. It has been shown[2] that for such a configuration there is an optimum grating width, W_{opt}, above which aberrations cause the resolution to deteriorate. The resolving power for a concave grating instrument with an optimum width grating is $R_{opt} = 0.92$ m W_{opt}/d (for an Eagle mount $R_{opt} = 0.95$ m W_{opt}/d). For W^{opt} either larger or smaller than optimum the theoretical resolution degrades. The value of W_{opt} depends on the angle of incidence θ_i, the wave length λ, the groove spacing and the radius of curvature of the grating R. Table I shows some values for a 3m radius instrument with a 1200g/mm grating.

Table I Optimum grating width and resolving power of 3m Rowland circle mount with 1200g/mm grating.[2]

$\lambda(\overset{o}{A})$	θ_i	W_{opt} (cm)	R_{opt}
1,000	7°	29	320,000
500	7°	23	254,000
100	80°	3.2	36,000
10	88°	1.1	12,600

The optimum grating width scales as $R^{3/4}$ so that the quest for higher resolution necessarily forces one to large radius of curvature instruments.

In order to achieve theoretical resolution it is necessary to coherently illuminate the grating. This require a slit width narrow enough that the grating is filled by the first diffraction maxima. For a grating with an aperture given by f = R/W, illuminated at wavelength λ, this requires a slit width w $\overset{<}{\sim}$ fλ. For an f/10 system at 1000Å, w $\overset{<}{\sim}$ 1μ while for f/60 at 500Å, w $\overset{<}{\sim}$ 3μ. The highest spectrographic resolution is obtained using photographic plates which have a grain size which limits the spatial resolution to $\approx$10μ. This is significantly less than the theoretical resolution of the best instruments so that in practice the spatial resolution of detectors is a major constraint in approaching theoretical resolutions at short wavelengths.

III. USEABLE RESOLUTION

The difficulty of reaching the diffraction limit when working at the short wavelength of the VUV is related to finite detector size, increased sensitivity to surface irregularities, low optical efficiencies with small slits and other factors. In this section we will consider just what resolution is achieved in practice in the VUV. We will discuss representative instruments and not claim to survey all available high resolution VUV facilities.

The 10.7m vacuum spectrograph at the National Bureau of Standards represents the highest resolution achievable by grating instruments in the VUV. This spectrograph is equipped with a 17.5cm grating with 1200g/mm and is an f/60 instrument. Used in second order at 1500Å the reciprocal dispersion is 0.4Å/mm. SWR plates have been used to resolve lines separated by 10μ giving $\Delta\lambda = 0.005$Å or $\Delta\bar{\nu} = 0.2\text{cm}^{-1}$ representing an achieved resolving power of 300,000.[3] If a 12" photographic plate is used, 30,000 data points are recorded giving a simultaneous coverage of 6000cm^{-1}. There are only a handful of such instruments in the world.

The highest resolution is achieved with the large gratings which can be used with modest angles of incidence, the so called normal incidence machines. However, the reflectivity of all materials drops rapidly as one moves to short wavelengths (high frequencies) and for wavelengths below ≈ 400Å it is necessary to go to grazing incidence optics. For short wavelengths the complex refractive index has a real component which is less than unity so that for very high angles of incidence the boundary from high index (vacuum) to low index (material) provides a high reflectivity.[4] The general guideline in the grazing incidence region is that if ϕ is the angle between the incident ray and the surface in degrees, then to have a useable reflectivity down to a wavelength λ(in Å) requires $\phi \lesssim 0.1\lambda$. The use of large angles of incidence greatly reduces the value of W_{opt}, and thus R_{opt}, as shown in Table I. The largest grazing incidence instrument is a 10m radius of curvature spectrograph at the National Bureau of Standards. This spectrograph has a 1200g/mm grating. Used at 100Å with $\phi = 10°$ one finds $W_{opt} = 8$cm and $R_{opt} = 88,000$. In actual practice at $\lambda = 304$Å using a 3μ slit, giving a 14μ image, a resolving power of 70,000 has been measured. Grazing incidence optics are much more sensitive to surface figure and alignment than normal incidence optics so that it takes considerably more care to achieve maximum resolution.

IV. H_2: AN EXAMPLE OF COMPLEMENTARY TECHNIQUES

The question we now want to address is just what resolution is necessary to extract the physics from spectroscopic problems. The molecular spectra of H_2 provide an excellent example of how different techniques can complement each other and an example of when conventional resolution is adequate and when it is not. The absorption

12

spectrum of H_2 between 765Å and 835Å has been recorded photographically by Herzberg and Jungen[5] using the 10.5m spectrograph at NRC in
Ottowa.   The resolution achieved was ≈ 1.5cm⁻¹ (0.01Å) at 800Å for a
resolving power of ≈ 100,000.   The spectrum was taken at 78°K to
observe para-hydrogen and the recorded spectrum covers 12,000cm⁻¹.
Many of the features are pre-dissociating or autoionizing and have
widths which are easily resolved by the 10.5m spectrograph.   For
these features additional resolution would not be useful.   The photographic plates provide excellent values for absolute wavelengths.
However, the photographic technique provides only crude intensity
measurements so that accurate line profiles cannot be obtained.   A
major advantage of the photographic plates is that the entire
spectrum can be viewed and those regions of special interest easily
identified.   Thus it was seen that in the region from 800Å to 805Å
there was considerable structure which was not resolved on the photographic plate.   Furthermore, the structure was complex and the photographic technique does not provide accurate intensity information so
the profiles and line widths could only be approximated.

   This second limitation was rectified by a second measurement of
H_2 made by Dehmer and Chupka[6] using a 3m scanning monochromator with
a helium continuum to observe photoionization of H_2 with a molecular
beam.   The resolution was ≈ 2.5cm⁻¹ (0.02Å) at 800Å giving R ≈ 40,000.
Coverage was from 715Å to 805Å or 14,000cm⁻¹.   Not only is the photoionization measurement complementary to photoabsorption in the
physics being probed but the measurement provides accurate intensities so that line profiles can be obtained.   In contrast to the
photographic technique, the different wavelengths are not recorded
simultaneously but rather scanned over ≈ 100 hrs.   This is not a
problem with a stable sample but has an obvious drawback for transient systems such as pulsed plasmas.

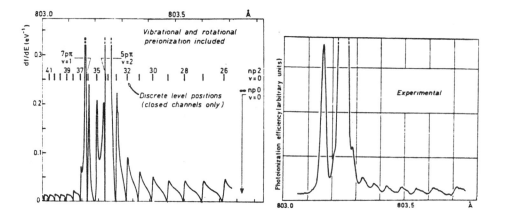

Figure 1.   Left spectrum shows theoretical calculations for
H_2 photoionization by Jungen and Dill.[7]   Right spectrum is
experimental results of Dehmer and Chupka[6] with 0.022Å resolution.

A portion of the experimental scan, along with theoretical calculations by Jungen and Dill,[7] are shown in figure 1. The experimental values agree very well with theoretical calculations when the instrumental resolution is included but it is clear that more resolution is desired. An expanded view of the theoretical calculations shows that the feature at 803.22Å has a complex shape with a peak having a width of 0.011cm⁻¹, clearly beyond the limiting resolution of any grating instruments.

A third experiment demonstrated the higher resolution which can be obtained using laser sources and upconversion techniques to reach the VUV. Figure 2 shows absorption spectra of H_2 near 830Å obtained by Rothschild et al[8] of University of Illinois at Chicago circle.

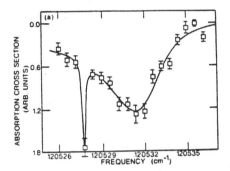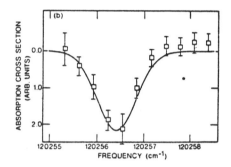

Figure 2.   Absorption in H_2 near 830Å using laser   upconverted source. Left spectrum shows asymmetric line profile while right spectrum is Doppler broadened Gaussian absorption. Note different scales.

These spectra provide accurate intensity measurements and very high resolution, $\Delta\nu \approx 0.01$cm⁻¹ or R $\approx 10^7$, but the coverage was only between 828Å and 833Å. It is seen that the very high instrumental resolution cannot be exploited because of the substantial Doppler width. Clearly the use of very high resolution laser sources will not only have to be applied to systems where the resolution is of intrinsic value but it will have to be matched by techniques to reduce other sources of line broadening. The limited coverage is characteristic of laser sources and is a significant limitation in studying new spectra.

Summarizing the H_2 data we see that photographic spectra provide an excellent overview of the H_2 absorption features, providing excellent wavelengths and a determination of areas of special interest. Monochromator scans provide intensity and profile information over a very broad region with adequate resolution for many, but not all, features. Finally the laser techniques provide extremely high resolution to probe those limited regions of the spectrum which require resolution beyond that of conventional instrumentation. It should be kept in mind that at higher intensities non-linear measurements using laser sources open up a much broader range of study than considered here.

V. ADDITIONAL STUDIES REQUIRING HIGH RESOLUTION

The rare gas spectrum in the region between the first ionization limit (mp^5 ^{2}P$_{3/2}$) and the second limit (mp^5 ^{2}P$_{1/2}$) shows two strong autoionizing series converging on the second limit. Figure 3 shows a photoionization spectrum obtained by Radler and Berkowitz[9] using a 3m scanning monochromator with a resolution of 0.045Å and also a photographic spectrum of the same region obtained by K. Yoshino[10] using a 6.7m spectrograph.

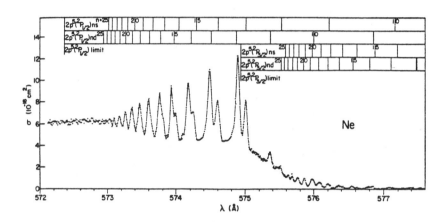

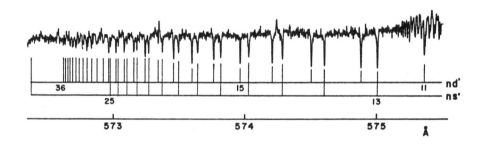

Figure 3. Spectrum of Ne between first and second ionization limits. Upper spectrum, photoionization with 0.045Å resolution. Lower spectrum, photoabsorption with 0.006Å resolution.

It is clear from these spectra that important information concerning line shapes and positions is lost because of inadequate resolution. Theoretical calculations by Johnson and Le Dourneuf[11] indicate that the inherent width of the autoionizing lines is considerably less than the resolution of Yoshino. This is a case where even with autoionization the spectrum is instrument limited.

It is not a sinple matter to anticipate when grating instruments will provide adequate resolution. Clearly the existence of autoionization is not sufficient to assure profiles which are broader than instrument limits. Figure 4 shows the absorption spectrum of Xe between the ionization limits recorded on the 6.7m spectrograph by Yoshino.[12] On the same figure for comparison is the spectrum of Ba[++] recorded by McIlrath and Lucatorto[13] with a 3m grazing incidence spectrograph. The spectra are isoelectronic and all levels above the first limit are autoionizing but while the nd' levels in Xe have a width comparable to the level separation, the nd' levels in Ba[++] and the ns' levels in both systems are too narrow to be resolved. Laser resolutions would be useful for these sharper transitions.

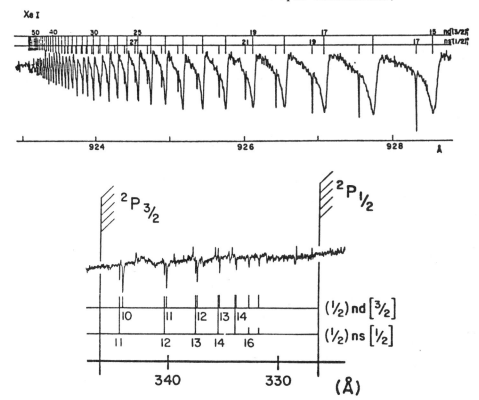

Figure 4. Upper spectrum: Xe recorded between the first and second ionization limits on a 6.7m spectrograph. Lower spectrum: Ba[++] recorded between the same limits on a 3m grazing incidence spectrograph.

Finally, as an example of the usefulness of Laser resolutions in locating new features, we show in figure 5 a high resolution scan of K vapor between 546Å and 542Å obtained by J.E. Rothenberg et al[14] of

Stanford. The spectrum was obtained photoelectrically in absorption using a tunable laser output which was upconverted to the VUV by anti-Stokes Raman scattering from metastable He atoms. The resolution was 1.2cm^{-1} giving a resolving power of 150,000 which exceeds all but the largest grating instruments. The broad features had been seen previously using conventional instrumentation but the sharp features were first seen in this experiment because they do not provide noticeable absorption in lower resolution spectra.

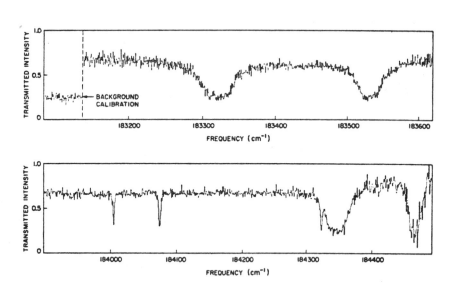

Figure 5. High Resolution absorption spectrum of K between 542Å and 546Å taken photoelectrically with an upconverted tunable laser source.

VI. <u>CONCLUSION</u>

It is clear from the above examples that for many problems in VUV spectroscopy conventional instrumentation provides more than adequate resolution to obtain line shape and strength information. This is largely due to the autoionization and predissociation which is ubiquitous in this region. Secondly, it is clear that the broad coverage and convenient instrumentation of conventional techniques provides an excellent way to identify regions of special interest and regions requiring higher resolution. Finally, however, it is clear that there are many problems which can only be studied using the high resolution achievable with laser sources. This is true not only for obtaining detailed line profiles but in observing features which are too narrow and weak to be seen with conventional resolutions. A comparison of the properties of conventional and laser sources, as summarized in Table 2, shows that grating based instrumentation provides a relatively easy and quick technique for

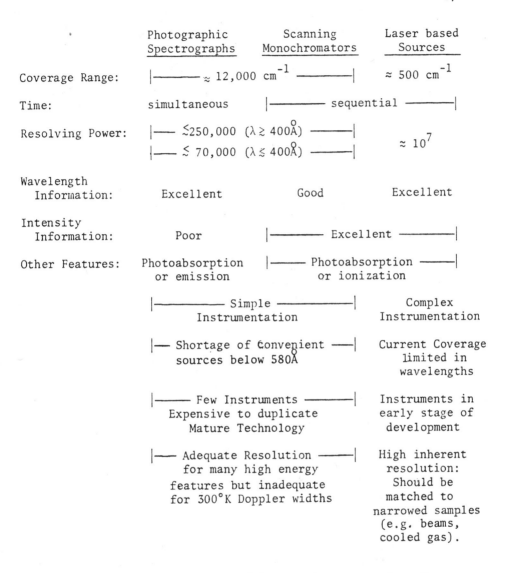

	Photographic Spectrographs	Scanning Monochromators	Laser based Sources
Coverage Range:	|————— $\approx 12{,}000$ cm^{-1} —————|		≈ 500 cm^{-1}
Time:	simultaneous	|————— sequential —————|	
Resolving Power:	|—— $\lesssim 250{,}000$ ($\lambda \gtrsim 400\overset{\circ}{A}$) ———| |—— $\leq 70{,}000$ ($\lambda \lesssim 400\overset{\circ}{A}$) ———|		$\approx 10^7$
Wavelength Information:	Excellent	Good	Excellent
Intensity Information:	Poor	|————— Excellent —————|	
Other Features:	Photoabsorption or emission	|——— Photoabsorption ———| or ionization	
	|————— Simple —————| Instrumentation		Complex Instrumentation
	|—— Shortage of Convenient ——| sources below 580$\overset{\circ}{A}$		Current Coverage limited in wavelengths
	|——— Few Instruments ———| Expensive to duplicate Mature Technology		Instruments in early stage of development
	|—— Adequate Resolution ——| for many high energy features but inadequate for 300°K Doppler widths		High inherent resolution: Should be matched to narrowed samples (e.g. beams, cooled gas).

Table 2: Summary of Comparative properties of high resolution systems.

covering very large spectral regions and for obtaining adquate information on much of the spectrum. Furthermore, photographic recording remains the only way to obtain simultaneous data over a large spectral region, a feature of great value for transient and unstable samples. The laser, however, has a resolution which is inherently greater than grating instruments and, when developed, will be the only tool for very high resolution in the VUV. Finally, it must be remembered that we have only considered the case of conventional absorption spectroscopy in this paper. The availability of high powered laser sources will also open the possibility of non-linear spectroscopy in the VUV region and this should be as fruitful here as it has been in the visible region.

This work has been supported by NSF grant CPE-81-19250: Useful discussions with M.L. Ginter, V. Kaufman, T.B. Lucatorto and W.C. Martin are greatfully acknowledged as is the assistance of K. Yoshino in providing unpublished data.

VII. REFERENCES

1. R.W. Ditchburn, Light (Blackie and Son, London, 1963).

2. J.A.R. Samson, Techniques of Vacuume Ultraviolet Spectroscopy (Pied Publications, Lincoln Neb., 1980).

3. V. Kaufman, Private Communication.

4. B. Edlén, Reports on Progress in Physics, 26, 181 (1963).

5. G. Herzberg and C.H. Jungen, J. Mol. Spectrose. 41, 425 (1972).

6. P.M. Dehmer and W.A. Chumpka, J. Chem. Phys. 65, 2243 (1976).

7. C.H. Jungen and D. Dill, J. Chem. Phys. 73, 3338 (1980).

8. M. Rothschild, H. Egger, R.T. Hawkins, J. Bokor, H. Pummer and C.K. Rhodes, Phys. Rev. A 23, 206 (1981).

9. K. Radler and J. Berkowitz, J. Chem. Phys. 70, 216 (1979).

10. K. Yoshino, Private Communication.

11. W.R. Johnson and M. Le Dourneuf, J. Phys. B: Atom. Molec. Phys. 13, L 13 (1980).

12. K. Yoshino, Private Communication.

13. T.J. McIlrath and T.B. Lucatorto in Laser Spectroscopy V, ed. A.R.W. McKellar, T. Oka and B.P. Stoicheff (Springer, Berlin, Heidelberg, New York 1981), p. 458.

14. J.E. Rothenberg, J.F. Young and S.E. Harris, Optics Lett. 6, 363 (1981).

TUNABLE, COHERENT SOURCES FOR HIGH-RESOLUTION VUV AND XUV SPECTROSCOPY*

B. P. Stoicheff, J. R. Banic, P. Herman,
W. Jamroz, P. E. LaRocque, and R. H. Lipson
Department of Physics, University of Toronto
Toronto, Ontario, M5S 1A7, Canada

ABSTRACT

A review of tunable laser sources for VUV and XUV spectroscopy is presented, with emphasis on the tuning ranges and resolution achieved to date. The most useful methods are based on four-wave frequency mixing in rare gases and in metal vapors. These sources provide tunable coherent radiation in the wavelength range of 200 to 106 nm, and in limited regions to shorter wavelengths. Typical resolving powers of 1 to 5×10^5 have been obtained, and values of $\sim 10^7$ are within reach. These results are illustrated with examples of atomic and molecular spectra.

INTRODUCTION

Just 20 years ago, great strides were made in high-resolution spectroscopy in the VUV, with the building of 6.4 m[1] and 10 m[2] grating spectrographs which provided resolving powers of $\sim 3 \times 10^5$ to 5×10^5 (that is, effective resolutions of ~ 0.2 to 0.3 cm^{-1} at 100 nm). To-day with holographic gratings, smaller spectrographs[3] achieve such resolution, but a resolving power greater than $\sim 5 \times 10^5$ has not been reported. The development of light sources for VUV (200 to 100 nm) and XUV (100 to ~ 20 nm) spectroscopy kept apace, so that modern laboratories are equipped with continuous or pulsed rare-gas discharges, and electron synchrotrons or storage rings. Typical wavelength coverage and relative intensities of some of these light sources[4] are illustrated in Fig. 1.

The availability of tunable dye lasers in the visible and near-infrared wavelength regions has had a profound effect on atomic and molecular spectroscopy. At the present time, in spite of considerable effort, the important wavelength region below 200 nm lacks tunable lasers. In fact, only a few lasers operate in this region, and these emit at discrete wavelengths with tunability limited to their bandwidths. However, there has been notable progress in generating tunable, coherent radiation below 200 nm by third order nonlinear processes (or frequency mixing of laser radiation) in rare gases and in metal vapors. Thus laser-driven sources of high brightness, and monochromaticity are now available for application to VUV and XUV spectroscopy.

BASIC THEORY OF NONLINEAR FREQUENCY MIXING

The process of optical frequency mixing is based on the nonlinear

*Research supported by the Natural Sciences and Engineering Research Council of Canada and the University of Toronto.

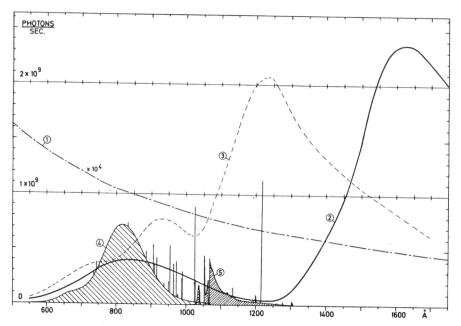

Fig. 1. Comparison of synchrotron and discharge light sources[4]. (1)
Calculated intensity of Stoughton storage ring: (2) measured intensity
after passing through 3 m monochromator. (3) Measured intensity of
DORIS storage ring after 3 m monochromator, (4) and (5) Measured intensities from continuum lamps of He and Ar, respectively.

susceptibility of atomic systems when irradiated by intense laser radiation[5]. It is well known that the polarization of a medium in the
presence of a monochromatic field $\bar{E}(r,t) = \sum_i E(\omega_i)$ can be written as

$$\bar{P}(\omega_i) = \chi^{(1)}(\omega_i) \cdot \bar{E}(\omega_i) + \sum_{j,k}\chi^{(2)}(\omega_i=\omega_j+\omega_k) \cdot \bar{E}(\omega_j) \cdot \bar{E}(\omega_k)$$

$$+ \sum_{jkl}\chi^{(3)}(\omega_i=\omega_j+\omega_k+\omega_l) \cdot \bar{E}(\omega_j) \cdot \bar{E}(\omega_k) \cdot \bar{E}(\omega_l) + \ldots \ldots \quad (1)$$

where $\chi^{(n)}$ are the susceptibility tensors of nth order. For third-
order processes (third harmonic generation, or 4-wave sum and difference mixing) we need be concerned only with $\chi^{(3)}$, whose principal term
may be written

$$\chi^{(3)}(\omega_0=\omega_1+\omega_2+\omega_3) = \frac{3e^4}{4\hbar^3} \frac{<g|\mu|a><a|\mu|b><b|\mu|c><c|\mu|g>}{(\Omega_{cg}-\omega_1-\omega_2-\omega_3)(\Omega_{bg}-\omega_1-\omega_2)(\Omega_{ag}-\omega_1)} \quad (2)$$

Here, $<g|\mu|a>$ is the electric dipole matrix element between the ground
state $|g>$ and an excited state $|a>$, having a lifetime Γ_a, and $\Omega_{ag} = \omega_{ag}-i\Gamma_a/2$ is the energy difference (Fig.2) between states $|a>$ and $|g>$,
e is the electronic charge and $\hbar=h/2\pi$, with h being Planck's constant.

Equation (2) shows that $\chi^{(3)}$ will be resonantly enhanced[6] whenever
the applied frequencies, ω_1, ω_2, ω_3 are such that the real part of the
resonance denominator vanishes, namely when $(\Omega_{ag}-\omega_1) = 0$, or $(\Omega_{bg}-\omega_1
-\omega_2)=0$, or $(\Omega_{cg}-\omega_1-\omega_2-\omega_3)=0$, corresponding to one, two, or three photon

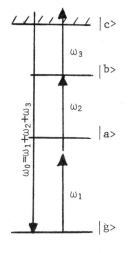

Fig. 2. 4-wave sum-mixing process, $\omega_0 = \omega_1 + \omega_2 + \omega_3$, with a 2-photon resonance transition ($\Omega_{bg} - \omega_1 - \omega_2 = 0$).

resonance, respectively. If any of $\omega_1, \omega_2, \omega_3$ is set equal to a resonance frequency (Ω_{ag}. etc.), $\chi^{(3)}$ will be enhanced but the incident radiation will be strongly absorbed. Similarly if $\omega_0 = \omega_1 + \omega_2 + \omega_3$ equals a resonance frequency, the generated radiation will be absorbed. If however, $\omega_1 + \omega_2$ is equal to a 2-photon transition (Ω_{bg}), the incident radiation at $\omega_1 + \omega_2$ is expected to be only weakly absorbed by the 2-photon transition, while the resonance enhancement of $\chi^{(3)}$ could be just as strong as for the 1-photon resonances.

For third harmonic generation (THG), $\chi^{(3)}$ simplifies to

$$\chi^{(3)}(\omega_0 = 3\omega) = \frac{3e^2}{4h^3} \frac{\langle g|\mu|a\rangle \text{ etc.}}{(\Omega_{cg} - 3\omega)(\Omega_{bg} - 2\omega)(\Omega_{ag} - \omega)} \qquad (3)$$

When 2ω approaches resonance, $\chi^{(3)}$ undergoes strong ($>10^4$) enhancement. For efficient THG, collinear phase-matching is necessary, that is, the refractive index $n(3\omega) = n(\omega)$ in order to yield a maximum effective interaction length. This may be obtained by varying the gas pressure or by a buffer gas. With focused incident radiation, THG can be observed only in negatively dispersive media[7]. Tunability is achieved by varying the incident frequency ω.

For generating tunable radiation by 4-wave sum mixing (4-WSM) or by 4-wave difference mixing (4-WDM), the process $\omega_0 = 2\omega_1 \pm \omega_2$ is of interest. $\chi^{(3)}$ then becomes

$$\chi^{(3)} = \frac{3e^2}{4h^3} \frac{\langle g|\mu|a\rangle \text{ etc.}}{(\Omega_{cg} - 2\omega_1 \mp \omega_2)(\Omega_{bg} - 2\omega_1)(\Omega_{ag} - \omega_1)} \qquad (4)$$

Strong enhancement is again achieved by tuning $2\omega_1$ to a parity - allowed 2-photon resonance, Ω_{bg}. Tunability (and further enhancement) is obtained by selecting ω_1 and ω_2, so that $2\omega_1 + \omega_2$ corresponds to the ionization continuum or to broad autoionizing levels above the ionization limit[8]. More detailed treatments of the relevant theory including phase-matching, saturation effects, and conversion efficiencies are given in Refs. 9 and 10.

COHERENT SOURCES AND REGIONS OF TUNABILITY

Second harmonic generation (SHG) of ruby laser radiation (694 nm) by Franken, Hill, Peters, and Weinreich[11] was a crucial step leading to the eventual production of coherent radiation in the VUV region. This was followed by third harmonic generation (THG) at 231 nm, demonstrated by Maker, Terhune, and Savage[12] in crystals, glasses and liquids. The major problem of generating even shorter wavelengths (due to the limited transparency of many nonlinear solids to the region above ~200 nm) was resolved when New and Ward[13] succeeded in producing THG in a number of gases. Harris and Miles then demonstrated that high conversion efficiency of THG and of sum-frequency mixing could be obtained by using phase-matched metal vapors as nonlinear media[14], and that efficiency could be improved further by resonance enhancement[6].

Frequency conversion into the VUV and XUV regions has been achieved by a variety of laser systems. Powerful pulsed lasers such as ruby, Nd:YAG, Nd in glass, flashlamp pumped dye (FPD), rare gas excimer, and rare gas halide exciplex lasers provide the primary coherent radiation. In some systems, tunable radiation from dye lasers (>320 nm) is used directly, and in others laser radiation (>400 nm) is doubled once or twice in nonlinear crystals to produce coherent radiation in the UV to about 200 nm. Subsequently, the coherent UV radiation is converted to coherent VUV and XUV by THG or frequency mixing in rare gases or metal vapors. Specific atomic (and molecular) systems are selected because of their large third order nonlinear susceptibility, negative dispersion for phase matching, suitability of energy levels for resonance enhancement, and low absorption at the desired VUV or XUV wavelength.

The nonlinear media and laser systems used to generate tunable, coherent, VUV and XUV radiation are summarized in Tables 1 and 2, along with their regions of tunability.

Table 1. Tunable Generation in Rare Gases

λ(nm)	Nonlinear Medium	Processes	Primary Laser	Ref.
206 – 160[a]	Xe	$2\lambda_{UV}-\lambda_L$	Nd:YAG-Dye	15
195 – 163	Xe	$2\times266\pm\lambda_s,\lambda_i$[b]	Nd:YAG and PO[b]	16
147 – 118	Xe	$2\times266\pm\lambda_s,\lambda_i$[b]	Nd:YAG and PO[b]	16
147 – 140	Xe:Kr	$3\lambda_{Dye}$	Nd:YAG-Dye	17
130 – 110[a]	Kr	$2\lambda_{UV}+\lambda_L$	Nd:YAG-Dye	15
123.6–120.3	Kr	$3\lambda_{Dye}$	KrF-Dye	18
123.5–120	Kr:Ar	$3\lambda_{Dye}$	Nd:YAG-Dye	17
121.6[c]	Kr	3×364.8	FPD	19
121.6[c]	Kr	3×364.8	Nd:YAG-Dye	20
106[c]	Xe	3×318	KrF-Dye	21
102.7[d]	Ar	3×308	XeCl	22
83[d]	Xe	3×248	cw Dye, Kr-Dye	23
79[c]	H_2	$2\times193.6+\lambda_D$	ArF, Dye	24
64[d]	Ar,H_2,Kr	3×193.6	cw Dye,ArF-Dye	25
57[d]	Ar	3×171	Xe_2	26

[a] λ_{UV} from a range of laser dyes.
[b] PO $\equiv$ parametric oscillator (signal and idler wavelengths λ_s,λ_i).
[c] Tunable over a small region.
[d] Tunable over the laser bandwidth.

Table 2. Tunable Generation in Metal Vapors

Nonlinear Medium (Ioniz. Limit in nm)	λ(nm)	Primary Laser	Ref.
Sr(217.8)	200 – 190	Nd:Glass-Dye	27
	195.7–177.8	N_2-Dye	28
Mg(162.1)	174 – 145	N_2-Dye	29
	160 – 140	N_2-Dye	30
	129 – 121	KrF-Dye	31
Be(133.0)	123 – 121	Nd:YAG-Dye	32
Zn(132.0)	140 – 106	XeCl/KrF-Dye	29,33
Hg(118.0)	125.1–117.4	Nd:YAG-Dye	34
	115.0–93.0	Nd:YAG-Dye	35

The two earliest systems used to generate coherent, tunable radiation in the VUV and XUV regions have set the pattern for recent work. In Fig. 3 is shown the method developed by Harris and his co-workers[36] at Stanford University. The primary laser source was a

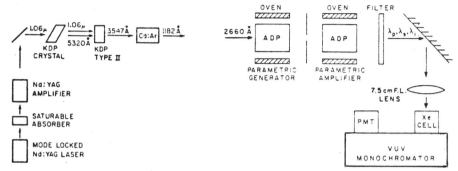

Fig. 3. Methods for generating harmonic and sum- and difference-mixing VUV radiation by third order nonlinear processes, at fixed wavelengths in a phase-matched metal vapor (Cd), and over a broadly tunable region in Xe gas[16,36].

mode-locked Nd:YAG laser and amplifier producing 1.06μm radiation in pulses of 50 psec duration, and peak power ∿20MW. This radiation was doubled to 532 nm in a KDP crystal, and then used directly, or doubled again, or mixed with 1.06μm radiation in a second KDP crystal, to produce radiation at 532, 266, or 355 nm with ∿10% efficiency. In this way, high-power radiation at several fixed wavelengths was available for mixing to shorter wavelengths in metal vapors and rare gases. Experiments in a Cd:Ar phase-matched mixture produced summing of 1064 and 2 x 354.7 to yield 152 nm, and tripling of 532 nm and of 354.7 nm to yield 177.3 and 118.2 nm radiation, respectively. In a phase-matched mixture of Xe:Ar, THG of 354.7 to 118.2 nm was achieved, and in Ar, THG of 266 to 88.7 nm.

Tunability was added to this coherent source by using the 266 nm radiation to pump an ADP parametric generator[16]. Variation of the crystal temperature from 50 to 105°C provided a tuning range of 420 to 720 nm, with >100μJ output in a bandwidth of 0.5 to 2 nm. The pump (266 nm) and resulting signal and idler radiation was then focussed into Xe gas (at pressures up to 1 atm). THG and sum and difference mixing produced VUV radiation tunable over portions of the region from 118 to 147 nm and continuously from 163 to 194 nm at peak powers of ∿1 W corresponding to ∿10^7 photons in a 20 psec pulse.

At about the same time as the above experiments on tunable VUV radiation, Hodgson, Sorokin and Wynne[28] of the IBM Thomas J. Watson Research Center reported a new 4-WSM technique providing a more extensive range of continuous tunability in the VUV, albeit at considerably lower peak power. Their experimental arrangement is shown in Fig. 4. A N_2 laser emitting ∿10 nsec pulses of ∿1 MW peak power at 337 nm was used to pump two dye lasers. These emitted up to 100 kW radiation of linewidth 0.1 to 1 cm^{-1}, tunable from 470 to 700 nm depending on the dye or dye mixture used. The two orthogonally polarized laser beams were spatially overlapped in a Glan-Thompson prism, and circularly polarized in opposed senses by a quarter-wave plate. These collinear beams of frequency ν_1 and ν_2 were focussed into a heated cell of Sr (at ∿800 to 900°C). Radiation of frequency ν_1 was tuned so that $2\nu_1$ was in resonance with various excited states of even parity, and ν_2 was swept to enable $2\nu_1+\nu_2$ to reach broad autoionizing levels in the ionization region. As already noted, additional resonance enhancement in $\chi^{(3)}$ arises when $2\nu_1+\nu_2$ corresponds to term values of autoionizing states.

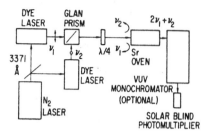

Fig. 4. Method of combining beams from two dye lasers (ν_1 and ν_2) to generate tunable coherent VUV radiation at $2\nu_1+\nu_2$ by 4-WSM in metal vapors[28].

The simple technique of using left and right circular polarization for ν_1 and ν_2 prohibits frequency tripling in isotropic media (since angular momentum is not conserved), and makes the 4-WSM process, $\nu=2\nu_1+\nu_2$, the exclusive nonlinear effect to be observed. In this way, Sr vapor was used to generate tunable VUV radiation over broad spectral ranges of ∿1200 cm^{-1} around each of the wavelengths 181.5 and 193 nm with the shortest wavelength reached being ∿155 nm. The extension of this method to other 2-electron atomic systems having broad autoionizing levels is described below.

Tunable Generation in Kr and Xe. Extensive wavelength tunability with rare gases has been achieved recently by Hilbig and Wallenstein[15]. They used non-resonant sum-frequency mixing ($\omega_{VUV} = 2\omega_{UV} + \omega_D$) in Kr and Xe to generate VUV radiation of ∿20 W power, tunable over most of the range 110 to 130 nm where these gases are negatively dispersive. These results are summarized in Fig. 5. They also generated VUV radiation of ∿50 W power at longer wavelengths in Xe by difference-frequency mixing. The process $\omega_{VUV} = 2\omega_{UV} - \omega_D$ resulted in radiation from 185 to 207 nm, and $\omega_{VUV} = 2\omega_{UV} - \omega_L$ at shorter wavelengths, from 160 to 190 nm. The frequencies ω_L, ω_D, and ω_{UV} refer to the output of

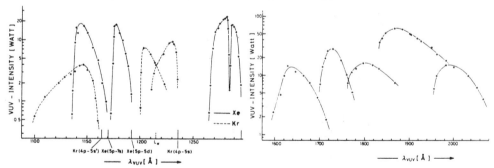

Fig. 5. Output power and tuning range of VUV generated by sum-frequency mixing in Kr and Xe[15].

Fig. 6. Output power and tuning range of VUV generated by difference-frequency mixing in Xe[15].

a Nd:YAG laser, a dye laser, and harmonic of the dye laser, respectively, used in these experiments[37]. For these processes, generation may take place in a medium with positive dispersion, so that efficient tripling in Xe was possible at $\lambda > 146.9$ nm. The tuning range and VUV output power are shown in Fig. 6. A conversion efficiency of $\sim 10^{-6}$ was obtained. With resonance enhancement[38], VUV radiation ~ 10 times as intense as that shown in Figs. 5 and 6 appears to be within reach.

<u>Tunable Generation in Mg and Zn Vapors.</u> The use of metal vapors for resonant 4-WSM, in particular Mg and Zn, has been pursued at the University of Toronto. The method is the same as that used by Hodgson et al[28]. Mg vapor was found to be a very efficient nonlinear medium by Wallace and Zdasiuk[30], who demonstrated a conversion efficiency of $\sim 0.2\%$ for THG[39]. They used 2-photon resonance enhancement and tuned ν_2 through the ionization continuum to obtain coherent VUV radiation tunable from 162 to 140 nm. An output of $\sim 10^{12}$ photons per pulse was obtained[40]. Recently a linewidth of ~ 0.02 cm^{-1} has been achieved.

The tuning range for Mg was extended to shorter wavelengths[31] (129 to 121 nm) by pumping UV dye lasers with a KrF laser. Tuning to longer wavelengths, as far as 174 nm, well beyond the ionization limit at 162 nm, has also been possible. An indication of the generated VUV signal intensity in the region 140 to 174 nm is given in Fig. 7. There is no dramatic change in intensity as $2\nu_1 + \nu_2$ is swept below the ionization limit (except for large enhancement at np resonances) indicating that the overlap of pressure and Stark broadened levels of high principal quantum numbers are as effective as autoionizing levels in the enhancement of 4-WSM.

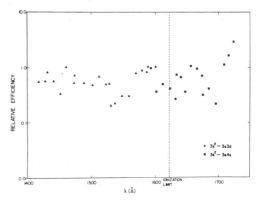

Fig. 7. Tuning curve of Mg vapor from 140 to 174 nm.

Recently, Zn vapor has been used to generate tunable VUV radiation from 139 to 106 nm (the LiF transmission limit). The VUV intensity in the vicinity of the ionization limit at 132 nm is shown in Fig. 8. It is seen that the intensity is essentially constant above the ionization limit; the intensity continues smoothly below the limit, with strong 3-photon resonant enhancement at np^{1}P levels, which provides conversion efficiencies $> 10^{-3}$. Finally, for the experiments with Zn and Mg, VUV generation occurs on both sides of np resonances and over broad regions between the resonances. This is in contrast to the results with Kr and Xe, where 4-WSM is limited to regions of negative dispersion (Fig. 5).

Considerable success has been obtained with VUV generation in Hg vapor[34,35], and will be reported at this Meeting.

<u>Tunable Generation in Molecular Gases.</u> As an alternative to metal vapors and in an effort to simplify the experiments, Innes, Stoicheff

26

and Wallace[41] explored the use of molecular gases for VUV generation. They found that NO was well-suited to the problem, with a strong 2-photon transition available for resonant enhancement. At 10 atm pressure, significant broadening of closely spaced rotational levels took place, and continuously tunable VUV radiation was generated in the γ bands (of breadth ∿600 cm⁻¹) at 151, 143, 136, and 130 nm. An estimated yield of ∿10⁷ photons/pulse was obtained for incident dye laser powers of 20 kW.

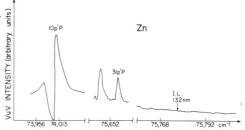

Fig. 8. Tuning curve of Zn vapor near the ionization limit at 132 nm.

Tunable Generation in the XUV. Rhodes and his colleagues have developed KrF and ArF laser systems of extremely high brightness, and tunable over the laser bandwidths, for use in THG and 4-WSM below 100 nm. A schematic outline of a typical multi-laser system for intense monochromatic radiation at 248 nm is shown in Fig. 9. The KrF radiation was tripled in flowing Xe to produce XUV radiation at ∿83 nm[23]. In similar experiments ArF radiation was focused in Kr, Ar, and H_2 to generate radiation at 64 nm[24].

A different type of laser-induced light source which is incoherent, yet tunable and of extremely high brightness has been developed by Harris and his colleagues[42,43]. It is based on spontaneous anti-Stokes scattering from He atoms stored in the $2s^1S$ metastable state (at ν = 166272 cm⁻¹). In these experiments, pulsed laser radiation at 1.06μm (ν_p∿9395 cm⁻¹) was focused in a He discharge, and spontaneously-emitted XUV radiation at $\nu \pm \nu_p$ or 56.9 and 63.7 nm was observed at right angles to the incident laser beam. This radiation exhibited several unique properties: linear polarization, narrow linewidth (1.3 cm⁻¹ at 56.9 nm and 1 cm⁻¹ at 63.7 nm, compared to the 5.6 cm⁻¹ of the 58.4 nm resonance line of the discharge) and high peak spectral brightness (with the 56.9 nm radiation being at least 100 times brighter than that of the resonance line emitted from the He discharge). Such spontaneous anti-Stokes radiation should be tunable over a range of ∿60,000 cm⁻¹ in the vicinity of 58 nm in He by pumping atoms in the metastable state

Fig. 9. A tunable, high spectral brightness XeF/KrF laser system for use in generating XUV radiation in Xe.

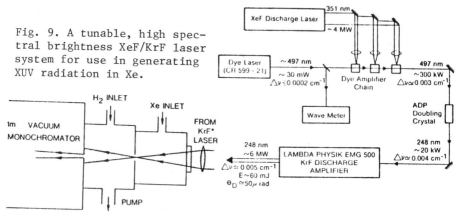

with tunable laser radiation.

Higher order frequency conversion has been used to generate radiation at fixed frequencies to wavelengths as short as 38 nm by Reintjes and co-workers[44]. They have used the fundamental, second, and fourth harmonics of a mode-locked Nd:YAG laser to generate XUV radiation in rare gases through fifth and seventh harmonic conversion and by 6-wave mixing ($\nu_3 = 4\nu_1 \pm \nu_2$). Fifth harmonic of 266.1 nm radiation produced radiation at 53.2 nm in He, Ne, Ar and Kr with a conversion efficiency of $\sim 10^{-5}$. Seventh harmonic at 38 nm was observed in He with peak power of ~ 100 W. While these sources are not yet tunable, such experiments are paving the way for tunable, coherent radiation in the XUV region.

MONOCHROMATICITY AND RESOLVING POWER

The laser-driven sources just described provide pulses of continuously tunable radiation over the region ~ 200 to 106 nm, and in limited regions at shorter wavelengths. The monochromaticity of these VUV and XUV sources is difficult to measure directly, but can be estimated from the bandwidths of the primary laser radiation used in frequency mixing in the rare gases or metal vapors. Linewidths as narrow as ~ 0.01 cm^{-1} have been obtained with various dye lasers and excimer lasers, leading to estimates of ~ 0.02 cm^{-1} for linewidths resulting from third order processes. Such monochromaticity corresponds to a resolving power (R.P.) of $\sim 5 \times 10^6$ at 100 nm.

SPECTROSCOPIC APPLICATIONS

Two classes of spectroscopic experiments have been reported using the sources listed in Tables 1 and 2. One is straightforward absorption spectroscopy at high resolution, making use of the narrow linewidths of coherent sources. The second is concerned with lifetime measurements of specific rotational levels of electronic states in simple molecules, making use of both frequency and time resolution. Some examples are briefly discussed.

In Figs. 10 and 11 are shown spectra of NO and HI in the region of 150 nm. For NO, the narrowest linewidth (FWHM) observed is ~ 0.25 cm^{-1}, corresponding to a convolution of Doppler breadth (0.15 cm^{-1}) and instrumental width of source and tuning mechanism (~ 0.2 cm^{-1}). This linewidth represents a R. P. of 2.7×10^5. The spectrum of HI shows an obvious predissociation, and although the Doppler breadth is 0.09 cm^{-1}, the narrowest line in this band system is 0.30 cm^{-1} representing a R. P. of $\sim 2 \times 10^5$. In Fig. 12, are shown two lines in the D←X spectrum of D_2[45]. The asymmetric broad (~ 4 cm^{-1} FWHM) line is the R(2) transition, and the sharp line is the Q(1) transition of the D(5)←X(0) system. Here the R. P. is $\sim 6 \times 10^5$. Further in the

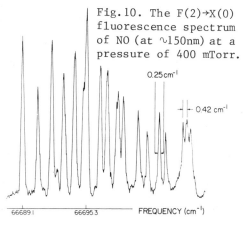

Fig. 10. The F(2)→X(0) fluorescence spectrum of NO (at ~ 150nm) at a pressure of 400 mTorr.

0.25 cm^{-1}

0.42 cm^{-1}

66689 l   666953   FREQUENCY (cm^{-1})

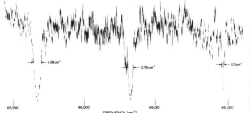

Fig. 11. Part of the $E^1\Sigma^+\leftarrow X$ and $d^3\Pi_2\leftarrow X$ absorption spectrum of HI (at ∿152 nm) at a pressure of 500 mTorr.

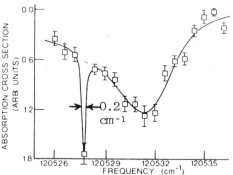

Fig. 12. D_2 absorption lines of $D(5)\leftarrow X(0)$ transitions at 83 nm[45].

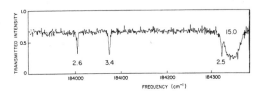

Fig. 13. Absorption spectrum of K at ∿54 nm[46].

XUV (at 54 nm), a spectrum of potassium is shown in Fig. 13, obtained using spontaneous anti-Stokes radiation[46]. The narrowest line shown is ∿2.5 cm^{-1}, and the narrowest observed is 1.9 cm^{-1}, corresponding to a R. P. of ∿1x10^5.

These examples show that the best resolving power achieved to date is ∿6x10^5. The main problem is the ubiquitous Doppler breadth which increases with frequency and is relatively large in the VUV and XUV, being ∿0.1 cm^{-1} even for the heaviest of atoms and molecules. Of course, the well-known Doppler-free laser techniques of saturated absorption and 2-photon absorption (with counter-propagating beams) could be used to improve the resolving power. While such experiments with the present VUV and XUV sources have not yet been reported, mention should be made of 2-photon absorption spectra which have used UV radiation (>200 nm) to excite atoms and molecules to levels of high energy corresponding to energies of VUV photons.

The highest effective resolution (∿300 MHz) was obtained by Hänsch and co-workers[47] in their experiment on the Doppler-free 2-photon spectrum of the 2S←1S transition in atomic hydrogen. They used SHG (at 243 nm) of dye laser radiation to excite H atoms to the 2S state, and monitored the process by observing fluorescence emission of Ly-α at 121.6 nm. The hyperfine doublet separation shown in Fig. 14 represents a R. P. of 8x10^6.

Also, Doppler-free, 2-photon absorption spectra of the $A^1\Pi\leftarrow X^1\Sigma^+$ bands of CO[48,49], and of the $a^1\Pi_g\leftarrow X^1\Sigma_g$ bands of N_2[49] have been obtained using radiation of wavelength 260 to 302 nm. Two-photon absorption with the present VUV and XUV sources would extend spectroscopic studies deep into the XUV region.

The short pulses (3 to 10 ns) and monochromaticity of present tunable coherent sources are ideally suited for selective fluorescence excitation and measurement of lifetimes of individual rovibronic levels. Such studies have been carried out with CO[50] and NO[51]. In a brief investigation with CO, rotational lifetimes of the $A^1\Pi(v=0)$ state

were measured with particular attention to
effects of degeneracies with rotational levels
of nearby, long-lived, $^3\Sigma$ and $^3\Delta$ states. Large
variations in lifetimes are apparent in Fig.15,
caused by strong perturbations for J'=12 (in
the Q branch) and for J'=16 (in the P branch)
which lead to a doubling of the unperturbed
lifetimes of these levels.

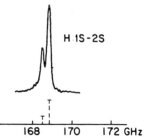

Fig.14. Hyperfine
doublet in Doppler-
free, 2-photon
absorption spect-
rum (2S←1S) of H[47].

A systematic study of radiative lifetimes
of the B'$^2\Delta$ state of NO has resulted in values
for rovibronic levels v=0,1,....8, and for
interacting levels of F$^2\Delta$ and N$^2\Delta$ Rydberg
states. A detailed study (at various pressures
of NO) of strongly perturbed levels of B'(v=7)
and N(v=0) has indicated significant dependence
of lifetimes on rotational quantum number, caused by collisions (Fig.16).

The above examples illustrate the possibilities for atomic and
molecular spectroscopy in the VUV and XUV regions, with the presently
available, tunable, coherent sources. There is no doubt that in the
near future, these monochromatic sources of high brightness will become
standard equipment in spectroscopic laboratories.

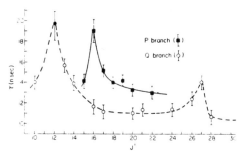

Fig.15. Radiative lifetimes of
selected rotational levels (J')
in the A$^1\Pi$ state of CO[50].

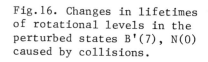

Fig.16. Changes in lifetimes
of rotational levels in the
perturbed states B'(7), N(0)
caused by collisions.

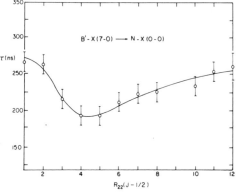

REFERENCES

1. P. G. Wilkinson, J. Mol. Spectroscopy 1, 288 (1957).
2. A. E. Douglas and J. G. Potter, Appl. Opt. 1, 727 (1962).
3. M. A. Baig, J. P. Connerade, J. Dagata, and S. P. McGlynn, J. Phys. B: At. Mol. Phys. 14, L25 (1981).
4. K. Radler and J. Berkowitz, J. Opt. Soc. Am. 68, 1181 (1978).
5. J. A. Armstrong, N. Bloembergen, J. Ducuing, and P. S. Pershan, Phys. Rev. 127, 1918 (1962).
6. R. B. Miles and S. E. Harris, IEEE J. Quantum Electron. QE-9, 470 (1973).
7. G. C. Bjorklund, IEEE J. Quantum Electron. QE-11, 287 (1975).
8. P. P. Sorokin, J. J. Wynne, J. A. Armstrong, and R. T. Hodgson, Annals N. Y. Acad. Sci. 267, 30 (1976).
9. C. R. Vidal, Appl. Optics, 19, 3897 (1980).
10. W. Jamroz and B. P. Stoicheff, in Progress in Optics, E. Wolf ed. (North-Holland Publ. Amsterdam) Vol. 20, Chapt. 5 (1982).
11. P. A. Franken, A. E. Hill, C. W. Peters, and G. Weinreich, Phys. Rev. Lett. 7, 118 (1961).
12. P. D. Maker, R. W. Terhune, and C. M. Savage, in Proc. Third International Congress in Quantum Electronics, P. Grivet and N. Bloembergen eds. (Dunod Éditeur, Paris and Columbia University Press, New York) 1559 (1964).
13. G. H. C. New and J. F. Ward, Phys. Rev. Lett. 19, 556 (1967).
14. S. E. Harris and R. B. Miles, Appl. Phys. Lett. 19, 385 (1971).
15. R. Hilbig and R. Wallenstein, Appl. Opt. 21, 913 (1982).
16. A. H. Kung, Appl. Phys. Lett. 25, 653 (1974).
17. R. Hilbig and R. Wallenstein, IEEE J. Quantum Electron. QE-17, 1566 (1981).
18. D. Cotter, Opt. Commun. 31, 397 (1979).
19. K. Mahon, T. J. McIlrath, and D. W. Koopman, Appl. Phys. Lett. 33, 305 (1978).
20. R. Wallenstein, Opt. Commun. 33, 119 (1980).
21. W. Zapka, D. Cotter, and U. Brackman, Opt. Commun. 36, 79 (1981).
22. J. Reintjes, Opt. Letters 4, 242 (1979).
23. H. Egger, R. T. Hawkins, J. Bokor, H. Pummer, M. Rothschild, and C. K. Rhodes, Opt. Letters 5, 282 (1980).
24. H. Egger, T. Srinivasan, K. Boyer, H. Pummer, and C. K. Rhodes, Present volume.
25. H. Egger, M. Rothschild, D. Muller, H. Pummer, T. Srinivasan, J. Zavelovich, and C. K. Rhodes, in Laser Spectroscopy V, A. R. W. McKellar, T. Oka and B. P. Stoicheff eds. (Springer-Verlag, Berlin) 446 (1981).
26. M. H. R. Hutchinson, C. C. Ling, and D. J. Bradley, Opt. Commun. 18, 203 (1976).
27. T. R. Royt and C. H. Lee, Appl. Phys. Lett. 30, 332 (1977).
28. R. T. Hodgson, P. P. Sorokin, and J. J. Wynne, Phys. Rev. Lett. 32, 343 (1974).
29. B. P. Stoicheff, J. R. Banic, P. Herman, W. Jamroz, P. E. La-Rocque, and R. H. Lipson, Present paper.
30. S. C. Wallace and G. Zdasiuk, Appl. Phys. Lett. 28, 449 (1976).
31. T. J. McKee, B. P. Stoicheff, and S. C. Wallace, Opt. Letters, 3, 207 (1978).

32. R. Mahon, T. J. McIlrath, F. S. Tomkins, and D. W. Kelleher, Opt. Letters, 4, 360 (1979).
33. W. Jamroz, P. E. LaRocque, and B. P. Stoicheff, Opt. Letters, 7, (1982) to be published.
34. F. S. Tomkins and R. Mahon, Present volume.
35. R. R. Freeman, R. M. Jopson, and J. Bokor, Present volume.
36. S. E. Harris, J. F. Young, A. H. Kung, D. M. Bloom, and G. C. Bjorklund, in Laser Spectroscopy I, R. G. Brewer and A. Mooradian, eds. (Plenum Press, New York) p. 59 (1974).
37. R. Wallenstein and H. Zacharias, Opt. Commun. 32, 429 (1980).
38. Y. M. Yiu, K. Bonin, and T. J. McIlrath, Present volume.
39. H. Junginger, H. B. Puell, H. Scheingraber, and C. R. Vidal, IEEE J. Quantum Electron. QE-16, 1132 (1980).
40. B. P. Stoicheff and S. C. Wallace, in Tunable Lasers and Applications, A. Mooradian, T. Jaeger, and P. Stokseth, eds. (Springer-Verlag, Berlin) p. 1 (1976).
41. K. K. Innes, B. P. Stoicheff, and S. C. Wallace, Appl. Phys. Lett. 29, 715 (1976).
42. S. E. Harris, Appl. Phys. Lett. 31, 498 (1977).
43. L. J. Zych, J. Lukasik, J. F. Young, and S. E. Harris, Phys. Rev. Lett. 40, 1493 (1978).
44. J. Reintjes, C. Y. She, R. C. Eckardt, N. E. Karangelen, R. A. Andrews, and R. C. Elton, Appl. Phys. Lett. 30, 480 (1977).
45. M. Rothschild, H. Egger, R. T. Hawkins, J. Bokor, H. Pummer, and C. K. Rhodes, Phys. Rev. A23, 206 (1981).
46. J. E. Rothenberg, J. F. Young, and S. E. Harris, Opt. Letters, 6, 363 (1981).
47. S. A. Lee, R. Wallenstein, and T. W. Hänsch, Phys. Rev. Lett. 35, 1262 (1975).
48. R. A. Bernheim, C. Kittrell, and D. K. Vries, Chem. Phys. Lett. 51, 325 (1977).
49. S. V. Filseth, R. Wallenstein, and H. Zacharias, Opt. Commun. 23, 231 (1977).
50. A. C. Provorov, B. P. Stoicheff, and S. C. Wallace, J. Chem. Phys. 67, 5393 (1977).
51. J. R. Banic, R. H. Lipson, T. Efthimiopoulos, and B. P. Stoicheff, Opt. Letters, 6, 461 (1981).

SPECTROSCOPIC MEASUREMENTS OF SUB-LASER LINEWIDTH AUTOIONIZATION RATES

W.E. Cooke, S.A. Bhatti, and C.L. Cromer
Deparment of Physics, University of Southern California
Los Angeles, Ca. 90007

ABSTRACT

We have simply determined autoionizing rates for states that have autoionization-induced linewidths smaller than the linewidth of our laser. By increasing our laser power to depletion-broaden the transition we have increased the apparent width of the transition beyond our laser's linewidth. The natural autoionization width can then be inferred from the laser power required to depletion-broaden the transition.

INTRODUCTION

Autoionizing states of the alkaline earth atoms in the 8eV range have been studied using primarily two techniques: VUV absorption[1] and a multi-step laser excitation process.[2] There are usually several advantages claimed for the laser technique, such as: (1) the laser technique produces simpler spectra, which are more easily analyzed; (2) the laser excitation technique is applicable to a wider range of states (such as high angular momentum states); and (3) the configuration of final states excited by the laser is usually easily identified.

Here, we will demonstrate that the laser technique, by virtue of the high spectral energy density of lasers, also provides a novel way to measure autoionization-broadened linewidths which are small even when compared to the laser's instrumental linewidth. We call this technique depletion broadening.[3]

Small autoionization linewidths are typical for highly excited, or Rydberg, states since the atom's volume increases dramatically with the principal quantum number and the electron-electron interaction which causes autoionization decreases accordingly. In many cases, the autoionization broadened linewidth of a state yields more information than its energy location does, and thus it often becomes important to measure small linewidths. The depletion broadening technique described here allows

ISSN:0094-243X/82/8900032-07$3.00 Copyright 1982 American Institute of Physics

the easy determination of small linewidths without
requiring excessive efforts to reduce the laser linewidth.

In the next section we will describe the depletion
broadening technique in a general way applicable to
any three level system for which states $|1>$ and $|2>$ can
be coupled by a laser field and state $|2>$ decays to
state $|3>$ with no return of population from $|3>$ to
either $|1>$ or $|2>$. For our studies, $|3>$ represents
the ion plus electron continuum, $|2>$ represents the
autoionizing Rydberg state and $|1>$ represents a bound
Rydberg state. The technique should be applicable to
other decaying systems, such as molecular predissociation, so we have left the formalism general.

In the final section we will demonstrate the tech-
nique as applied to autoionizing $6P_{\frac{1}{2}}ns$ states of barium
for values of n=20, 30, 35 and 40.

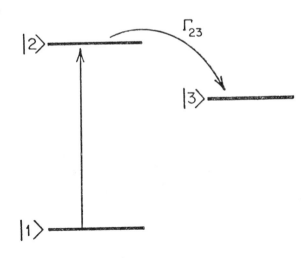

Fig. 1. Three-level system in which $|1>$ and $|2>$ are
strongly coupled by a laser and $|2>$ decays to $|3>$, with
neglibigle return from $|3>$ to either $|1>$ or $|2>$.

THEORY

Figure 1 illustrates the three levels under consideration: an initial state $|1>$ which is coupled by a laser to a state $|2>$ which decays at a rate Γ_{23} into state $|3>$. Γ_{23} is the decay rate that we will measure, assuming we have some way to monitor the population in state $|3>$.

The laser's electric field ε couples states $|1>$ and $|2>$ through their dipole moment μ, so that

$$\Omega = \tfrac{1}{2}\mu\varepsilon \tag{1}$$

defines the Rabi transition frequency. Then population will be transferred from state $|1>$ through state $|2>$ to state $|3>$ at a rate[4]

$$\frac{dN_3}{dt} = \frac{\Omega^2}{\Delta\omega^2+(\Gamma/2)^2}\Gamma_{23}N_1 \tag{2}$$

where the laser is detuned from resonance by $\Delta\omega(\Delta\omega>\Omega)$, Γ is the total decay rate of state $|2>$ in the absence of the laser and N_i is the population in state $|i>$. If the population N_1 does not change appreciably during the laser pulse, then the integral of equation (2) gives a normal Lorenztian lineshape for the N_3 excitation as a function of laser tuning.

However, for a sufficiently long laser pulse time, τ, it is possible to transfer all the population to state $|3>$, thus depleting N_1. In this case one does not obtain a simple Lorentzian. Consider the case where negligible total population is ever kept in state $|2>$. Then $(N_1+N_3)=N$, where N is the initial population in state $|1>$. Equation (2) can be now integrated:

$$N_3 = N[1-\exp-(\frac{\Omega^2\Gamma_{23}\tau}{\Delta\omega^2+(\Gamma/2)^2})] \tag{3}$$

When τ is sufficiently large,

$$\Omega^2\Gamma_{23}\tau>>\Gamma^2 \tag{4}$$

then equation (3) produces a lineshape with Full Width Half Maximum:

$$\Gamma_{FWHM} = 2\Omega(\Gamma_{23}\tau/\ln2)^{\frac{1}{2}} \tag{5}$$

The shape of equation (4) depends on the laser power, because in the tails $(\Delta\omega >> \Gamma_{FWHM})$ the signal is small and still exhibits a linear dependence on $\Omega^2\tau$, whereas near center $(\Delta\omega << \Gamma_{FWHM})$, the signal has saturated at its maximum size. This spectrum looks qualitatively like a power broadened lineshape where the dipole coupling Ω has been effectively enhanced by the factor $(\Gamma_{23}\tau)^{\frac{1}{2}}$. In order to use this depletion broadening to measure Γ_{23}, one needs only to increase the laser power or pulse length sufficiently to broaden the transition to a measurable linewidth (i.e., one that is greater than instrumental effects, or inhomogeneous broadenings).

Although equation (4) determines the required power to depletion-broaden the 1→2 transition, its derivation required $\Delta\omega$ to be greater than Ω in equation (2). This additional condition can be written as

$$\Gamma_{23}\tau > 1 \tag{6}$$

Equation (6) requires that the laser pulse be long enough that population transferred to $|2>$ can decay to $|3>$. Equation (6) also limits the ultimate resolution of this technique to the Fourier transform limit of the laser pulse.

If equation (6) is satisfied, then the power required to depletion-broaden the lineshape to an amount Γ_I, equal to the instrumental linewidth, can be obtained from equation (4) by replacing Γ with Γ_I. This gives us a limit on the total laser pulse energy density, ε, which is proportional to $\Omega^2\tau$:

$$\varepsilon > 2\times10^{-10} \left(\frac{\Gamma_I}{f\Gamma_{23}}\right) \omega\Gamma_I \tag{8}$$

where ε is measured in joules/cm^2, Γ_I, Γ_{23} and ω are measured in cm^{-1} and f is the 1→2 transition oscillator strength.[5] For a strong optical transition, a power density of $\sim\mu J/cm^2$ will double the natural width of a 1cm^{-1} broad line.

EXPERIMENT

We used the technique described above to measure the autoionization linewidth of Ba6P$_{\frac{1}{2}}$ns state for n=20, 30, 35 and 40. The lowest state, n=20, was measured two ways: by using the depletion broadening technique and by directly scanning a low power laser (since its linewidth is 3.1cm^{-1}, which is greater than our laser resolution).

The apparatus consists of an effusive atomic beam of Ba that passes between two parallel plates where it is crossed by three pulsed (5nsec-duration) dye lasers. The first two lasers excite the $6s^2 \to 6s6p$, ^1P, (5536Å) and $6s6p \to 6sns$, ^1S_o (~4200Å) transitions.[6] The 6sns state is our state $|1>$. The third laser drives the transition $6sns \to 6P_{\frac{1}{2}}ns$ (4935Å) corresponding to the $|1> \to |2>$ transition in the above discussion. The doubly excited state $6P_{\frac{1}{2}}ns$, decays into an ion electron pair (state $|3>$). After the third laser, a small positive-voltage pulse is applied to the bottom plate, forcing the ions through a grid in the top plate and into a particle multiplier. Data is obtained by scanning the frequency of the third laser, while monitoring the total number of ions detected. This procedure is described in detail elsewhere.[2]

In order to eliminate the Rabi frequency, Ω, from equation (5), we used pairs of measurements: one to a low n state for which Γ_{23} can be directly measured, and one for the higher n state where Γ_{23} is too small.

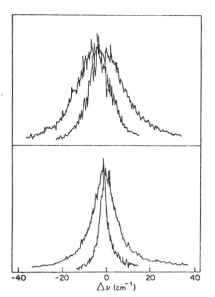

-40   -20   0   20   40
$\Delta\nu$ (cm^{-1})

Fig. 2. Excitation spectra of the $6P_{\frac{1}{2}}16s$ (upper) and $6P_{\frac{1}{2}}20s$ (lower) states. At high power the line shapes are depletion broadened; at low powers the linewidth represents the autoionization decay rate.

Figure two shows typical data comparing the n=16 case to the n=20 state. For both cases figure 2 shows a high power ($I \sim 10kW/cm^2$) trace and a low power trace. By measuring the natural width (11 cm^{-1}) and the broadened width (24cm^{-1}) for the n=16 case, we could evaluate $\Omega^2\tau=11$ for the high power sweep, using equation (3). Then for the n=20 case, where the depletion broadened width was only 11.3cm^{-1}, we again used equation (3) to obtain $\Gamma_{23}=2.2$cm^{-1}, assuming the same value of $\Omega^2\tau$. This is to be compared to the directly measured value of 3.1cm^{-1}. From different traces we obtained values from 2.1 to 3.9 cm^{-1} suggesting that the 30% uncertainty is primarily due to power variations and errors in measuring the width of a slightly assymetric line.

Our assumption that $\Omega^2\tau$ would be the same for the $6s16s \rightarrow 6P_{\frac{1}{2}}20s$ transition has been shown to be valid if the states are not severely configuration mixed and if the effective quantum number of the Rydberg electron does not change much (<0.1) during the transition.[7] Under those circumstances, the core electron is being excited, while the Rydberg electron is just a spectator, not affecting the transition.

By making similar measurements for the n=30, 35 and 40 states, using the n=20 state as a reference, we obtained the values shown in Table 1, for the $6P_{\frac{1}{2}}ns$ states' autoionizing rates. These rates show a continuation of the expected $(n^*)^{-3}$ scaling with the effective quantum number (n^*).

TABLE I

n	$\Delta(cm^{-1})$	$\Gamma(cm^{-1})$	$(n^*)^3\Gamma(a.u.)$
20	9.7	3.1a	5.5×10^{-2}
30	3.7	0.41b	3.2×10^{-2}
35	3.0	0.27b	3.6×10^{-2}
40	2.5	0.18b	3.7×10^{-2}

aMeasured at low power; uncertainty, ±10%
bCalculated using Eq. (6); uncertainty, ±30%

ACKNOWLEDGEMENTS

This work was supported by the National Science Foundation under grant PHY79-16444. Some of the equipment was provided by the Research Corporation. W.E. Cooke acknowledges a grant from the Alfred P. Sloan Foundation.

REFERENCES

1. C.M. Brown and M.L. Ginter, J. Opt. Soc. Am. 68, 817 (1978).
2. W.E. Cooke and T.F. Gallagher, Phys. Rev. Lett. 41, 1648 (1978).
3. W.E. Cooke, S.A. Bhatti and C.L. Cromer, Opt. Lett. 7, 69 (1982).
4. J.L.F. de Meijere and J.H. Eberly, Phys. Rev. A 17, 1416 (1978).
5. H.A. Bethe and E.E. Saltpeter, Quantum Mechanics of One and Two Electron Atoms (Academic Press,

38

New York, 1957) p. 250.

6. J.R. Rubbmark, S.A. Borgstrom, and K. Bockasten, J. Phys. B. $\underline{10}$, 421 (1977).

7. S.A. Bhatti, C.L. Cromer and W.E. Cooke, Phys. Rev. A $\underline{24}$, 161 (19 1).

LASER-INDUCED IONIZATION OF Na VAPOR

C. Y. Robert Wu and D. L. Judge
Department of Physics & Space Sciences Center
University of Southern California, Los Angeles, CA 90007

F. Roussel, B. Carré, P. Breger, and G. Spiess
Centre d'Etudes Nucleaires de Saclay
Service de Physique Atomique
91191 Gif-sur-Yvette Cedex
France

ABSTRACT

The production of Na_2^+ ions by off-resonant laser excitation in the 5800-6200Å region mainly results from two-photon absorption by the Na_2 molecule to highly excited *gerade* states followed by (a) direct ionization by absorbing a third photon or (b) coupling to the molecular Na_2 $D^1\Pi_u$ Rydberg state which is subsequently ionized by absorbing a third photon. This mechanism, i.e., a two-photon resonance three photon ionization process, explains a recent experimental observation of Roussel et al. It is suggested that the very same mechanism is also responsible for a similar observation reported by Polak-Dingels et al in their work using two crossed Na beams. In the latter two studies the laser-induced associative ionization processes were reported to be responsible for producing the Na_2^+ ion. From the ratio of molecular to atomic concentration in the crossed beam experiment of Polak-Dingels et al we estimate that the cross section for producing Na_2^+ through laser-induced associative ionization is at least four orders of magnitude smaller than ionization through the two-photon resonance three photon ionization process in Na_2 molecules.

INTRODUCTION

Recently laser-induced associative ionization[1-3] has been proposed as a mechanism for producing Na_2^+ ions by utilizing a resonant (referring to the excitation of Na D lines) and/or off-resonant two-photon laser excitation of the Na vapor. The two-photon ionization of the Li_2 molecule through a triplet van der Waals state has also been proposed.[4] The molecular structure, i.e., the yield of molecular ions as a function of laser wavelength, is in general very weak and complex and has often been attributed to the vibrational structure of the ground electronic state of the molecular ion. However, multiphoton ionization through the dimer molecules, which could amount to a few percent of the atomic concentration at elevated temperature[5], is believed not important. In this paper we shall show that it is the Na_2 molecules that are responsible for producing the Na_2^+ ions and the mechanism is a two-photon resonance three photon ionization process.

The expected properties of the various proposed mechanisms are briefly summarized as follows:

ISSN:0094-243X/82/900039-06$3.00 Copyright 1982 American Institute of Physics

A. Associative ionization through resonant excitation of the Na atoms

    (a) Using one laser beam at a wavelength tuned to the Na D line

$$Na(3s) + 5897Å \rightarrow Na(3p)$$

$$Na(3p) + Na(3p) \rightarrow Na_2^+ + e$$

$$i(Na_2^+) \approx [Na(3p)]^2 \approx n^2 \cdot I^2 \qquad (1)$$

    (b) Using two laser beams with one of them tuned to the Na D line

$$Na(3s) + 5897Å \rightarrow Na(3p)$$

$$Na(3p) + Na(3s) + h\nu \rightarrow Na_2^+ + e$$

$$i(Na_2^+) \approx [Na(3p)] [Na(3s)] \cdot I \approx n^2 \cdot I \cdot I' \qquad (2)$$

B. Associative ionization through off-resonant excitation of Na atoms

$$Na(3s) + Na(3s) + 2 h\nu \rightarrow Na_2^+ + e$$

$$i(Na_2^+) \approx n^2 \cdot I^2 \qquad (3)$$

C. Multiphoton ionization through Na_2 molecules

$$Na_2 + m h\nu \rightarrow Na_2^+ + e$$

$$i(Na_2^+) \approx [Na_2] \cdot I^m \qquad m \leq 3 \qquad (4)$$

where $i(Na_2^+)$ is the ion signal of Na_2^+, $n=[Na]$ is the atomic sodium vapor density, and $[Na_2]$ is the molecular sodium vapor density. I is the laser power density and m is the number of photons involved in the process.

The main differences between processes (1)-(3) and process (4) are threefold: (a) three photons are required to ionize Na_2 molecules, (b) in the process (4), the yield of Na_2^+ ions should exhibit a linear dependence on the dimer vapor density, but in processes (1)-(3) it should be a quadratic dependence on the atomic vapor density, and (c) the complex molecular spectrum due to process (4) should include the characteristics of the <u>bound</u> molecular constants of both ground and excited states whereas that due to processes (1)-(3) should exhibit only the molecular constants of the excited states since the lower state involved in the associative transition is a quasi-molecular state with a repulsive potential.

RESULTS AND DISCUSSIONS

In the following discussion it is argued that the proposed mechanisms of Roussel <u>et al</u>[1], i.e., process (3), and of Polak-Dingels <u>et al</u>[3], i.e., process (2), are probably not as important as the multiphoton ionization process through Na_2 molecules, i.e., process (4). Furthermore, we shall point out that it is desirable to use the lowest Na atomic vapor density possible in the measurement of resonant associative ionization cross sections.[6-8]

The off-resonant complex molecular spectrum obtained by Roussel et al[1] is shown in Figure 1. Peaks A and B correspond to the Na atomic D_1 and D_2 lines while peak C corresponds to the Na 3s-5s two-photon transition. In trying to analyze the complex spectrum we find surprisingly that the spectrum matches several of the recently reported two-photon *gerade-gerade* transitions[9-12] and that of the $Na_2 D^1\Pi_u - X^1\Sigma_g^+$ transition.[13] The peak positions of the $F^1\Sigma_g^+ - X^1\Sigma_g^+$ transitions are taken directly from the rotationless (J=0) vibrational levels reported by King et al.[11] Those of the $D^1\Pi_u - X^1\Sigma_g^+$ transitions are calculated band head positions using the molecular constants of Barrow et al.[14] Other transitions involving *gerade* states shown in Figure 1 have been calculated in the present work (for J=0 only) using the reported molecular constants of Taylor et al[9] and Carlson et al.[10] The ground state molecular constants of Demtröder et al[15] are used in the above calculations. The calculated positions are, of course, half the laser wavelengths shown in Figure 1. Every single peak (even the weakest one) can be accounted for. From this spectroscopic evidence it is clear that the Na2 molecule is responsible for producing the observed complex structure.

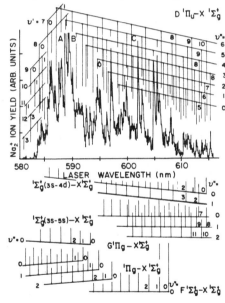

Figure 1. The yield of Na_2^+ ions as a function of laser wavelength in the 5800-6150Å region. The sodium density and the laser power used are $4.3 \times 10^{12} cm^{-3}$ and $\sim 10^5 W/cm^2$, respectively. Peaks A and B correspond to the Na atomic D_1 and D_2 resonance lines while peak C corresponds to the Na 3s-5s two-photon resonance line. The vibrational assignments indicated are due to the Na2 molecular transitions (see text for details). It is to be noted that the wavelengths of the transitions are half of the laser wavelength indicated.

The *gerade* states populated by the two-photon excitation are evidently ionized by absorbing a third photon. However, those *gerade* states also couple to the *ungerade* states, i.e., $D^1\Pi_u$, $C^1\Pi_u$, $B^1\Pi_u$, and $A^1\Sigma_u^+$, through radiative and/or nonradiative processes with/without the collisional quenching effects. The $Na_2(D^1\Pi_u)$ molecules thus formed are also subsequently ionized to produce the Na_2^+ ion by absorbing a third photon. The energy deficit of the $^1\Sigma_g^+$, $^1\Pi_g$, and $D^1\Pi_u$ states relative to the v=0 level of the Na_2^+ ion is typically less than 1 eV. Thus, the third photon can be provided by the laser, the fluorescence radiations of the $D^1\Pi_u \rightarrow X^1\Sigma_g^+$

(2860-3360A), $C^1\Pi_u \rightarrow X^1\Sigma_g^+$ (3150-3800Å), $B^1\Pi_u \rightarrow X^1\Sigma_g^+$ (4600-5900Å), $A^1\Sigma_u^+ \rightarrow X^1\Sigma_g^+$ (5900-8500Å), and even by radiations from transitions of the excited *gerade* states ($^1\Sigma_g^+$ and $^1\Pi_g$) to the *ungerade* states ($C^1\Pi_u$, $B^1\Pi_u$, $A^1\Sigma_u^+$). (However, these transitions are not known at the present time.)

King et al[11] have made use of the fluorescence radiation in the 3600-4000Å range as the monitor of two-photon transitions. Fluorescence in this wavelength range could be due to $C \rightarrow X$ and/or one or more of the above-mentioned excited *gerade* $\rightarrow$ *ungerade* transitions. At the 6021.7Å two-photon excitation Woerdman[16] observed emissions at wavelengths longer than 3120Å which strongly suggests that transitions such as $D \rightarrow X$, in addition to the $C \rightarrow X$ and $B \rightarrow X$, must be involved. These results support the present assertion that two-photon excitation of Na$_2$ molecules in the 5800-6150Å region will lead to the population of the $D^1\Pi_u$ and other *ungerade* states of Na$_2$. Further study to unravel the nature of coupling of the *gerade* states to the $D^1\Pi_u$ is warranted.

The striking similarity between the complex structure reported by Polak-Dingels et al[3] and that by Roussel et al[1] strongly suggests that the mechanism is the same in both cases despite the fact that the experimental techniques and methods employed are completely different. The ratio of molecular to atomic concentration was estimated to be ~ 0.03% in the cross beam experiment of Polak-Dingels et al.[3] We can therefore conclude that the cross sections for producing Na$_2^+$ through the laser-induced Na(3p) - Na(3s) associative ionization process proposed by Polak-Dingels et al[3] is at least four orders of magnitude smaller than that through the multi-photon ionization process in the Na$_2$ molecule.

Now, let's examine the pressure dependence. The Na$_2$ ion signal as a function of Na$_2$ and Na vapor density is shown in Figures 2(a) and (b), respectively. This figure is drawn based on the published results of Roussel et al.[1] The dimer vapor density is calculated by extrapolating the tabulated percent concentration of Na atomic density.[5] A "quenching" effect is evident in the region of high dimer vapor densities. The slope at low dimer density region is estimated to be ~0.6 as indicated in Fig. 2(a). This suggests that the yield of Na$_2^+$ is approximately linearly dependent on Na$_2$ density. Most recently, Roussel et al[17] have found that the yield of Na$_2^+$ ions is indeed dependent on the dimer concentration which was controlled by separately heating the double-chamber oven, a well-known technique for suppressing the number of dimer molecules.

As mentioned above, there is strong competition between ionization (Na$_2^+$), dissociative ionization (Na$^+$ + Na), and fluorescence processes. Thus, it may be reasonable to say that the branching ratio for producing Na$_2^+$ ions is about 0.6 at a dimer vapor density $< 1.5 \times 10^{10}$cm^{-3} (or at a Na atomic vapor density $< 5 \times 10^{12}$cm-3). It should be interesting to study the branching ratios for the other processes.

In Figure 2(b), the yield of Na$_2^+$ ions produced through associative ionization by resonant excitation of the Na D lines, i.e., process (1), exhibits an approximately quadratic dependence on n

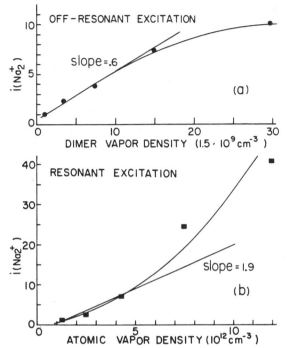

Figure 2. The yield of Na_2^+ ions as a function of Na density. 2(a) is due to the off-resonant excitation and 2(b) is due to the Na D line resonant excitation.

which is what one would expect. However, as the vapor density increases, the yield of Na_2^+ ions sharply increases and significantly deviates from the n^2 dependence. This may result from effects[18] such as collisional heating of electrons, energy pooling collisions, inelastic collisions, etc. Thus, it becomes clear why the literature cross section data for the Na(3p)-Na(3p) associative ionization process differ by two orders of magnitude.[6-8] Evidently, it is desirable to carry out the cross section measurements in the low vapor density condition.

Finally, it is important to investigate the dependence of Na_2^+ yield on pressure, dimer concentration, and laser power. In addition, time-resolved spectroscopy would be a significant aid in understanding the processes taking place in laser-induced ionization of Na vapor.

REFERENCES

1. F. Roussel, B. Carré, P. Breger, and G. Spiess, J. Phys. B 14, L313 (1981).
2. J. Weiner and P. Polak-Dingels, J. Chem. Phys. 74, 508 (1981).
3. P. Polak-Dingels, J.-F. Delpech, and J. Weiner, Phys. Rev. Lett. 44, 1663 (1980).
4. M. E. Koch, W. C. Stwalley, and C. B. Collins, Phys. Rev. Lett. 42, 1052 (1979).

5. A. N. Nesmeyanov, Vapour Pressure of the Elements (New York, N.Y., 1963).
6. A. deJong and F. van der Valk, J. Phys. B $\underline{12}$, L561 (1979).
7. A. Klyucharev, V. Sepman, and V. Vuinovich, Opt. Spectrosc. $\underline{42}$, 336 (1977).
8. G. H. Bearman and J.J. Leventhal, Phys. Rev. Lett. $\underline{41}$, 1227 (1978).
9. A.J. Taylor, K.M. Jones and A.L. Schawlow, Opt. Commun. $\underline{39}$, 47 (1981).
10. N.W. Carlson, A.J. Taylor, K.M. Jones, and A.L. Schawlow, Phys. Rev. A $\underline{24}$, 822 (1981) and references therein.
11. G.W. King, I.M. Littlewood, and N.T. Littlewood, Chem. Phys. Lett. $\underline{80}$, 215 (1981).
12. R. Vasudev. T.M. Stachelek, W.M. McClain, and J.P. Woerdman, Opt. Commun. $\underline{38}$, 149 (1981).
13. S.P. Sinha, Phys. Soc. (London) Proc. $\underline{59}$, 610 (1947).
14. R.F. Barrow, N. Travis, and C.V. Wright, Nature $\underline{187}$, 141 (1960).
15. W. Demtröder, M. McClintock, and R.N. Zare, J. Chem. Phys. $\underline{51}$, 5495 (1969).
16. J.P. Woerdman, Chem. Phys. Lett. $\underline{43}$, 279 (1976).
17. F. Roussel, B. Carré, P. Breger, and G. Spiess (unpublished results).
18. T.B. Lucatorto and T.J. McIlrath, Appl. Opt. $\underline{19}$, 3948 (1980).

OPTICAL-OPTICAL DOUBLE RESONANCE MULTIPHOTON IONIZATION SPECTROSCOPY OF NO

M. Seaver, W.Y. Cheung, D. Gauyacq, W.A. Chupka and S.D. Colson
Sterling Chemistry Laboratory, Yale University, P.O. Box 6666
New Haven, CT 06511

ABSTRACT

Double resonance studies using two pulsed lasers were performed on NO yielding highly simplified Rydberg and valence state spectra showing effects of internal and external perturbations.

INTRODUCTION

Various double resonance techniques including optical-optical double resonance (OODR) have been found to be extremely useful in many areas of spectroscopic research. These techniques have usually been confined to regions of the spectrum above the vacuum ultraviolet in wavelength. Using multiphoton techniques we have extended the method to study molecular states of NO at energies up to and above the ionization limit. Of particular concern in this study are the spectroscopic properties of Rydberg states, perturbations of these states by external electric and magnetic fields and collisions, perturbation by valence states and decay by predissociation and autoionization.

EXPERIMENTAL

The apparatus and experimental conditions will be described in detail in a forthcoming paper. Briefly, the apparatus consists of two tuneable dye lasers (Molectron DL200 and Lambda-Physic FL2002) simultaneously excited by a XeCl excimer laser (Lambda Physic EMG102). The two dye laser beams counterpropagate through a common focus within the ionization cell.

This cell is depicted in Fig. 1. The standard configuration is a cylinder held at -500 volts and a wire. This configuration produces electric fields of 100- 200 volts/cm at the laser focus depending on the focal position. When more careful control of the electric field is needed an insulated grid is inserted. As the voltage on the grid varies from -490 to -350 volts fields of $\sim$10-150 volts/cm occur reproducibly.

The current generated at the collection wire is amplified (Kiethley 427) and signal averaged with a linear gate. The output of the linear gate connects directly to a strip chart recorder.

46

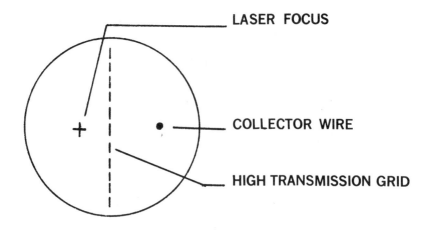

LASER FOCUS

COLLECTOR WIRE

HIGH TRANSMISSION GRID

Fig. 1 Double Chamber Ionization Cell.

Common experimental conditions are 1) a photon flux of $\sim 10^{21}$ photons/cm^2-s for each laser, 2) a ~ 50 μm diameter focal "spot", and 3) about 7 torr of NO. The optics are arranged to provide $\sim 90\%$ time overlap between the two 10 ns pulses.

GENERAL RESULTS

We have pumped, by both two-photon and three-photon processes, the $A^2\Sigma^+(3s\sigma)$, v=0-3 and $C^2\Pi(3p\pi)$, v=0 states of NO. From these states we probed a large number of Rydberg states with values of $\ell=0$, 1, 2, 3 and n up to about 35. Since the resolution in the pump transition is sufficient to select a single rotational level, the rotational structure of the transition is highly simplified and lends itself to very simple analysis.

ROTATIONAL STRUCTURE

As an example of the simplicity of rotational structure obtained in studies of lower Rydberg states, Fig 2 shows an OODR spectrum of the transition 4f-A $^2\Sigma^+(1,1)$ obtained by pumping the N=6 level of the A state. The nine lines correspond to the expected nine rotational branches. By pumping successive values of N ranging upward from zero, a Fortrat diagram for this transition is constructed simply by plotting the energies of the observed lines as a function of N as shown in the figure. Such data has been found to agree well with the results of Miescher and co-workers[1] where their data were available. In

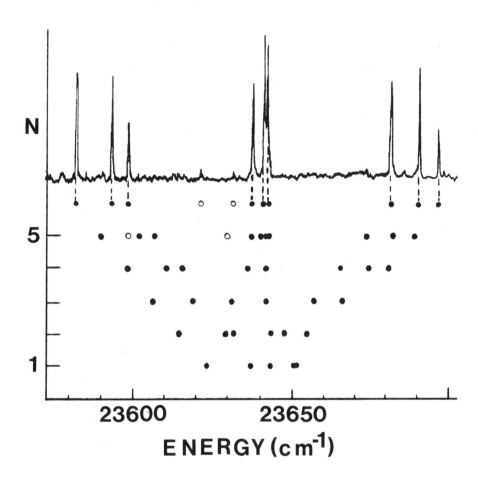

Fig. 2 Rotational structure of the 4f-A $^2\Sigma^+$ (1,1) OODR and illustration of the formation of the Fortrat diagram.

addition, because of the selectivity of the OODR technique some states which could not be investigated by Miescher et al. due to spectral congestion yielded readily to analysis. One such example is the strongly perturbed v=8 level of the L valence state which will be treated in detail elsewhere.

HIGH RYDBERG STATES

By pumping the C state with three photons, it was possible to observe the OODR spectrum of high Rydberg states converging

to the ionization limit. A small portion of this spectrum is
shown in Fig. 3. The dominant features of the spectrum are the

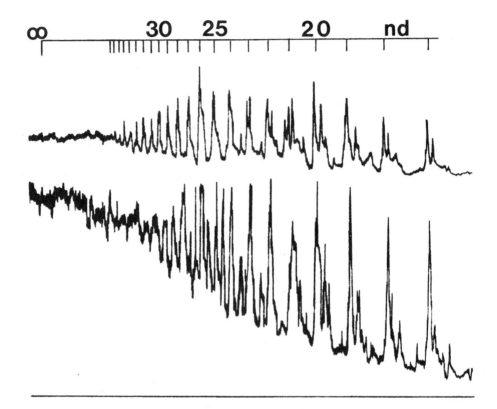

Fig. 3. Rydberg series converging to the first IP of NO. The
upper trace was taken at an electric field strength of 10 V/cm
while the lower trace was taken at about 100 V/cm.

members of the nd series. The mechanism of ionization for the
higher memebers is predominantly collisional ionization. The
upper part of the spectrum taken at 10 V/cm electric field
strength is seen to be resolved up to about n=37, while the
lower part, taken at about 100 V/cm is seen to be very considerably
broadened by the Stark effect. A more detailed study of this
Stark effect on the series shows that the broadening is semi-
quantitatively accounted for by the hydrogenic formula for the

linear Stark effect.[2] The dominant ℓ=2 series is very nearly
degenerate with all ℓ>2 levels and develops a hydrogenic Stark
manifold whose width increases approximately linearly with
electric field strength. At lower values of n at which the s and
p levels are well resolved from the d levels, the p levels, which
have a quantum defect far from integer value and hence are not

degenerate with $\ell \geq 2$ levels, show a strong quadratic Stark effect. In the region of n ≈ 10 the p levels are unobserved at fields of 10 V/cm, while at fields of about 100 V/cm they are of intensity comparable to that of the strong nd levels from which they borrow oscillator strength by the Stark effect. In this respect, the gross behavior of the NO series strongly resembles that of the Rydberg series of Na observed by Zimmerman et al.[3] (A careful examination of the lower trace of the figure also appears to reveal the emergence of some strong np levels but this effect is much clearer at lower values of n. It is of interest that the effect does not appear to behave monotonically with n.)

The effects of NO pressure on shifting line positions has been investigated and shown to be negligibly small for the pressures used in the experiments of Fig. 3. An extrapolation of the dominant nd Rydberg series yields a preliminary value for the ionization potential of NO of 74, 716.5 cm^{-1} in good agreement with the spectroscopic value of 74, 721.5 cm^{-1} obtained by Miescher[1] but in disagreement with the value of 74, 664.8 cm^{-1} obtained by Ono et al.[4] by photoionization. It is estimated that the probable cause of the error in the latter results is field ionization. Ionization by an electric field of about 70 V/cm would account for the discrepancy.

OODR PEAK SHAPES

While D.C. Stark broadening of high-n Rydberg peaks was readily seen, transitions to most lower n Rydbergs were found to have line widths determined by the laser bandwidth at lower powers of both pump and probe lasers. (Higher laser powers readily led to broadening by A.C. Stark effects, but those effects could be avoided in most cases). However, some predissociating states could be probed and the resulting transition observed in some cases as a dip in the background signal and in other cases as a positive signal but with a width consistent with the predissociation lifetime. One such example is shown in Fig. 4. The observed dip with a half-width of about 7 cm^{-1} corresponds to the 5pσ state which is known to pre-dissociate but whose width cannot be determined by standard methods due to spectral congestion.

A particularly unusual behavior of peak shapes is shown in Fig. 5. In this experiment the A $^2\Sigma^+$(3sσ) state is pumped and the probe laser scanned in the region of the relatively strong transition to the E $^2\Sigma$(4sσ) state. The v=20 level of the B $^2\Pi$ valence state lies nearby and at low pump powers appears as a group of positive peaks in the OODR spectrum. However, as the probe laser power is increased and the E-A transition becomes saturated and broadens, the B state peaks become dips in the increasingly intense tail of the E-A transition. A plausible explanation invokes an interference between the two dominant amplitudes of the two photon process from the A state to the ionization continuum. The effect may

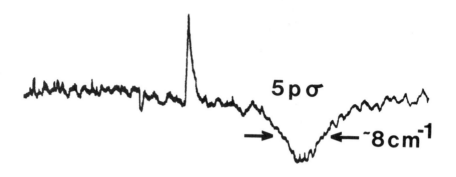

Fig. 4 The OODR signal in the region of the 5pσ, v=1 state. The pumped level was N=0 of the A, v=1 state.

be analogous to the Fano profile[5] common in one photon processes in which the parameter q has a value much less than unity when the amplitude via the E state is very large. The analogous situation in the present case may result from the circumstance that, while the E state with configuration $(\pi)^4$ (4sσ) has a large cross section for single photon ionization to the NO^+ ground state with configuration $(\pi)^4$, the B state must have a very low single ionization cross section since it has the double excited configuration $(\pi)^3 (\pi*)^2$ and in the single configuration approximation cannot be photoionized by one photon to form the NO^+ ground state. Further work is necessary to elucidate the details of this process. For instance, it will be important to see if the phenomenon persists unchanged when pump and probe pulses do not overlap in time.

VIBRATIONAL AUTOIONIZATION

By pumping the v=3 level of the A state we have been able to probe the v=3 level of the 4f, 4d, 6s and 7s Rydberg states. All these levels lie above the v=0 ionization limit and one of them (7s) lies above the v=1 limit. Thus all are subject to vibrational autoionization, the first three states via a $\Delta v=-3$ process and the 7s state via a $\Delta v=-2$ process. (For technical reasons we have not yet tried to produce states which can autoionize via the faster $\Delta v=-1$ process but this will shortly be feasible.) All these observed transitions have line widths which are laser band width limited ($\le .5$ cm^{-1}). Thus the auto-ionization rates are less than about 10^{11} sec^{-1}. This rather small value (for such low values of n) is in agreement with our estimates of autoionization rates made using an approximation

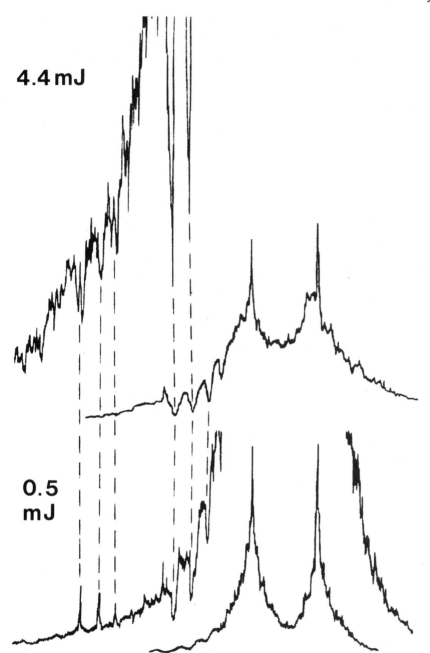

Fig. 5. OODR signal obtained by pumping A, v=1, N=3 and probing the region of E(4sσ), v=1 and B, v=20, showing peak shape change as a function of probe laser power.

developed by Herzberg and Jungen[6] in a study of the same process occuring for highly excited states of H_2. These calculations indicate that measurable widths should be observed for certain Δv=-1 processes and especially for autoionization of np states which however are also strongly predissociated and which we have not yet observed. The use of etalon-narrowed band widths will be very fruitful in this research area.

CONCLUSION

Many transitions to states normally observable in the vacuum ultra-violet are now accessible by multiphoton OODR techniques. High Rydberg states of molecules can now be studied with detail comparable to that attainable previously only far atoms. Many transitions obscured by spectral congestion so common in the vacuum ultra-violet region are clearly observed by this technique. Decay processes such as predissociation and preionization can be studied with laser band widths which can be made much narrower than those available in conventional spectroscopy.

REFERENCES

1. E. Miescher and K.P. Huber, in International Review of Science, Physical Chemistry Series Two, Vol. 3, Spectroscopy (ed. Dr. Ramsay (Butterworths, 1976). Chap. 2.
2. H.A. Bethe and E.E. Salpeter, Quantum Mechanics of One- and Two-Electron Atoms (Plenum, New York, 1977).
3. M.L. Zimmerman, M.G. Littman, M.M. Kash, and D. Kleppner, Phys. Rev. A20, 2251 (1979).
4. Y. Ono, S.H. Linn, H.F. Prest, C.Y. Ng and E. Miescher, J. Chem. Phys. 73, 4855 (1980).
5. U. Fano, Phys. Rev. 124, 1866 (1961).
6. G. Herzberg and Ch. Jungen, J. Mol. Spectrosc. 41, 425 (1972).

FREQUENCY CONVERSION PROCESSES IN HYDROGEN

Y.GONTIER and M. TRAHIN

Service de Physique des Atomes et des Surfaces
Centre d'Etudes Nucléaires de Saclay
91191 Gif-sur-Yvette Cedex, France

ABSTRACT

Frequency conversion probabilities for resonant and non-resonant multiphoton Raman-like processes taking place above and below
the ionization threshold of Hydrogen are compared.

1. INTRODUCTION

During the interaction of a radiation field with an atom, the
optical electron can make bound-bound transitions depicted in Fig(1)

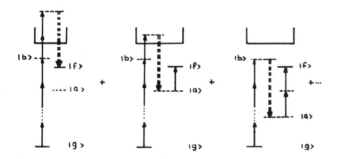

Fig(1): First few contributions to fifth-order frequency conver-
    sion process.

For a N-th order transition, the electron absorbs N-1 photons
of energy E_p and reemits a photon of energy $E_R = E_g + (N-1)E_p - E_g$; E_g
and E_f being the energies of the initial and final states
respectively. Such Raman-like process is said to take place above
or below the ionisation threshold according to whether the quantity
$E_g + NE_p$ is positive or not. When the initial and the final states
are identical one deals with harmonic generation. In general, the
light emitted in such process do not coincide with a characteristic
frequency of the vapor. Thus, at moderate pressure the radiations
cannot be resonantly reabsorbed by the medium. In addition, for
large absorption orders, the emission spectrum can be extended towards arbitrarily high frequencies. Lowest-order ionization i.e.

process where the electron absorbs just the number of quanta required to go through the ionization limit, gives rise to losses. It is thus of central interest to determine the most favourable situations for observing the emitted signal.

2. THEORY

To lowest-order of the perturbation theory, the probability for the absorption of N-1 photons of energy E_p and the emission of a photon of energy E_R in the solid angle $^P d\Omega$ is given by

$$\frac{dP^{(N)}}{d\Omega} = \frac{\alpha^4}{2\pi} \frac{c}{a_o} (\frac{I}{I_o})^{N-1} E_R \sum_i |M_i^{(N)} (E_p E_R)|^2, \qquad (1)$$

where α is the fine structure constant, a_o is the radius of the first Böhr orbit, I is the intensity in w/cm^2, $I_o = 14.03810^{16} w/cm^2$ and the sum runs over the two independent polarization states of the emitted photon. The matrix element $M_i^{(N)}$ is the sum of the N contributions obtained by considering the N possible ways of emitting the UV photon (see fig(1)) i.e.

$$M_i^{(N)} = \frac{1}{E_p^{(N-1)}} \sum_{\nu=1}^{N} M^{(N, \nu)} (E_p, E_R) \qquad (2)$$

The N-1 sums contained in each Matrix element $M^{(N,\nu)}$ are performed exactly by using an implicit summation technique /1/.

3. NUMERICAL RESULTS

a) Non resonant processes.

The probabilities for processes taking place above and below the ionization threshold have been compared when no multiple of the photon frequency coincides with any characteristic frequency of the atom. Table (1) shows the ratios of 4th-, 6th- and 8th- order above threshold frequency conversion probabilities over the corresponding ones below the ionization limit (harmonic generation at 1060 nm).

Transitions	order	λ_p (nm)	λ_R (nm)	Ratio
===========	======	=========	=========	==========
1S - 2S	4	271.2	353.3	1.85 (+5)
	6	286.1	212	1.06 (+6)
	8	471.8	151.4	1.06 (+5)

Table 1 : Ratios of probabilities for generating a radiation at λ_R through processes taking place above and below the ionization threshold.

The results shown in Table (1) illustrate a general rule according to which the generation of a well-defined frequency is more efficient for process taking place above than below the ionization threshold.

b) E-Resonances

By definition an E-resonance (Emission resonance) appears in frequency conversion processes when any two atomic levels, different from the initial level, are resonantly coupled by an integer number of photons. For example, in Fig (1), such a resonance occurs when the state $|a\rangle$ is a real state of the atom. Such a resonance do not affect ionization because it does not couple the state $|g\rangle$ to anyone. By this way one can enhance radiative process with respect to ionization. In hydrogen, such a resonance can be induced in a 7th-order process at 433.9 nm where the n=5 and n=2 levels are coupled. For a resonance detuning of 4 cm^{-1} (λ_R = 434 nm) an interaction time of 10 ns and an intensity value equals to 0.95 x I_s, I_s being the saturation intensity of 5th-order ionization process, the resonant frequency conversion probability is $P_{RFC}^{(7)}$ = 0.068, the corresponding 7th-order non-resonant frequency conversion probability (calculated at 530 nm) is $P_{NRFC}^{(7)}$ = 1.3×10^{-9} and the 5th-order non-resonant ionization probability is $P_{NRI}^{(5)}$ = 0.77 .

The resonance provides a considerable enhancement with respect to non-resonant frequency conversion, ($P_{RFC}^{(7)}$ / $P_{NRFC}^{(7)}$ = 5.2×10^7) and non-resonant 5th - order ionization $(P_{RFC}^{(7)}$ / $P_{NRI}^{(5)}$ = $8.8 \times 10^{-2})$

c) "E-Plus I"-Resonances

When an excited state like the state $|b\rangle$ of Fig(1) is resonantly coupled to the initial state ($|g\rangle$), one is faced with a I-resonance. It is called Ionization resonance because it is the familiar resonance of ionization processes.
E- and I-resonances can occur simultaneously in hydrogen where the energy separation between the n=1 and n=3 levels is fourfold that of the n=2 and n=6 levels. Thus, at λ_p = 410 nm, 7th-order frequency conversion becomes a doubly resonant process (first-order 2S-6P and fourth-order 1S-6S) which we compare to the corresponding 5th-order resonant (single) ionization . As before, with a pulse duration of 10ns and for a detuning E_p = 2 cm^{-1} of the 2S-6P transition, the seventh-order doubly resonant frequency conversion probability is $P_{DRFC}^{(7)}$ = 2.2×10^{-5} and the 5th-order resonant ionization is $P_{RI}^{(5)}$ =0.77. Although the ratios $P_{DRFC}^{(7)}$ / $P_{RFC}^{(7)}$ = 3.06×10^2 and $P_{DRFC}^{(7)}$ / $P_{NRFC}^{(7)}$ = 1.5×10^{10} increase compared to the preceding case , the ratio $P_{DRFC}^{(7)}$/ $P_{RI}^{(7)}$ = 2.8×10^{-5} is smaller than $P_{RFC}^{(7)}$ / $P_{NRI}^{(7)}$ =8.8×10^{-2}.

4. CONCLUSION

In non resonant processes, the most favourable condition for producing energetic photons involve processes taking place above the ionization threshold whereas it is preferable to induce E-resonances only to obtain significant enhancements of radiative processes with regard to lowest.order ionization.

REFERENCES

/1/   Y. Gontier, M. Poirier and M. Trahin
        J. Phys. B. Atom. and Molec. Phys. $\underline{13}$, 1381-87 (1980).

ATOMIC H AND D CONCENTRATIONS AND VELOCITIES MEASURED WITH HARMONICALLY GENERATED LYMAN-α (1215Å) RADIATION

R. W. Dreyfus[*] and P. Bogen
Institut für Plasmaphysik, Assoc. EURATOM-KFA
D-517 Jülich, GERMANY

H. Langer[**]
Max-Planck-Institut für Plasmaphysik, EURATOM Association
D-8046 Garching, GERMANY

ABSTRACT

A resonance fluorescence scattering system remotely senses H- and D-atom concentrations in the $\geq 10^9 \text{cm}^{-3}$ range in a tokamak. The Doppler broadening measures temperatures down to $100°\text{K}$.

INTRODUCTION

The present work aims at measuring H- and D-atom densities by the resonance fluorescence technique. As presently utilized, this technique involves the generation, scattering, collection and detection of Lyman-α (1215Å) photons. Equipment for carrying out this measurement is shown in Fig. 1. An important point is that this equipment represents a complete, working assembly. Furthermore, since the present Lyman-α source displays the usual narrow linewidths of laser sources, not only are H-atomic densities measured, but also the Doppler broadening gives the density as a function of velocity, i.e. temperature. The present equipment is in its final stages of refinement with the ultimate objective being the measurement of H-atomic behavior during a high current tokamak discharge. This portion of the measurement has progressed to the point of demonstrating practicality. In the meantime, the following significant goals have been attained: demonstration of high sensitivity (10^9 H-atoms/cm^3), selective excitation to give H-atom velocities down to 10^5cm/sec, i.e.

[*] Permanent Address: IBM Thomas J. Watson Research Center, Yorktown Heights, New York 10598

[**] Present Address: Fa. Carl Baasel Lasertechnik, Sandstr. 21, München 2, GERMANY

100°K, and demonstration of energy and wavelength calibration techniques that permit quantitative measurements.

APPARATUS

The operation of the assembled equipment is described with the help of Fig. 1. An excimer/dye laser pair[1] deliver ~2MW of tunable UV near 3646Å. The dye laser pulses are ≤10nsec FWHM. The available repetition rate is ~20Hz although stable operation is usually limited to 2 to <10Hz range. In any case, multiple shots are possible during a single 2sec tokamak discharge.

Lens L focuses the 3646Å input into a phasematched mixture of 1 Kr:2.5 Ar in order to generate third harmonic energy. The UV to VUV power conversion efficiency is commonly in the 10^{-5} range, i.e. one obtains ~100 Watts of VUV power; however, the pulse length is shortened to about 2 to 3nsec in the third harmonic generating process. The output energy is also decreased by the fact that the 2mm MgF_2 exit window on the Kr/Ar gas cell only transmits 60% of the energy.

The third harmonic gas cell plus the other VUV optical components have all been assembled into the 2.5m long vacuum chamber of the Cassegrain telescope. This 2.5m length is advantageous in avoiding x-rays and magnetic fields from interfering with the photomultiplier and also helps avoid plasma damage to the optical components.

Our seemingly simple choice of utilizing a single focusing lens, L, with a 15cm focal length, masks the numerous and important effects of this choice. The direct effect on the focal region is that the diffraction limited beam waste diameter is ~100μ, while the confocal parameter[2], b, is unusually long, ~1.5cm. As can be deduced from Eq. (11) of Ref. 2, the efficiency of generating VUV is independent of b when first order Kerr effect limits the phasematching. Going beyond this point, however, one sees two advantages of using a large focal area. First, one normally increases the Kr/Ar gas pressure as high as possible (≳3.4 bar in the present case) until gas breakdown becomes a limitation. With a shorter focal length lens, gas breakdown would have occurred at even lower pressures and thereby resulted in less VUV output. In the other direction, the focal length of L cannot be increased significantly further because the spectral linewidth of the 3646Å laser will not sustain phasematching over much larger values of b. Also increasing the focal length increases the optical power density on the windows of the high pressure cell; these windows are already operating near the breakdown power density. VUV absorption[2] in the Kr/Ar gas mixture limits the length of the gas cell and hence limits our ability to reduce optical power on its windows.

A second advantage of the 15cm focal length is that no additional lens is required to produce the desired 1cm^2 illuminated area in the plasma boundary. The laser and telescope axes (see Fig. 1) intersect at 5°, thus the usefully excited area has a depth of ~5cm. It is from this 5cm^3 that hydrogen atoms resonantly scatter Lyman-α radiation into the 30cm dia. Cassegrain telescope.

One disadvantage of the 15cm length is that phasematching to efficiently generate VUV energy is limited to a 0.6Å tuning range. Consequently, very accurate, ±0.5%, Kr/Ar gas mixtures are necessary in order to center this phasematch maximum at the Lyman-α wavelength.

After the signal beam exits from the telescope, unwanted wavelengths are rejected by a system of an O_2 gas cell[3] and/or a Lyman-α filter.[4] A fast solar-blind photomultiplier detects the photons with a CsI or KBr cathode[5]; the total system response time is 5nsec. This last number is important because it is the background light within this 5nsec window that ultimately determines the sensitivity.

Accurate calibration of the dye-laser wavelength is important for two reasons. First, calibration makes it possible to determine Doppler shifts on an absolute basis; hence H densities can be measured as a function of absolute velocity, i.e. net H flow should be detectable. Second, while the H and D lines are reasonably broad, ~1cm^{-1}, the lasers must be tuned within this linewidth for purposes of adjusting the other optical components. The wavelength is calibrated by recording the optogalvanic effect in a hollow cathode Ar discharge lamp. The ease of using this technique plus its accuracy are illustrated in Fig. 2. The accuracy is ±0.02Å in the UV, and hence, ±0.007Å (±0.5cm^{-1}) in the VUV. As a point of comparison, note that normal changes in ambient barometric pressure can contribute a similar uncertainty in the wavelength calibration.

One problem that has to be solved in making calibrated scattering measurements is the accurate measurement of pulsed VUV power and energy. A reasonable secondary standard is an ionization chamber filled with a gas (nitric oxide in the present case) of known ionization cross section.[3,6] Such a chamber both measures the VUV energy directly and provides the basis for calibrating photodiodes. The latter then measure power with subnanosecond resolution. The photodiode in the present case is of conventional planar construction (cathode to anode gap is ~3mm) with a gold plated cathode. The cathode photoefficiency is just under 3%, i.e. ~2.7mA/watt at Lyman-α and in good agreement with previous data.[3]

Since the 3646Å pump beam also falls on these detectors, it is interesting to note their resistance to spurious radiation. The photoionization chamber has a UV sensitivity $>10^8$ times smaller than the VUV sensitivity. This small UV sensitivity apparently comes from stray light generating photoelectrons directly on the charge collecting wire. With respect to the photodiode, the cathode is solid metal, thus there is no damage from the $\sim 2MW/cm^2$ of background UV light. The latter, however, generates a spurious signal equivalent to a UV sensitivity $\leq 10^{-9}$A/watt (at present power levels). Illumination with various intensities indicates that this signal arises from multiphoton photoemission.

Returning to the photoionization chamber, it is interesting to comment upon previous observations[7] that anomalously low output is obtained with nsec laser pulses. The problem was previously noted to arise at >1 Torr fill gas pressures.[7] The reproducibility of the present VUV source has now permitted direct observation of this saturation effect, see Fig. 3. Note that as the NO gas pressure increases above 1 Torr, the measured signal rapidly falls below the anticipated level. Based on known ion-electron recombination coefficients, this sublinear response can be shown to arise from gas-phase charge recombination occurring faster than the charge collection by the wire electrode. The problem has been known for some time in the fields of high intensity x-ray and accelerator sources,[8] but is only presently being analyzed for the case of nsec laser pulses and low (~1 Torr) fill gas pressures.[9]

SCATTERING RESULTS

Having established that measured energies ($\leq 10^{11}$ photons) of known wavelength VUV can be generated, the next question is the appearance of signals when this energy is scattered off of H or D atoms. Two types of atomic sources were utilized in the laboratory and two sources within the Asdex Tokamak.

Thermal and Microwave D-Atom Sources

The two laboratory sources were a calibrated thermal source[10] and a conventional microwave discharge source. While both sources scattered satisfactorily with H, results with D are given here because of the 30% narrower Doppler profile.

Figure 4 gives an example of an absorption measurement result. The geometry shown in Fig. 5 was utilized for this measurement. Using the linewidth values given in Table I, a calculated curve has been added to Fig. 4; the agreement with the experimental curve is apparent. The fact that a room-temperature line shape is obtained with the geometry shown in Fig. 5, indicates that the D atoms are rapidly (<3 collisions) thermalized by collisions with the stainless steel walls but primarily without recombining to form D_2. The total absorption agrees to within 10% with the values predicted on the basis of an independent calibration[10] of this same source and thus provides a cross check on the previous calibration.

Results, such as Fig. 6, are the basic output of the present resonance fluorescence measurement. These results again indicate the close approach of the measured line shape to the 300°K calculated curve. H-atom concentrations from a number of such curves are listed in the first three rows of Table II. For the laboratory sources summarized in row 1, direct comparison with calculated signals is possible; the latter are about 30% larger than the measured intensities. One reason for this difference is that the mirror reflectivity could only be estimated as 50%/reflection by comparison with similar mirrors. *In situ* calibration of the reflectivity has not yet been carried out. In order to visualize the meaning of these intensities, the photon signals have been inserted in brackets, [ ], into Fig. 5.

Scattering from H-Atoms Within a Tokamak

Using the experimental arrangement for a typical case shown in Fig. 1, the resonance fluorescence signal is obtained from H atoms within the Asdex Tokamak, see Fig. 7 and also Ref. 11. The H atom density and temperature for the two curves are given in Table II. Note that indeed the higher current has produced a higher H atom density plus an increased kinetic energy. An important point is that this is apparently the first experiment in which laser generated VUV has been successfully scattered from H atoms in a Tokamak. Thus the ability to measure H atom velocities (with laser resolution) and densities during even the highest current discharges is at hand.

Minimum Detectable H-Atom Density

The question now arises as to what constitutes the minimum detectable H atom density in each type of discharge. For the two sources (thermal and microwave) in the laboratory this minimum is set by scattered light due to both the 3646Å fundamental and 1215Å VUV radiation. In the initial laboratory tests, only minimal light shielding was installed, without a *bona fide* beam dump, and the spurious signal was about 10^4 photons/pulse. While this is a finite number of photons, it should be recalled that the input VUV signal was between 10^{10} and 10^{11} photons; hence less than a fraction of 10^{-6} spurious photons were reaching the photomultiplier. It is this small signal that now limits the H atom detectability to $\gtrsim 10^9 \text{cm}^{-3}$. Continuing this question of minimum detectable H-atom density, we note that results with the equipment installed on the Asdex Tokamak give <1 spurious photon/pulse. In this case, an efficient beam dump on the inner tokamak wall absorbed nearly all the direct beams. The small amount of scattered light was then efficiently dissipated by the large volume of the vacuum vessel. *With such a low background signal, the minimum detectable H atom concentration is* $\sim 3 \times 10^5 \text{cm}^{-3}$.

The detectability of H atoms in very high current, $\sim$300kA tokamak pulses, is limited by two additional effects. First, the atom temperature is usually estimated at 10^5 to $10^{6\circ}$K. (Note that it is precisely this temperature in the plasma boundary that the present technique is designed to measure.) In this case, the scattered intensity is equivalent only to $\sim 2 \times 10^8$ room-temperature H atoms, hence the signal is small. The second limitation is that this small signal is masked by the plasma light. The tokamak current directly accounts for $\sim$400 Lyman-α photons (60 photoelectrons) at the photomultiplier cathode during the 5nsec response time of the system. The shot noise from these 60 photoelectrons competes strongly with the resonantly scattered laser light. Steps to maximize the signal and override this background light are underway.

SUMMARY

To summarize, a complete VUV resonance fluorescence system for detecting H and D atoms has been successfully operated. The system is compatible with environments as hostile as tokamak plasmas. The sampling time is in the nsec range, and has a $\sim$10Hz repetition rate. Initial results have detected as few as 10^9 atoms/cm^3 and established their velocities to better than 10^5cm/sec, i.e. 100°K. The present results also indicate that as few as 3×10^5 H atoms/cm^3 should be detectable with the present apparatus. Furthermore, hydrogen atom temperatures and concentrations have been detected for low and intermediate current tokamak discharges. Background light measurements have indicated that similar results should be forthcoming for high energy tokamak discharges.

ACKNOWLEDGEMENTS

The contribution of Y. T. Lie in supplying and operating the thermal H source is gratefully acknowledged as is the help and interest of Dr. Keilhacker and the Asdex Tokamak Group (Max Planck I.P.P., Garching). The authors wish to thank the following people for their many helpful suggestions and contributions: E. Hintz, R. Kaufmann, H. Rohr, K.-H. Steuer, H. Scheingraber, C. Vidal and H. Wannier. The experiment received extensive technical assistance from K. Huber, H. Reichert, W. Beck and W. Wiegmann.

REFERENCES

1.   Models EMG 201 and FL 2002, respectively, Lambda Physik Corporation, Gottingen, Germany.

2.   H. Langer, H. Puell and H. Rohr, Opt. Comm. **34**, 137 (1980).

3.   J. A. R. Samson, **Techniques of Vacuum Ultraviolet Spectroscopy**, (Wiley, New York, 1967) and references therein.

4.   Acton Research, Cambridge, MA.

5.   EMI Gencom, Plainview, NY.

6.   K. Watanabe, F. M. Matsunaga and H. Sakai, Appl. Opt. **6**, 391 (1967).

7.   R. T. Hodgson and R. W. Dreyfus, Phys. Rev. **9**, 2635 (1974).

8.   P. Langevin, Comptes Rendus **134**, 533 (1902); J. Boag and J. Currant, Br. J. Radiol. **53**, 471 (1980), and R. Ellis and L. Read, Phys. Med. Biol. **14**, 293 (1969).

9.   R. Dreyfus, H. Langer and P. Bogen, to be published.

10. P. Bogen and Y. Lie, Appl. Phys. **16**, 139 (1978). The source calibration involved a comparison with radiation scattered from Kr.

11. P. Bogen, R. W. Dreyfus, H. Langer and Y. T. Lie, Proc. Int. Conf. on Plasma-Wall Interaction, submitted to J. Nucl. Materials.

TABLE I: LINEWIDTHS (cm^{-1})

Laser ($0.4 cm^{-1}$ at $3646 Å$)	0.7

H, D	
$^2P_{1/2}$-$^2P_{3/2}$ Splitting	0.37
H → D Shift	22.4

DOPPLER ($300°K$)	
H	1.0
D	0.7

TABLE II: CONCENTRATIONS AND

TEMPERATURES OF H-ATOM SOURCES

Source	$N(cm^{-3})$	$T(°K)$
Thermal Dissoc. and Microwave Discharge	10^9 to 7×10^{10}	300
Asdex Tokamak 2 A. Cleaning Discharge	5×10^9	300
Asdex Tokamak 600 A, 50Hz	2×10^{10}	6,300
Goals: Tokamak, 300 KA, Pulse	10^{10}	3×10^5
No Scattered Light	3×10^5	300

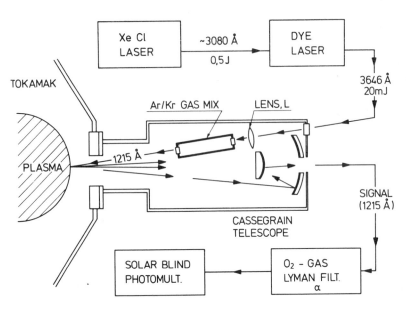

Figure 1: Schematic representation of the remote sensing equipment for H-atom measurements by the resonance fluorescence technique at Lyman-α (1215Å). This wavelength is generated by having lens L focus the 3646Å fundamental into a phasematched mixture of 1 Kr:2.5 Ar at 2 to 4 bar pressure. A solar blind gold photodiode can be remotely inserted immediately after the tripling gas cell in order to accurately establish the VUV power.

Figure 2: The optogalvanic wavelength calibration concept and signals from two lines in the Ar filled hollow cathode discharge.

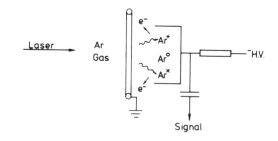

The calibration accuracy is about one half the laser linewidth, as is to be expected. The base line fluctuations, particularly near the Lyman-α driving frequency, may be ignored.

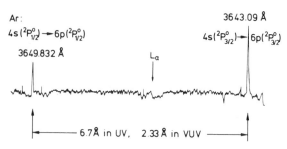

Figure 3: Charge collected vs. NO fill pressure in the photoionization chamber with high and low intensity 1215Å excitation. The theoretical lines presume linear response and the usual exponential absorption[3,6] law. An important point is that with lower excitation intensity near theoretical response is retained up to higher NO pressure, i.e. 2.5 vs. 1.3 Torr for 50% undershooting.

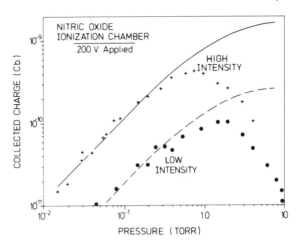

Figure 4: Absorption signal from a 26cm path of D atoms. The dashed theoretical curve is fitted only at the maximum absorption, the line shape is derived from the linewidth values given in Table I.

Figure 5: A schematic illustration of the experimental configuration used for the present laboratory atom sources. Photon fluxes have been inserted in brackets, [ ], at critical points and are based on 10^9 H atom/cm^{-3} and 300°K atom temperature.

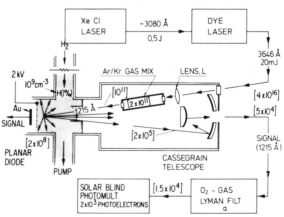

68

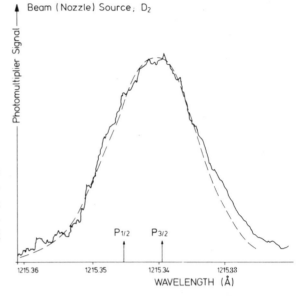

Figure 6: The resonance fluorescence signal obtained from a deuterium atom source. The theoretical curve is again based on Table I.

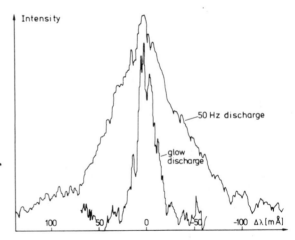

Figure 7: The resonance fluorescence signal obtained with a low current, 2A, glow discharge and an intermediate (600A) current, 50Hz, discharge within the Asdex Tokamak. Note the increased current has produced a 3.5 times broader Doppler profile.

LASER INDUCED PROCESSES IN HIGHLY
EXCITED STATES IN MOLECULES

J. Lukasik, S. C. Wallace[*], W. R. Green[**] and F. Vallée
Laboratoire d'Optique Quantique du CNRS, Ecole Polytechnique
91128 Palaiseau, France

ABSTRACT

Multiphoton excitation techniques are used to demonstrate laser induced processes involving high energy (E>11ev) electronic states in such molecular gases as nitrogen or carbon monoxide. The method utilizes high peak power, tunable dye lasers and produces vacuum ultraviolet radiations through direct multiphoton excitation, laser assisted collisional processes or nonlinear optical effects such as sum-frequency mixing.

INTRODUCTION

Although the study of laser induced effects in free molecules is now almost 20 years old, the field is, at present, experiencing a new wave of activity and a particular interest is focused on the vacuum ultraviolet (VUV) region[1-4]. Variety of molecules, with their rich spectrum of high-energy states, easily available large number densities and experimental handling simplicity, are attractive candidates for VUV radiation sources. Molecular spectroscopy in the region below 2000 A provides important data on the properties of those highly excited electronic states. Laser field effects in this region play an important part in developing and understanding models in molecular dynamics and provide the basis for exploring the rich field of dressed state chemistry in other areas such as photofragmentation and even reactive scattering. Recent interest in laser molecular studies in the VUV region also stems from the idea, first considered for atoms, of exploiting two-photon resonant processes. Another stimulus has been the increasing availability of high-power, tunable, frequency doubled visible dye lasers making such processes feasible.

This paper summarizes recent progress in our laboratory on experimental demonstration of processes involving high-energy valence or Rydberg molecular states using multiphoton excitation techniques. It will describe VUV emissions from laser-excited $b'^1\Sigma_u^+$ valence state of molecular nitrogen[5], observation of laser assisted collisional energy transfer between Rydberg states of carbon monoxide[6] and generation of coherent and tunable VUV radiation in the 1150 A region also in carbon monoxide[7].

Permanent address:
[*] Departement of Chemistry and Physics, University of Toronto, Toronto, Ontario M5S1A1, Canada
[**] Watkins-Johnson Co., 3333 Hillview Ave., Palo Alto, Ca. 94304, USA

ISSN:0094-243X/82/900069-10$3.00 Copyright 1982 American Institute of Physics

EXPERIMENTAL

The experiments are carried out with an excitation system consisting of two independently tunable and synchronized in time 5320 Å-pumped visible (rhodamine) dye lasers and a VUV detection setup. The dye lasers operate in a usual oscillator-amplifier mode, use transversely pumped capillary cells, work at a repetition rate of 5 Hz and deliver typically 40 mJ of energy in about 10 nsec. Tuning is accomplished in the oscillator by a system consisting of a fixed-grazing incidence grating and a movable 100 % reflecting mirror. A prism beam expander inserted in the dye oscillator provides a linewidth of the order of 0.08 cm^{-1}. One of the lasers is frequency doubled in a KD*P crystal to create a tunable 8 nsec UV source with more than 5 mJ of energy. These pulses can be spatially overlapped and focused into a gas cell mounted in front of a 1-meter vacuum Seya-Namioka monochromator to a measured area of about 10^{-4} cm^2. The monochromator is followed by a solar blind photomultiplier with a CsI cathode and an amplifier. A boxcar or a photon counting system after the photomultiplier provide additional gain and signal processing.

NITROGEN
VUV EMISSIONS FROM LASER-EXCITED b$'^1\Sigma_u^+$ STATE

We have observed the first selective excitation of the b$'^1\Sigma_u^+$ valence state of molecular nitrogen using, in this particular experiment, a single laser source. The energy level diagram of Fig. 1 describes the excitation process. The first step in the two-step excitation is a two-photon transition from the ground molecular state (v"=0) to the a$^1\Pi_g$ metastable state (v=1). The final step is a single-photon transition from the a$^1\Pi_g$ (v=1) state to the b$'^1\Sigma_u^+$ (v'=3) target state induced by a third photon of the pumping laser. After excitation to the final state, the molecule can then make a dipole transition down to an upper vibrational level of the ground state with a corresponding emission of a VUV photon.

This two-step technique, besides conveniently avoiding the traditional Franck-Condon-factor limitations associated with multiphoton excitation, offers excellent potential for inversion of high-energy levels of N$_2$. The reasons are, at least, threefold: the excitation is com-

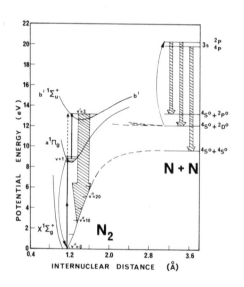

Fig. 1. Potential energy curves of N$_2$ (Ref. 8)

pleted within a laser pulsewidth, the laser provides selective excitation, and the upper vibrational levels of the molecular ground state
are not thermally populated. If sufficient population is stored in an
excited electronic state in a time comparable to the spontaneous emission time then the gain may be high enough to produce a pulse of stimulated emission even without mirrors[9].

Fig. 2 shows a monochromator scan of the cell emission detected
at right angles to the excitation axis. The pump laser at 2832 A is
tuned to the two-photon transition exciting $v=1$ level of the $a\,^1\Pi_g$ state. An additional pump photon, nearly resonant ($\Delta\omega \simeq 1\,\mathrm{cm}^{-1}$) with the

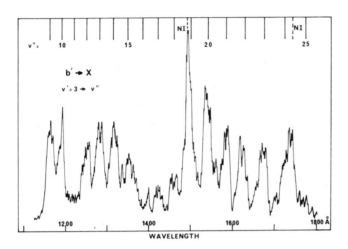

Fig. 2 Nitrogen VUV emission spectrum produced by the two-step laser
excitation of the $b'\,^1\Sigma_u^+$ state. Resolution is 15 A.

$v'=3$, $J'=8$ level of the $b'\,^1\Sigma_u^+$ state[10], then induces a transition to
this state. The b' state then emits to the ground state. A progression of over 10 lines is identified corresponding to transitions from
the $v'=3$ level of the $b'\,^1\Sigma_u^+$ state to the $X\,^1\Sigma_g^+$ ground vibrational levels.
Also evident in the scan are two atomic nitrogen emission lines cor-
responding to the 3s ^2P-2p^3 $^2D^0$ and 3s ^2P-2p^3 $^2P^0$ transitions at
1493 A and 1744 A respectively.

In the experiment repeated with the pump laser tuned to the two-
photon transition corresponding the the excitation of the $v=2$ level
of the $a\,^1\Pi_g$ state the vibrational progression observed in Fig. 2
clearly disappeared. The only features remaining in the scan were atomic nitrogen lines at 1493 A and 1744 A. We also observed a line at
1200 A which may correspond to the 3s ^4P-2p^3 $^4S^0$ atomic transition
as well as some weak and unidentified emissions around 1300 A.

It is interesting to note that there is not enough energy in any
of the excited N_2 states to dissociate into the emitting 3s state of
atomic nitrogen. We therefore expect the excitation process to be two-
step. It may be speculated that after excitation of the $a\,^1\Pi_g$ state a
third pumping photon crosses one of many dissociative states[8] of N_2

producing the $^4S^0$ and the $^2D^0$ states of atomic nitrogen (see Fig. 1). This is then followed by a dipole-quadrupole collision between a nitrogen molecule in the $a^1\Pi_g$ metastable state and a nitrogen atom in the metastable ^2D state. The molecule makes a quadrupole transition to ground, inducing a dipole transition in the atomic nitrogen from the $2p^3$ ^2D to the 3s state (curved arrows in Fig. 1) which then emits. It has been experimentally demonstrated that dipole-quadrupole interactions can play a significant role in energy transfer processes[11]. This particular electric quadrupole transition in N_2 is thought to have a strength of 2.6 atomic units[8].

We estimate the total population in the v'=3 level of the $b'^1\Sigma_u^+$ state to be greater than 10^{13} cm^{-3} while working with p_{N_2} =2 atm. If one takes into account the rotational distribution in v'=3 level this magnitude of population would be of the order of the threshold population required to observe H_2 lasing on Werner bands[12]. The near laser threshold populations we obtained suggest the future application of feedback mirrors or picosecond pulses to achieve laser action. A separate tunable laser for each step in the two-step process will dramatically enhance the flexibility of the excitation scheme described here and such experiments are to be undertaken in the near future.

CARBON MONOXIDE
LASER-ASSISTED COLLISIONAL ENERGY
TRANSFER IN THE VUV BETWEEN RYDBERG STATES

Laser-field effects in collisions now constitute a well documented class of phenomena and have recently been shown to occur with significant cross-sections[13] and in a wide range of atomic media[14]. We describe here the first observation of analogous processes in a molecular system, where we have demonstrated some striking features of laser-assisted intermolecular energy transfer involving Rydberg states in carbon monoxide (CO).

The types of laser-assisted energy transfer processes which we studied in CO are as follows :

$$CO(1)\{A^1\Pi, v_1=5\} + CO(2)\{X^1\Sigma^+, v_2=0\} + \hbar\omega \rightarrow$$

$$\rightarrow CO(1)\{X^1\Sigma^+, v_1=1\} + CO(2)\{B^1\Sigma^+, v_2=0\} \qquad (1a)$$

$$\rightarrow CO(1)\{X^1\Sigma^+, v_1=0\} + CO(2)\{B^1\Sigma^+, v_2=1\} \qquad (1b)$$

and these are indicated schematically in Fig. 3. A laser pulse at t=0 creates a population ($>10^{13}$ cm^{-3}) in a rovibronic level (v=5, J=12) of the $A^1\Pi$ state via two-photon absorption of a frequency doubled dye laser (λ=2784 A). After a 12 nsec delay the transfer laser, counterpropagating with respect to the pump laser is focused into the CO cell where it prepares a dressed state (indicated by the dotted line) which can transfer its energy to another molecule if it simultaneously makes a collision.

Using the known oscillator strengths[15], Franck-Condon factors[16] and rotational line strengths for B-A and B-X transitions we estimate that we are in the strong field regime for our laser intensities and

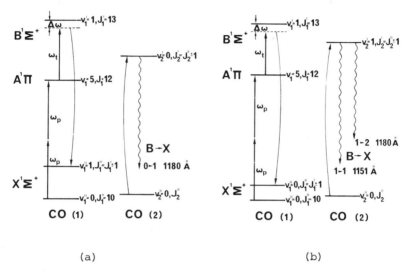

(a)   (b)

Fig. 3. Energy level schemes for laser-assisted intermolecular energy transfer in CO. Fig.(a) and (b) correspond to processes (1a) and (1b) described in text.

therefore the cross-section may be expressed by[17]

$$\sigma = 1.4 \; \frac{3\pi}{\sqrt{3}} \left(\frac{\mu_1 \; \mu_2 \; \mu_3}{h^2 \; \bar{v} \; \Delta\omega} \right) E \tag{2}$$

where μ_1, μ_2 and μ_3 are matrix elements correspondig to $B \leftarrow A(1,5)(CO(1))$, $B \rightarrow X(CO(1))$, $B \leftarrow X(CO(2))$ transitions respectively, $\Delta\omega$ is detuning from a real rovibronic level in the B state and E is the transfer laser field strength. Allowing for a $10\,cm^{-1}$ detuning from a rovibronic level in the $B^1\Sigma^+$ state and a factor of 1/3 for the orientational average with respect to the linearly polarized transfer laser[17] ($P/A = 8 \times 10^9 \, W/cm^2$), we estimate $\sigma = 1.7 \times 10^{-15} \, cm^2$ for (1a) and $\sigma = 0.3 \times 10^{-15} \, cm^2$ for (1b).

VUV B-X fluorescence is detected at 90° through our detection system described previously. At low intensities of the transfer laser ($10^5 \, W/cm^2$), we observe only the direct (intramolecular) excitation of the $B^1\Sigma^+(v'=1) \leftarrow A^1\Pi(v''=5)$ transitions, with line positions as observed by conventional spectroscopy[18]. As can be seen from the portion of the R branch shown in the lower trace of Fig. 4 ($p_{CO}=30$ torr) we are populating primarily J=12 in the $A^1\Pi$ state with a small contribution from J=13, but with no J=11 (both J=12 and J=13 are excited simultaneously because the bandwidth of the pump laser is comparable to their separation at the S branch A-X bandhead). Even for pressures as high as 100 torr, there is no evidence from these excitation spectra for any rotational relaxation in this relatively unperturbed[19] vibrational level of the $A^1\Pi$ state.

Under conditions of high transfer laser intensity ($8 \times 10^9 \, W/cm^2$)

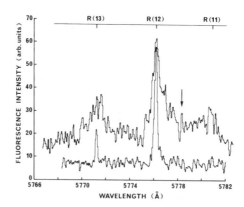

Fig. 4. Excitation spectrum of CO, detecting all B-X emissions (0,1)+
(1,1)+(1,2). Lower trace P/A=10⁵W/cm², upper trace P/A=8×10⁹W/cm². The
signals registered between the R lines correspond to the background
level, observed without the transfer laser (see lower trace).

we observe a dramatic change in the excitation spectrum. In the upper
portion of the Fig. 4 we show the excitation spectrum taken while simultaneously monitoring the fluorescence from both v'=0 and v'=1 levels
of the $B^1\Sigma^+$ state. This choice was made because (0,1) emission lies
too close (Δλ=0.5 Å) to the (1,2) emission to be resolved with our monochromator and because we also wish to emphasize that both processes
(1a) and (1b) are occuring simultaneously. As well as the asymmetric
lineshape around the real transition noted above, there is also an additional set of peaks of lower intensity but reproducible ir position
as for example indicated by the arrow in Fig. 4. Furthermore, the emission spectra obtained when exciting at 5776 Å and 5778.2 Å give different intensity ratios at 1151 Å ((1,1) emission) and 1180 Å ((0,1) +
(1,2)emission) namely 4:1 vs 2:1, corroborating the additional presence
of (0,1) emission resulting from process (1a).

A detailed theoretical analysis of the rotational state dependence
(taking into account the Boltzman distribution of acceptor states of
CO(2)) of processes (1a) and (1b) reveals an extremely dense spectrum
of switched collisions with Δω extending ± 200 cm⁻¹. We have in fact
observed similar quasi-continuous spectral features, well above the
background level, tuning the high intensity transfer laser as far as
150 cm⁻¹ from any real B-A (1,5) transition.

The intensity dependence of the observed fluorescence for λ_t=5778Å
(indicated by the arrow in Fig. 4) is given in Fig. 5. It is initially
linear in P/A and then goes over to a $(P/A)^{1/2}$ dependence at high intensities as is predicted by theory[17] in eq.(2). This behavior was observed at several different wavelengths, except in a 0.4 cm⁻¹ region
around the intramolecular transitions shown in the low intensity scan
in Fig. 4.

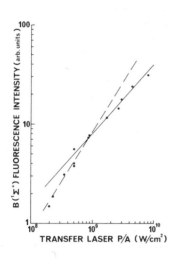

It is possible to make an estimate of the cross-section observed experimentally for the sum of processes (1a) and (1b). We calculate that 2 % of the $A^1\Pi$ state is transferred to the $B^1\Sigma^+$ state via this laser assisted collisional process at 8×10^9 W/cm^2. This corresponds to a total cross section of $\sigma_{a+b}=0.5\times10^{-15}$ cm^2 as compared to our initial estimate of $\sigma_{a+b}=2\times10^{-15}$ cm^2. Agreement is reasonable considering the total uncertainties in the input parameters. It should be noted that the large cross sections achieved in these experiments establish the feasibility of future applications, as for example exciting VUV laser transitions via laser-induced collisions.

Fig. 5. Intensity dependence of the switched collision process, p_{CO}=100 torr, λ_t=5778.2 A. Solid and broken lines represent slopes of 1/2 and 1 respectively.

TUNABLE COHERENT VUV GENERATION IN THE 1150 Å RANGE

In this experiment we demonstrate the usefulness of a high-energy Rydberg state $B^1\Sigma^+$ in carbon monoxide to produce coherent and continuously tunable over 500cm^{-1} VUV radiation in the vicinity of 1150 A by resonantly enhanced four-wave sum frequency mixing.

For optical frequency mixing processes which have provided an important source of coherent radiation in the VUV[20] and are currently the only source of coherent tunable radiation in the XUV[21], the use of rich in rovibronic structure molecules seems to be advantageous over other nonlinear media such as metal vapors or rare gases. So far, only Wallace and al.[22] have reported in detail such experiments in a molecular system where nitric oxide was used to produce tunable radiation in a few regions between 1300 A and 1520 A.

Our experiments use two- and three-photon resonant enhancement effects due to the $A^1\Pi$ and $B^1\Sigma^+$ electronic states and the pertinent energy diagram is schematically shown in Fig. 6. In these experiments, contrary to the switched collisions work in CO, the ω_1 and ω_2 beams are propagating colinearly. Their careful spatial overlapping is assured by the use of a dichroic mirror. Also any time delay between the two beams is now suppressed. The two beams are focused by a 190 mm fused silica lens (non-corrected for chromatic abberations) into a CO cell at pressures of a few torr and the resulting, highly directional VUV radiation, is viewed in the forward direction.

Fig. 7 shows a portion of our results from a double resonance type experiment where the ω_1 radiation at 2893 A is tuned to the band-head of the S branch of the A-X transition and ω_2 is chosen such that

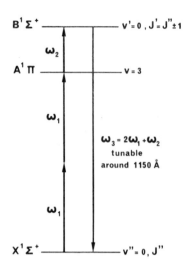

Fig. 6. Resonantly enhanced sum frequency mixing process in carbon monoxide.

the sum frequency $\omega_3 = 2\omega_1 + \omega_2$ probes the $v'=0$ level of the $B^1\Sigma^+$ state. Tuning the ω_1 frequency to the R or Q branches bandheads results in different and less intense spectra due to the fact that different selection rules on rotational transitions must be considered. Additional to the B-X branch structure clearly and reproducibly observed as in Fig. 7 indicates possible influence of various rovibronic levels in the vicinity of $v'=0$ of the B state. This structure has not yet been further explored or analyzed but it may be pointed out a likely contribution of the A-X(19-0) transition which at least partially overlaps B-X(0-0) transition[23].

The absolute measurement of conversion efficiency for the process reported here is difficult. It is only crudely estimated that for the ω_1 dye laser of about 600 kW the lower limit for the sum frequency

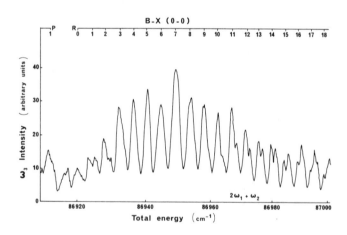

Fig. 7. Portion of double resonant enhancement spectrum showing the R branch of the B-X(0-0) transition in CO. The $2\omega_1$ frequency is fixed and set to the S(7) two-photon A-X transition at 69119cm^{-1} while ω_2 is scanned. CO pressure is 1.8 torr. Total tunability range is 500 cm^{-1}.

mixing process is 10^{-6}. This appraisal may, however, be pessimistic by 1 or perhaps even 2 orders of magnitude mainly due to the present deteriorated reflectivity of our grating in the VUV. Moreover, the efficiency of the process should also increase by the use of a focusing lens corrected for chromatic abberations.

Since all the input parameters for the nonlinear susceptibility $\chi^{(3)}$ $(\omega_3 = 2\omega_1 + \omega_2)$ (see, for example, eq. 4 of the Ref. 22) are available for CO[15,16], its value can be calculated and under doubly resonant conditions is 4×10^{32} esu. Power created at ω_3 can be evaluated using focused beam configuration analysis[24,25]. Under our experimental conditions, with the intensities $I(\omega_1) \approx 10^9$ W/cm^2 and $I(\omega_2) \approx 10^{10}$ W/cm^2 and taking the Boltzman rotational distribution at $p_{CO} = 3$ torr into account, we find $P(\omega_3)$ to be of the order of 20 W, in reasonable agreement with the value obtained in the experiment.

As predicted by the theory[25] our experiments show that the ω_3 radiation intensity is quadratically dependent on CO pressures, at least up to 3 torr. A deviation from this behavior at higher pressures suggests either optical depth effect of the double resonance radiation or it may demonstrate phase matching destruction of the process.

Further extension of tunable VUV regions in CO may be obtained by exploring other vibrational levels of the B state as well as reaching even higher Rydberg states such as $E^1\Pi$ or $C^1\Sigma^+$. It should be noted that high-resolution nonlinear spectroscopy through $\chi^{(3)}$ (ω_3) can become a powerful tool in clarifying many details of electronic structure in high lying energy levels of molecular systems.

REFERENCES

1. S. V. Filseth, R. Wallenstein and H. Zacharias, Opt. Commun. <u>23</u>, 231 (1977)
2. J. Bokor, J. Zavelovich and C. K. Rhodes, J. Chem. Phys. <u>72</u>, 965 (1980)
3. D. J. Kligler, J. Bokor and C. K. Rhodes, Phys. Rev. <u>21</u>, 607 (1980)
4. K. K. Innes, B. P. Stoicheff and S. C. Wallace, Appl. Phys. Lett. <u>29</u>, 715 (1976)
5. W. R. Green and J. Lukasik, Opt. Lett. <u>5</u>, 537 (1980)
6. J. Lukasik and S. C. Wallace, Phys. Rev. Lett. <u>47</u>, 240 (1981)
7. F. Vallée, S. C. Wallace and J. Lukasik (to be published)
8. A. Lofthus and P. H. Krupenie, J. Phys. Chem. Ref. Data <u>6</u>, 113 (1977)
9. R. T. Hodgson, J. Chem. Phys. <u>55</u>, 5378 (1971)
10. K. P. Huber and G. Herzberg, Constants of Diatomic Molecules (Van Nostrand Reinhold, N.Y., 1979)
11. W. R. Green, M. D. Wright, J. Lukasik, J. F. Young and S. E. Harris, Opt. Lett. <u>4</u>, 265 (1979)
12. R. T. Hodgson and R. W. Dreyfus, Phys. Rev. Lett. <u>28</u>, 536 (1972)
13. W. R. Green, J. Lukasik, J. R. Willison, M. D. Wright, J. F. Young and S. E. Harris, Phys. Rev. Lett. <u>42</u>, 970 (1979)
14. S. E. Harris, J. F. Young, W. R. Green, R. W. Falcone, J. Lukasik, J. C. White, J. R. Willison, M. D. Wright and G. A. Zdasiuk, Laser Spectroscopy IV (Springer-Verlag, N.Y., 1979)

78

15. P. H. Krupenie, The Band Spectrum of Carbon Monoxide, U. S. NBS, National Standards Reference Data Series 5 (U. S. GPO, Washington, D.C., 1966)
16. D. L. Albritton, Franck-Condon factors, personal communication
17. L. I. Gudzenko and S. I. Yakovlenko, Zh. Eksp. Teor. Fiz. 62, 1686 (1972)(Sov. Phys. JETP 35, 877 (1972))
18. J. Danielak, R. Kepa, K. Ojczyk and M. Rytel, Acta Phys. Pol. A39, 29 (1971)
19. D. Grimbert, M. Lavolée, A. Nizan and A. Tramer, Chem. Phys. Lett. 52, 45 (1978)
20. See, for example, a review by S. C. Wallace, Photoselective Chemistry, Part 2, ed. by J. Jortner (John Wiley and Sons, N.Y. 1981)
21. H. Egger, R. J. Hawkins, J. Bokor, H. Pummer, M. Rothschild and C. K. Rhodes, Opt. Lett. 5, 282 (1980)
22. S. C. Wallace and K. K. Innes, J. Chem. Phys. 72, 4805 (1980)
23. J. D. Simmons, A. M. Bass and S. G. Tilford, Astrophys. J. 155 345 (1969)
24. R. B. Miles and S. E. Harris, IEEE J. Quantum Electron. QE-9, 470 (1973)
25. G. Bjorklund, IEEE J. Quantum Electron. QE-11, 287 (1975)

OVERVIEW OF CORE-EXCITED Li SPECTROSCOPY: CALCULATIONS AND EXPERIMENTS

Carlos F. Bunge[*]

Instituto de Física, Universidad Nacional Autónoma de México,
Apartado 20-364, Delegación Alvaro Obregón, 01000 México D.F., México

Abstract

A close collaboration between experimentalists and calculists has recently led to a detailed interpretation of the core-excited spectrum of neutral lithium. Some doublets, which may be reached from a close-lying energy-rich metastable quartet, have attracted considerable attention as candidates for emission of laser radiation in the 200 Å region. The more salient results are used to describe the methods and philosophy inherent to accurate atomic ab-initio calculations.

1. Introduction

In principle, quantum mechanical calculations of atomic properties can be carried out far enough to verify the existing theory against experiments believed to be reliable.[1] It is our purpose here to illustrate another interesting use of ab-initio structure calculations, namely, the determination of accurate absolute term values by combining *partially* accurate calculations and experimental data.

We assume that the theory is essentially correct. The starting point is Schrödinger's equation:

$$H\Psi = ih(\partial/\partial t)\Psi , \tag{1}$$

where H is the Breit-Pauli effective Hamiltonian:[2]

$$H = H_{nr} + H_{rel} , \tag{2}$$

and H_{rel} is the relativistic correction, which can be conveniently partitioned into a fine-structure term H_{fs} and the rest:

$$H_{rel} = H_{fs} + (H_{rel} - H_{fs}) . \tag{3}$$

Many of the core-excited states of neutral lithium are embedded in a continuum with the same L-S symmetry and parity. When this happens, autoionization[3] usually takes place in $\tau \simeq 10^{-13} - 10^{-15}$ sec through the coulomb interelectronic repulsion (the half lifetime τ may increase considerably in the presence of highly excited orbitals).

The core-excited emission spectrum is produced by states which are metastable against autoionization, lying below the onset of the continuum with the same L-S symmetry and parity. Narrow line spectra result from transitions between quartet states,[4-6] and between doublets in the

---

[*]John Simon Guggenheim Memorial Fellow 1981-82

ISSN:0094-243X/82/900079-11$3.00 Copyright 1982 American Institute of Physics

2000–5500 Å region[5-7] and in the 200 Å region.[7-10] Intercombination lines have not been observed.[7] Slightly widened transitions between bound and resonance doublets have been observed both in emission[5-7] and in absorption.[11-13]

The fine-structure terms H_{fs} of Eq.(3) - the spin-orbit, spin-other orbit and spin-spin interactions - are responsible for L-S mixing, for the splitting of L-S terms into LSJ multiplets, and eventually for the mixing of the latter with a possible adjacent continuum characterized by the same J and parity. Thus the lowest quartet Li 1s2s2p $^4P°$ is metastable, autoionizing slowly in about 5 μsec,[3] due to the mixing of the $^4P°_{1/2}$ and $^4P°_{3/2}$ wave functions with the adjacent $1s^2kp_{1/2}$ and $1s^2kp_{3/2}$ continuum, respectively (spin-orbit induced decay). Also, the $^4P°_{5/2}$ wave function mixes with the $1s^2kf_{5/2}$ continuum, producing a still slower spin-spin induced autoionization.

2. Two ways to calculate energy levels

The calculation of energy levels may follow either of two lines:

a) Calculate first nonrelativistic energy levels E_{nr}:

$$H_{nr}\Phi = E_{nr}\Phi , \tag{4}$$

and then approximate the total energy E by $\tilde{E}$ as:

$$E \simeq \tilde{E} = E_{nr} + E_{rel} + \tilde{E}_{rad} , \tag{5}$$

where E_{rel} and $\tilde{E}_{rad}$ are relativistic and radiative corrections, respectively. In the past, E_{rel} and $\tilde{E}_{rad}$ were often neglected. This is the traditional way embodied, in the course of time, in the approaches of A.Weiss,[14] A.Bunge,[15] Ch.Froese-Fischer,[16] and many others.

b) Calculate first a zeroth-order relativistic energy E_R:

$$(H_D + V_{ee})\Psi = E_R\Psi , \tag{6}$$

where H_D is a sum of one-electron Dirac Hamiltonians and V_{ee} is the interelectronic coulomb repulsion. Then E is approximated by E* as:

$$E \simeq E^* = E_R + E_{TB} + E^*_{rad} , \tag{7}$$

where E_{TB} is the transverse Breit correction and E^*_{rad} is the radiative term. This more rigorous approach, pioneered by I.P.Grant[17] and others becomes indispensable when treating moderate and high Z species. For low nuclear charge, however, the traditional approach is more accurate, on account of powerful methods to deal with the nonrelativistic correlation energy problem.

3. Four styles to produce numbers

Let us consider the calculation of a quantity like E_{nr}:

$$E_{nr} = E_{nr}(\text{calc.}) + \Delta . \qquad (8)$$

For many-electron systems, $\Delta \neq 0$ always. Four different attitudes towards Δ may be distinguished:[18]

a) Δ is calculated ab-initio.[19] A few significant calculations[20] have exhausted the present possibilities of current theories of bounds to expectation values. In general, today's rigorous error bounds are too large to be of much use in atomic spectroscopy. (Rigorous calculations therefore, belong to the future.)

b) Δ is ignored. Neat in appearance but certainly incomplete, these calculations should be interpreted as numerical probes to test particular methods. (Other interpretations may cause a lot of confusion.)

c) Several Δ's corresponding to various levels of accuracy are first deduced by comparison with selected experimental results. Comparable Δ's are assumed to be valid in similar but new situations. This approach,[14] which accounts for the core of contemporary ab-initio activity, may be called the *pragmatic* ab-initio approach. A large amount of useful numerical results for atomic properties and most high-quality molecular calculations fall in this category.

d) Δ is calculated by means of empirically developed formulas, without resort to experiment. This *error-conscious* approach was pioneered by Pekeris[21] in his first work on He. Our own work falls between styles c) and d). More about this in Secs. 6-8.

4. Methods available

Grossly speaking, the methods to approximate Schrödinger's equation for stationary states may be divided into "r_{ij}" (or Hylleraas-type) and orbital methods.

r_{ij} methods are largely variational. Two-electron energy levels can be calculated with spectroscopic accuracy.[22] For three-electron systems, the calculations[23] look very promising because Δ is certainly small, but unfortunately, no one knows *how* small. This is the weakness of the r_{ij} approach: formulated as it is in a nonorthogonal basis, it does not yield itself to a convergence analysis (except for two electrons). In 1971, Sims and Hagstrom[24] came up with a combined orbital-r_{ij} strategy which may give excellent results for Li and Be.[18] Beyond four or five electrons, the r_{ij} method founders in conceptual difficulties (no patterns of convergence, $\Delta = ?$) as well as taxing computational demands (too many indices).

Orbital methods are either perturbation or variationally oriented.

Many-body perturbation theory (P.T.) to infinite order in certain diagrams,[25] coupled cluster P.T.,[26] the hierarchy of Bethe-Goldstone equations,[27] and Padé approximants to the Rayleigh-Schrödinger P.T. expansion complete up to a given order,[28] are certainly amenable to error conscious treatments. Unfortunately, these methods have not yet been exploited in quantitative studies of atomic spectroscopy, mainly because a comprehensive strategy to calculate Δ is still lacking.

While P.T. methods are often implemented with nonoptimized orbitals, the variational calculations are generally carried out either with self-consistent-field (S.C.F.) orbitals,[16] or with brute-force energy-optimized orbitals.[14,15] Explicit or implicit use is made of some sort of expansion of the approximate Ψ in terms of Slater determinants $D_{K\alpha}$:

$$^{YZ}\Psi = \sum_{K,p} {}^{YZ}\phi_K^{(p)} a_{Kp} \tag{9}$$

$$^{YZ}\phi_K^{(p)} = \sum_{\alpha} D_{K\alpha} c_{K\alpha}^{(p)} \tag{10}$$

In Eqs.(9,10), YZ stands for LS (nonrelativistic expansions) or JM (relativistic calculations). The Clebsch-Gordan coefficients $c_{K\alpha}^{(p)}$ in (10) insure that $^{YZ}\phi_K^{(p)}$ has proper Y-Z symmetry, whereas the a_{Kp}'s are linear coefficients. In principle, the superscript p runs over all degenerate elements (all possible internal couplings between the electrons) corresponding to configuration K and symmetry Y-Z. In practice, conveniently small partitions[15] of these degenerate spaces are achieved, resulting in crucial simplifications for all open-shell states.

In nonrelativistic calculations, the spinorbitals $\psi_{i\ell m s}$ are necessarily restricted as indicated below:

$$\psi_{i\ell m s} = R_{i\ell}(r) \cdot Y_{\ell m}(\theta,\phi) \cdot \sigma_s \tag{11}$$

as otherwise the imposition of Y-Z symmetry becomes intractable.[29]

In S.C.F. methods,[16,30] the radial orbitals $R_{i\ell}$ are obtained as solutions of integrodifferential equations. In the standard configuration interaction (C.I.) method, on the other hand, orthonormal $R_{i\ell}$'s are expanded in a primitive basis $P_{j\ell}(r)$:

$$R_{i\ell}(r) = \sum_j P_{j\ell}(r) \cdot b_{j\ell i} \tag{12}$$

In Eq.(12) the $P_{j\ell}$'s are defined in terms of nonlinear parameters which are usually energy optimized by brute force. The coefficients $b_{j\ell i}$ are eventually chosen to optimize the convergence of Eqs.(9,10).[31]

At present, relativistic S.C.F.C.I. methods[17] are unable to achieve spectroscopic accuracy for low-Z atoms, or for any system where correlation effects are important. The nonrelativistic counterpart[16] can go much further than R.S.C.F.C.I. but it cannot give Δ's as accurate as one may get by using a standard C.I. approach.[32] (The extant reasons will become evident in Secs. 7 and 8.)

5. <u>Cancellation of E_{rel} and $\tilde{E}_{rad}$</u>

Although in principle E_{rel} can be calculated straightforwardly and $\tilde{E}_{rad}$ can be estimated somehow or other, it is more convenient, from the point of view of accuracy, to look for transition energies ΔE where

$\Delta E_{rel} + \Delta \tilde{E}_{rad}$ is negligible. In these cases $\Delta E = \Delta E_{nr}$. For example, let us consider the calculation of the absolute term value of Li 1s2s2p $^4P^\circ_{3/2}$, illustrated in Fig. 1. The Li 1s2s3s ^4S state has a 1s2s ^3S

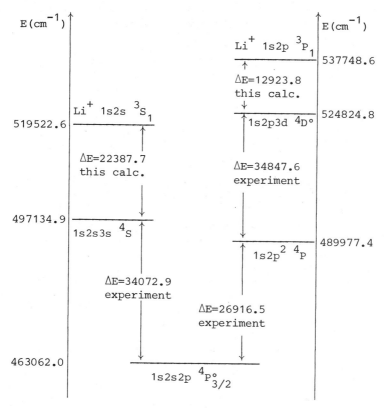

Fig. 1. Two different paths leading to Li 1s2s2p $^4P^\circ_{3/2}$, taking advantage of $E_{rel} + \tilde{E}_{rad}$ cancellations.

inner core very similar to the one in Li$^+$ 1s2s ^3S. Because the contribution of the outer electron to $E_{rel} + \tilde{E}_{rad}$ is insignificant, we assume $\Delta E = \Delta E_{nr}$ which we calculate[33] to be equal to 22387.7±0.7 cm^{-1}. A similar argument is used to calculate ΔE between Li$^+$ 1s2p ^3P and Li 1s2p3d $^4D^\circ$. The Li 1s2s2p $^4P^\circ_{3/2}$ energies calculated by the two paths differ only in 1.3 cm^{-1}, which is smaller than the experimental uncertainty in the $^4D^\circ \rightarrow ^4P$ transition. Furthermore, $E_{rel} + \tilde{E}_{rad}$ is equal to 118 and 108 cm^{-1} in Li$^+$ ^3S and ^3P, respectively,[34] clearly showing that these effects cannot be neglected. This strategy also works when the cores are linear combinations of L-S states.[35]

6. Exact expression for Δ

One can always write E_{nr} as a sum of a variational upper bound E_u, plus the error due to the truncation of the one-electron basis, T_1, plus T_N, the error due to truncations in the N-electron basis:

$$E_{nr} = E_u + T_1 + T_N . \tag{13}$$

We now write T_1 as:

$$T_1 = T_1^\circ + \tilde{T}_1 \pm \delta T_1 . \tag{14}$$

T_1° contains errors usually localized in the inner shells and it is expected to *cancel* with another T_1° from a level with a similar inner shell. $\tilde{T}_1 \pm \delta \tilde{T}_1$ is actually computed as described in the next section.
Similarly, T_N may be written as:

$$T_N = T_N^\circ + \tilde{T}_N \pm \delta \tilde{T}_N \tag{15}$$

In Eq.(15), the T_N° term, which one expects to cancel, involves *deep* inner shells. Interactions between the outer shell and the peel of the core are usually included in E_u or in $\tilde{T}_N$.

7. Estimate of $\tilde{T}_1$

Twenty years ago, in a landmark paper, Sinanoglu[36] suggested that C.I. expansions could be calculated "by parts." Here we will use C.I. by parts as an *intermediate step* to develop the primitive basis $P_{j\ell}$, Eq.(12), and to estimate $\tilde{T}_1$.

Let us consider an expansion of Φ in terms of a reference configuration Φ_o, singly excited configurations Φ_i^a (reference orbital i substituted by orbital a), and doubly excited configurations Φ_{ij}^{ab}:

$$\Phi = \Phi_o + \sum_{ia} \Phi_i^a c_i^a + \sum_{ij} \sum_{ab} \Phi_{ij}^{ab} c_{ij}^{ab} . \tag{16}$$

The terms ij and jk in (16) are visualized in Fig. 2. A ray in Fig. 2,

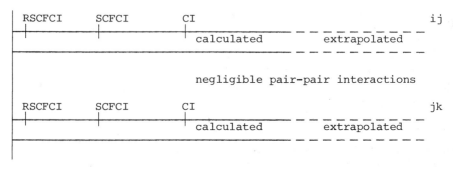

Fig. 2. Schematic representation of the terms ij and jk in Eq.(16).

like the one denoted by "ij", represents a pair-expansion $\sum_{ab} \Phi_{ij}^{ab} c_{ij}^{ab}$ ordered according to decreasing importance of its terms. We note the following:

a) RSCFCI cannot go very far in the pair-expansion because of S.C.F. convergence problems.

b) The same S.C.F. convergence problems arise in nonrelativistic SCFCI at a much later stage. Yet the impracticality of adding further terms to pair-expansions prevents the calculation of accurate truncation errors.

c) Standard C.I.[14,15,37] allows for larger C.I. expansions. After a certain point, when pair-pair interactions become negligible, it does not make sense to keep adding $P_{j\ell}$'s to expand the *whole* Ψ. We just calculate explicitly the energy effect T_1^{ij}(calc.) using energy-optimized $P_{j\ell}$'s which are subsequently ignored, thus keeping the size of the one-electron basis at a reasonable size. The truncation energy $\tilde{T}_1^{ij}$ for the pair-expansion "ij" is written as:

$$\tilde{T}_1^{ij} = T_1^{ij}(\text{calc.}) + T_1^{ij}(\text{extrap.}) \pm \delta T_1^{ij} \tag{17}$$

where $T_1^{ij}(\text{extrap.}) \pm \delta T_1^{ij}$ is estimated from patterns of convergence[38] uncovered in the calculations of $T_1^{ij}(\text{calc.})$.

d) The total truncation error $\tilde{T}_1$ may then be written as:

$$\tilde{T}_1 = \sum_{ij} T_1^{ij} + \sum_{ijkl} T_1^{ijkl} + \dots \tag{18}$$

The terms T_1^{ijkl} may be estimated by perturbation theory.[18,32] In Be ground state, for example, even with a large 10s9p8d7f5g3h1i basis of radial orbitals, it amounts to 50 cm^{-1}. Since these terms are tedious to calculate, one should avoid ground states and other situations where such terms do not cancel. For example, the calculation of ionization potentials should proceed through highly excited states, which are easy to calculate and which are tied to the ground state through experimentally known transitions.

e) The most accurate and economical way to calculate $\tilde{T}_1$ should be by means of coupled-cluster P.T.[26]

8. Estimate of $\tilde{T}_N$

Given even a moderate-size one-electron basis, the number of N-electron functions that can be generated is prohibitively large to be included in a single variational calculation. Thus the need to truncate the C.I. expansion. The following strategy is employed:

a) A master list including all single and double excitations and a *well defined* selection of triples and quadruples is subjected to C.I. by parts. The corresponding truncation error T_N(list) is calculated as:[32]

$$T_N(\text{list}) = \sum_s \Delta T_s . \tag{19}$$

ΔT_s is the difference between two variational energies for stage "s" in the search through the configuration list, and it is given by:

$$\Delta T_s = E_{bt} - E_{at} \tag{20}$$

where E_{bt} and E_{at} are the energies before and after truncation, respectively. The selection of deleted terms is guided by approximate formulas for partial energy contributions.[39] Also, the selection of configurations to be included at each stage involves a delicate judgment.[18]

b) The effect of all unlinked triples and quadruples not included in the master list, T_N(no list), is estimated by perturbation theory.[18]

c) All *linked* triples and quadruples not present in the master list are assumed to give negligible contributions to $\tilde{T}_N$. This is a reasonable hypothesis when triples and quadruples are ordered according to a hierarchy involving the occupation numbers of the excited orbitals.[18]

d) Finally, $\tilde{T}_N$ is obtained as:

$$\tilde{T}_N = T_N(\text{list}) + T_N(\text{no list}) \tag{21}$$

9. An application: the 3490 Å line of core-excited negative Li

A relatively intense $\lambda = 3490$ Å line in the beam-foil spectrum of core-excited Li defied a correct interpretation until recent calculations[40] of energy levels and transition probabilities, *together* with previous experimental findings[41] led to the conclusion[42] that this line was due to an electric dipole transition between two bound states of core-excited negative lithium. Three subsequent experiments[43] confirmed this *numerically-guided* discovery. The relevant energy levels are shown in Fig. 3.

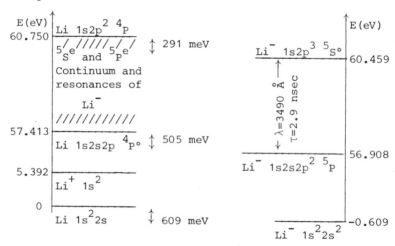

Fig. 3. Energy diagram of bound Li⁻ ions and photodissociation thresholds.

In order to cancel $E_{rel}+\tilde{E}_{rad}$ as much as possible, we first calculate the electron affinities of Li $^4P°$ and ^4P, which define the photodissociation thresholds of Li⁻ ^5P and $^5S°$, respectively. A detailed analysis of the calculations is given in Table 1.

Table 1. Calculation of the electron affinities of core-excited Li, all energies in a.u.(Li). 1 a.u.(Li)=219457.48 cm⁻¹=27.20953(10) eV.

	^5P		$^5S°$
E_u	-5.386 346		-5.255 890
T_1	-0.000 147(33)		-0.000 148(25)
T_N	-0.000 040(5)		-0.000 009(1)
E_{nr}	-5.386 533(38)		-5.256 047(26)
E_{nr} ($^4P°$)	-5.367 992(37)[a]	(^4P)	-5.245 351(37)[a]
E.A.	0.018 541(76)		0.010 696(63)
	=504.5±2.0 meV		=291.0±1.8 meV

a. Ref. 44.

Because these are small systems, T_1 and T_N, Eqs.(14,15), have been calculated entirely (no $T_1°$ or $T_N°$ cancellations needed).

If produced in good yield, metastable Li⁻ 1s2s2p² ^5P ions might provide an alternative energy reservoir[42] in Harris' recent proposal[45] for an x-ray "laser."

10. The core-excited Li doublets

These states have been studied in emission[5⁻10] and in absorption[11⁻13] and there is currently a strong interest[10] in nonautoionizing (1s2p)¹Pnℓ levels in connection with the development of a laser in the 200 Å region. As expected, the (1s2p)¹Pnℓ doublets decay to 1s²nℓ doublets in the 200 Å region more intensely than the (1s2p)³Pnℓ doublets.[7] The latter also decay appreciably into core-excited 1s2snℓ and 1s2p² doublets.[7] Although the branching of (1s2p)¹Pnℓ states into the core-excited system is very small, Harris and coworkers[46] succeeded to observe quite neatly the λ=2845 Å photons from { (1s2p)¹P3d}²D°→1s2p² ²P.

Li {1s(2s2p)³P}²P° is the lowest core-excited doublet that can be observed by absorption from the 1s²2s ground state. The absorption experiment[11] placed its absolute position at 475150 cm⁻¹. This resonant state can also be reached from above through the following steps:[7] Li⁺ 1s2p ³P₁°→{(1s2p)³P3d}²D° by calculation, {(1s2p)³P3d}²D°→1s2p² ²P by experiment (λ=3660.9 Å)[6] and calculation, the former being the more accurate one, and 1s2p² ²P→{1s(2s2p)³P}²P° also by experiment (λ=4590 Å)[5] and calculation. The second path yields an absolute term value of 475147 cm⁻¹, in good agreement with the absorption experiment. A similar analysis can be applied to the 1s2p² ²D resonance.

Finally, let us remark that the third and fourth ²P states are well represented by:[9] $(1s2p)^3 P4p \pm (1s2p)^1 P3p$, offering the first example of mutual perturbations between two series which converge to two different L-S ionic states represented by the same set of orbitals.

11. Quick comments on Li quartets

After many years of confusion, calculations[47] prompted a reinterpretation of transitions involving excited $^4P^\circ$ states, which was later confirmed by experiment[6] and by more accurate calculations.[48] The determination of empirical bounds on lifetimes[44] permitted to predict unambiguously[44] the existence of cascades (confirmed by experiment[6]) and of incipient autoionization in $1s2p3d$ $^4D^\circ$. The knowledge of quartet spectra is quite satisfactory after Mannervik's excellent review.[6]

12. Computational prospects

The most obvious developments ahead seem to be: a) use of nonrelativistic SCFCI to drive RSCFCI more efficiently in the calculation of major orbitals and nearly degenerate ones, b) relativistic pseudo-potentials for inner cores, c) incorporation of the above in a relativistic C.I., d) automatization of many intermediate steps to reduce human errors and boredom, and e) implementation of more efficient techniques of calculation, especially from the field of quantum chemistry.

Acknowledgment

I have received all sorts of enlightenment and friendly stimuli from Gordon Berry, Annik Vivier Bunge, Manuel Galán, Steve Harris, Rocío Jáuregui, Tom Lucatorto, Sven Mannervik and Andy Weiss, to whom I give my thanks.

References

1. C.F.Bunge, Chem.Phys.Lett. 42,141(1976).
2. H.A.Bethe and E.E.Salpeter, *Quantum Mechanics of One and Two Electron Atoms*, Academic, New York (1957).
3. P.Feldman and R.Novick, Phys.Rev. 160,143(1967).
4. W.S.Bickel et al., Phys.Rev. 178,118(1969); J.Opt.Soc.Am. 59,830 (1969); J.P.Buchet et al., Phys.Lett. 28A,529(1969).
5. H.G.Berry et al., Phys.Scr. 3,63(1971); J.Opt.Soc.Am.62,767(1972).
6. S.Mannervik, Phys.Scr. 22,575(1980); J.Bromander et al., J.Phys. (Paris) Colloq. C1, Supplément 40,6(1979).
7. R.Jáuregui and C.F.Bunge, Phys.Rev. A23,1618(1981).
8. J.P.Buchet et al., Phys.Rev. A7,922(1973).
9. C.F.Bunge, Phys.Rev. A19,936(1979).
10. J.R.Willison et al., Phys.Rev.Lett. 44,1125(1980); S.E.Harris et al., G.L.Reports 3289 and 3335 (1981), to be published.
11. D.L.Ederer et al., Phys.Rev.Lett. 25,1537(1970) and unpublished calculations by A.W.Weiss.
12. T.J.McIlrath and T.B.Lucatorto, Phys.Rev.Lett. 38,1390(1977) and unpublished calculations by A.W.Weiss.
13. A.M.Cantú et al., J.Opt.Soc.Am. 67,1030(1977).
14. A.W.Weiss, Phys.Rev. 162,71(1967); Adv.At.Mol.Phys. 9,1(1973).
15. A.Bunge, J.Chem.Phys. 53,20(1970); C.F.Bunge and A.Bunge, Int.J. Quant.Chem. 7,927(1973).

16. Ch.Froese-Fischer, *The Hartree-Fock Method for Atoms*, Wiley, New York (1977).
17. I.P.Grant et al., Comp.Phys.Commun. 21,207,233(1980); ibid., 17 149(1979), and references therein.
18. C.F.Bunge, Phys.Scr. 21,328(1980).
19. P.-O.Löwdin, J.Chem.Phys. 43,S175(1965); F.Weinhold, Adv.Quant. Chem. 6,299(1972).
20. M.T.Anderson and F.Weinhold, Phys.Rev. A9,118(1974); J.S.Sims et al., ibid., A8,2220,2231(1973); A13,242(1976); C.F.Bunge, Int.J.Quant.Chem. 12,343(1977).
21. C.L.Pekeris, Phys.Rev. 112,1649(1958).
22. Y.Accad et al., Phys.Rev. A4,516(1971); J.S.Sims et al., J.Phys B 15,327(1982).
23. S.Larsson, Phys.Rev. 169,49(1968); J.Muszynska et al., Chem.Phys. Lett. 76,136(1980).
24. J.S.Sims and S.A.Hagstrom, Phys.Rev. A4,908(1971).
25. H.P.Kelly, Adv.Chem.Phys. 14,129(1969), Adv.Theor.Phys. 2,75(1968)
26. J.Cizek and J.Paldus, Phys.Scr. 21,251(1980); I.Lindgrén and S.Salomonson, ibid., 21,353(1980).
27. R.K.Nesbet, Adv.Chem.Phys. 14,1(1969).
28. S.Wilson and D.M.Silver, Phys.Rev. A14,1949(1976).
29. C.F.Bunge, Phys.Rev. 154,70(1967).
30. J.Hinze and C.C.J.Roothaan, Prog.Theor.Phys.Supp. 40,36(1967).
31. P.-O.Löwdin, Phys.Rev. 97,1474(1955); C.F.Bunge, ibid., A7,15 (1973).
32. C.F.Bunge, Phys.Rev. A16,1965(1976).
33. C.F.Bunge, Phys.Rev. A23,2060(1981).
34. C.F.Bunge and A.V.Bunge, Phys.Rev. A17,816(1978).
35. M.Galán and C.F.Bunge, Phys.Rev. A23,1624(1981).
36. O.Sinanoglu, J.Chem.Phys. 36,706(1962).
37. F.Sasaki and M.Yoshimine, Phys.Rev. A9,17,26(1974).
38. C.F.Bunge, Theor.Chim.Acta 16,126(1970); D.P.Carroll et al., J.Chem.Phys. 71,4142(1979).
39. I.Shavitt, in *Modern Theoretical Chemistry* (H.F.Schaefer III, ed.) Vol. 3, pp 129-188, Plenum, New York (1977).
40. C.F.Bunge, Phys.Rev. A22,1(1980).
41. H.G.Berry et al., Nucl.Instr.Meth. 90,269(1970); Phys.Scr. 3,63 (1971).
42. C.F.Bunge, Phys.Rev.Lett. 44,1450(1980).
43. S.Mannervik et al., J.Phys.B 13,L441(1980); R.L.Brooks et al., Phys.Rev.Lett. 45,1318(1980); A.Denis and J.Desesquelles, J.Phys. Lett. (Paris), 42,L57(1981).
44. C.F.Bunge and A.V.Bunge, Phys.Rev. A17,822(1978).
45. S.E.Harris, Opt.Lett. 5,1(1980).
46. S.E.Harris et al., this issue.
47. S.Lunell and N.H.F.Beebe, Phys.Scr. 15,268(1977); S.Lunell, ibid., 16,13(1977); T.Ahlenius et al., Phys.Lett. 63A,270(1977).
48. C.F.Bunge, J.Phys. B 14,1(1981).

AUTOIONIZING AND HIGH-LYING RYDBERG STATES
OF LUTETIUM ATOMS

C. M. Miller and N. S. Nogar
Group CNC-2, Mail Stop 738
Los Alamos National Laboratory, Los Alamos, NM 87545

ABSTRACT

Two color multiphoton ionization has been used to examine the spectroscopy of Lu atoms near the ionization threshold of 5.43 electron volts (43800 cm^{-1}).

INTRODUCTION

Sensitive elemental analysis of the rare earth elements, particularly with respect to isotope ratios, is often complicated by the presence of isobars which interfere with mass spectrometric analysis. Resonance or multiphoton ionization spectroscopy offers the opportunity to alleviate these interferences by selectively ionizing the species of interest, using the electronic transitions characteristic of that element. Two-photon, one-color photoionization of lutetium, using the $^2D^0_{3/2}$ intermediate level at 22125 cm^{-1}, has been shown to discriminate against isobaric ytterbium by a factor of at least 50000, while accurately reproducing the lutetium isotope ratio.[1] The use of secondary resonances, such as Rydberg or autoionization levels, offers the possibility of further selectivity, through the simultaneous matching of two resonances, and increased sensitivity, through increased cross sections for absorption near or above the ionization threshold. The intent of the current effort is to elucidate those levels in lutetium which may promote such detection techniques.

EXPERIMENTAL

A XeCl* excimer laser was used to simultaneously pump two dye laser systems for these photoionization experiments. The first, operating with pulse energies of 20 to 100 microjoules, was used to saturate the $5d6s^2 \; ^2D_{3/2} \rightarrow 5d6s6p \; ^2D^0_{3/2}$ transition of lutetium at 22125 cm^{-1}. The second dye laser, with pulse energies up to 2 millijoules, was tuned through the spectral region corresponding to transitions from the $^2D^0_{3/2}$ state to highly excited Rydberg levels and autoionizing states. In both cases, the dye coumarin-460 was utilized, with dye laser bandwidths on the order of 0.2 cm^{-1} over the 4380 to 4780 Å tuning curve.

The dye laser beams were directed antiparallel through the ionization region of a time-of-flight mass spectrometer. Both beams were unfocussed and 3 mm in diameter. Ions thus created were extracted into the drift tube of the apparatus with a pulsed field synchronized to follow the firing of the laser. When high potentials were required for field ionization of Rydberg states, DC extraction was used. Signal processing electronics recorded the lutetium ion signal arriving at the spectrometer's electron multiplier.

Lutetium atoms for photoionization were provided by resistively heating a thin foil of the metal. Source temperatures ranged from 900 to 1200°C.

RESULTS

Figure 1 shows the observed one- (lower) and two- (upper) color, two photon photoionization spectra of lutetium. In the lower trace, a single laser was scanned over the indicated wavelength region; in the upper trace, the second laser at $22\,125$ cm^{-1} (4519 Å) was added. Both traces are shown on the same scale and are uncorrected for the tuning curve of the laser dye. At the laser powers used, optical saturation of several transitions was observed to occur.

Table 1 lists the observed photoionization peaks, with the origin of each peak listed, and cross references to Figure 1. Four autoionization levels are observed, three of which have been previously noted.[2] In two cases, both single- and two-color transitions to the same autoionizing level are seen. Several single-color resonances are also found, corresponding to known lutetium transitions.[3]

A number of single-color resonances are found which have no origin in known lutetium transitions (peaks A-E). Some have significantly enhanced ionization probability with the addition of the 4519 Å laser ("augmented single-color"), and two (peaks D and E) have previously been assigned as autoionizing resonances.[2] The true origin of these signals is currently under investigation.

Transitions to Rydberg levels from the intermediate $^2D^0_{3/2}$ level were also observed, but were few in number due to field ionization limitations in our apparatus.

DISCUSSION

Our results indicate that the use of autoionizing levels in conjunction with resonant intermediate levels can lead to a significant increase in the ionization probability of lutetium. Power studies indicate that this increase may be as large as two orders of magnitude, compared to single-color resonance ionization under similar laser conditions.

In further efforts to increase the analytical sensitivity of the resonance ionization process, calculations have been performed using rate equations modeling.[4] These have confirmed the beneficial effect of the use of autoionizing resonances found experimentally. Results indicate that the applicable absorption cross sections are sufficient for the use of continuous wave laser excitation, with the accompanying increase in the ionization duty cycle.

REFERENCES

1. C. M. Miller, N. S. Nogar, A. J. Gancarz, and W. R. Shields, submitted for publication.
2. G. I. Bekov and E. P. Vidalova-Angelova, Sov. J. Quantum Electron. 11, 137 (1981).
3. W. C. Martin, R. Zalubas, and L. Hagen, Atomic Energy Levels-- The Rare-Earth Elements, NSRDS-NBS 60, NTIS.
4. C. M. Miller and N. S. Nogar, manuscript in preparation.

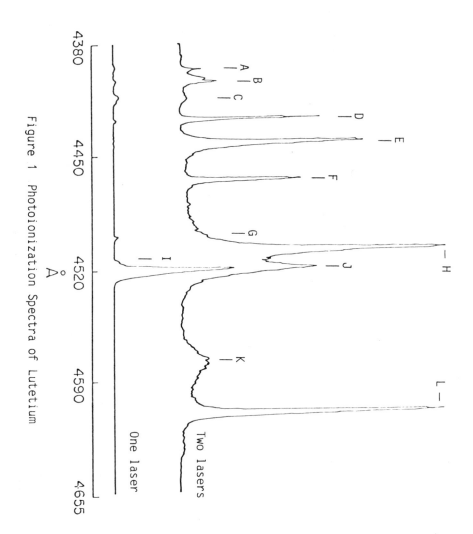

Figure 1 Photoionization Spectra of Lutetium

Table 1 Observed Photoionization Transitions

Label	Wavelength (Å)	Origin
A	4397	Augmented single-color (doublet)
B	4404	Augmented single-color (doublet)
C	4415	Single-color
D	4428	Augmented single-color
E	4443	Augmented single-color
F	4465	Two-color transition to auto-ionizing level at 44515 cm^{-1}
	4480	Single-color resonance for thermally excited lutetium
G	4499	Single-color resonance
H	4508	Two-color transition to auto-ionizing level at 44302 cm^{-1}
I	4513	Single-color transition to auto-ionizing level at 44302 cm^{-1}
J	4519	Single-color resonance (both lasers at same wavelength)
	4562	Single-color transition to auto-ionizing level at 43832 cm^{-1}
K	4576	Two-color transition to auto-ionizing level at 43973 cm^{-1}
L	4606	Two-color transition to auto-ionizing level at 43832 cm^{-1}

VACUUM UV LASER INDUCED FLUORESCENCE OF THE NO MOLECULE

H. Scheingraber and C.R. Vidal
Max-Planck-Institut für Physik und Astrophysik
Institut für extraterrestrische Physik
8046 Garching, W.-Germany

ABSTRACT

Tunable vuv radiation of narrow linewidth in the region between
190 and 200 nm was generated by two-photon resonant sum frequency
mixing of two flashlamp pumped dye lasers. Using this radiation individual rotational vibrational levels around the dissociation limit
of the NO molecule were excited. The subsequent fluorescence was
spectrally resolved using a 1 m vuv grating spectrometer. This technique allows a direct measurement of the Franck-Condon factors, which
were found to be strongly J dependent.

INTRODUCTION

In recent years, coherent tunable narrow bandwidth light sources
have become available in the spectral region below 200 nm[1]. Thus a
lot of the powerful methods of high resolution molecular laser spectroscopy developed in the infrared and visible regime can be extended
into the vacuum UV region. As a first example, we have investigated
some levels of the A and C electronic states of NO around the dissociation limit at 52 400 cm^{-1}.

So far, individual rotational vibrational levels could be excited only by exploiting an accidental coincidence of a strong atomic
emission line and a molecular absorption line. This method was applied to the NO molecule by Broida and Carrington[2] using the resonance
line of ionized cadmium at 214.44 nm and in the vuv region by Hikida
et al.[3] using the 184.9 nm resonance line of a low pressure mercury
lamp. The disadvantage of these experiments was the lack of tunability of the light source. The work of Guest and Lee[4], who measured
absorption and fluorescence of NO between 106 and 200 nm using the
easily tunable synchrotron radiation, suffered from intensity and
bandwidth limitations of the synchrotron.

These difficulties were overcome in the experiments of Banic et
al.[5], who measured radiative life times of several levels of the NO
B'$^2\Delta$ state. They used narrow bandwidth tunable vuv radiation generated by four-wave sum frequency mixing in Mg vapor to measure the
total fluorescence intensity as a function of the excitation frequency.

In our experiment, however, the fluorescent light coming from
a single rotational vibrational level was spectrally resolved by
means of a 1 m vuv grating monochromator. This technique allows a
direct measurement of the transition probability of the observed
fluorescent transitions.

EXPERIMENTAL

The coherent vuv radiation in our experiment was generated by two-photon resonant up-conversion of two flashlamp pumped dye lasers. The output power of the lasers was 20 kW per pulse with a pulse duration of 600 nsec and a linewidth of 0.3 cm^{-1}. As a nonlinear medium strontium vapor was used while xenon served as the phase-matching medium. This mixture was prepared in a concentric heat pipe oven in the crossed tube design[6]. The tuning range of the system using rhodamin 19 and rhodamin 6G dyes was 191-197 nm. This tuning range covered the whole (3,0) band of the $A^2\Sigma - X^2\Pi$ system and most parts of the (0,0) band of the $C^2\Pi - X^2\Pi$ system of the NO-molecule.

The signal intensity at the exit slit of the spectrometer is typically four orders of magnitude lower than the intensity of a simple excitation spectrum. Furthermore, due to the rather strong self-quenching of NO[7], the fluorescence measurements had to be performed at pressure below 0.1 Torr. In order to exploit a sufficient amount of energy in the fluorescent sample, the effective optical pathlength had to be raised by using a multiple reflection cell in front of the entrance slit of the spectrometer. In this manner the signal intensity was increased by a factor of five with respect to a simple fluorescence cell.

RESULTS

In Fig. 1 a fluorescence spectrum originating from a single rotational level of the $A^2\Sigma$ state with v'=3 is shown. The whole v"-progression up to v"=11 can be seen. Each transition consists of two lines corresponding to the doublet components $^2\Pi_{1/2}$ and $^2\Pi_{3/2}$ of the NO ground state. Since the spectral sensitivity of the system was calibrated, the Franck-Condon factors of these transitions can be directly extracted from such spectra. The comparison of the measured Franck-Condon factors with calculated values from Spindler et al.[8] is shown in Fig. 2. The significant discrepancies are due to the R-dependence of the electronic transition moment, which has been neglected in the calculations.

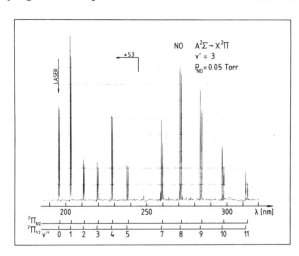

Fig. 1. Spectrally resolved fluorescent light originating from a single rotational level of the $A^2\Sigma$ state with v'=3 of the NO-molecule.

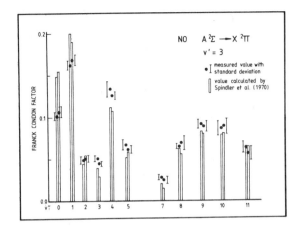

Fig. 2. Comparison of measured and calculated   Franck-Condon factors for the
v" progression with v'=3 of the $A^2\Sigma$ -
$X^2\Pi$ system. The deviations are due to an
R-dependence of the electronic transition moment, which was not taken into
account in the calculations.

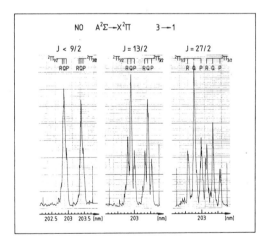

Fig. 3. Fluorescence spectrum of the
(3,1) band of the $A^2\Sigma$ - $X^2\Pi$ system for
different total angular momenta J. The
change from Hund's coupling case (a)
to Hund's coupling case (b) with increasing values of J gives rise to a change in
the relative rotational line strength.

For the (3,1) band
of the $A^2\Sigma$ - $X^2\Pi$ system
spectra for different J
values are recorded. Such
spectra are shown in Fig.
3. The total angular mo-
mentum J was determined
using infrared measure-
ments of Amiot et al.[9].
With increasing angular
momentum J the rotation-
al splitting becomes com-
parable with the fine
structure splitting of the
ground state. This causes
a transition from Hund's
coupling case (a) to
Hund's coupling case (b)
for increasing values of
J. The change in the
coupling of the angular
momenta can be seen by
the change of the rota-
tional line strength for
the doublet components in
Fig. 3.

Pumping the $C^2\Pi$ state
with v'=0 the fluorescence
spectrum obtained, looks
very similar to the spec-
trum reported by Groth et
al.[10]. They populated
levels with v'=0 of the
NO $C^2\Pi$ system by an inverse predissociation
process[11]. Such an experiment, however, is not se-
lectiv with respect to
the rotational vibration-
al levels, and the reso-
lution of this experiment
was not sufficient to re-
solve the multiplet com-
ponents of the ground
state. The measured intensity distribution, there-
fore, is the sum of several different rotational
vibrational levels.

In the selective op-
tical excitation of our
experiment a sufficient

fluorescence signal was obtained only for small values of J. Furthermore, the intensity distribution of the doublet components was extremely sensitive to the particular J value. Since the range of the
angular momenta J was very small, these intensity variations cannot
be due to a change in Hund's coupling case but originate from perturbations[11,12]. The perturbing state is known to be the $B^2\pi$ state
of NO. For higher values of J, a strong predissociation via $a^4\pi$
takes places[11] and lowers the fluorescence intensity.

The fluorescence spectrum obtained by pumping the $C^2\pi$ state with
v'=0 shows besides the δ-bands due to the $C^2\pi - X^2\pi$ system also the
γ-bands due to the $A^2\Sigma - X^2\pi$ system. The $A^2\Sigma$ state with v'=0 is populated by an infrared transition from the $C^2\pi$ state to the $A^2\Sigma$ electronic state[11,12]. Due to the almost identical internuclear distances
of these two potential energy curves only transitions between levels
with identical vibrational quantum numbers take place. As a result
the vuv fluorescence spectrum is simple and the branching ratio for
the $C^2\pi - X^2\pi$ and $C^2\pi - A^2\Sigma$ transitions could be determined.

REFERENCES

1. C.R. Vidal, Appl. Opt. 19, 3897 (1980).
2. H.P. Broida and T. Carrington, J. Chem. Phys. 38, 136 (1963).
3. T. Hikida, N. Washida, S. Nakajima, S. Yagi, T. Ichimura, and
    Y. Mori, J. Chem. Phys. 63, 5470 (1975).
4. J.A. Guest and L.C. Lee, J. Phys. B, 14, 3401 (1981).
5. J.R. Banic, R.H. Lipson, T. Efthimiopoulos, and B.P. Stoicheff,
    Opt. Letters 6, 461 (1981).
6. H. Scheingraber and C.R. Vidal, Rev. Sci. Instrum. 52, 1010
    (1981).
7. A.B. Callear and I.W.M. Smith, Trans. Faraday Soc. 59, 1720
    (1963).
8. R.J. Spindler, Jr., L. Isaacson, and T. Wentink, Jr.,
    J. Quant. Spectr. Rad. Transf. 10, 621 (1970).
9. C. Amiot, R. Bacis, and G. Guelachvili, Can. J. Phys. 56, 251
    (1978).
10. W. Groth, D. Kley, and V. Schurath, J. Quant. Spectr. Rad.
    Transf. 11, 1475 (1971).
11. F. Ackermann and E. Miescher, J. Mol. Spectrosc. 31, 400 (1969).
12. T.W. Dingle, P.A. Freedman, B. Gelernt, W.J. Jones, and
    I.W.M. Smith, Chem. Phys. 8, 171 (1975).

HIGHLY ROTATIONALLY EXCITED CO SPECTRUM[*]

A. V. Smith & P. Ho
Sandia National Laboratories, Albuquerque, NM 87185

ABSTRACT

The rotational spectrum of the $X^1\Sigma^+(v=0) - A^1\Pi(v=1)$ transition of CO is measured and analyzed for $25 \leq J \leq 62$. The rotationally hot CO is generated by photolysis of formaldehyde (H_2CO).

Formaldehyde (H_2CO) irradiated with 355 nm light is known to dissociate with high quantum yield into H_2 and CO ([1]). A substantial portion of the 3.5 ev excess can appear as rotational energy of the CO molecule, producing ground state $X^1\Sigma^+$ (v=0) CO in rotational levels with J greater than 60.

To obtain the rotational spectrum for $25 \leq J \leq 62$ of the $X^1\Sigma^+$ (v=0) - $A^1\Pi$ (v=1) transition, we photolyze .3 torr formaldehyde in a static gas cell with 25 mJ of 355 nm third harmonic of one Nd: YAG laser. A second Nd: YAG laser, delayed 250 ns, pumps two dye lasers whose outputs are mixed in Mg vapor to produce coherent light with a line width of 1.0 cm^{-1} tunable from 151 to 155 nm ([2],[3]). A sample spectrum is shown in Fig. 1.

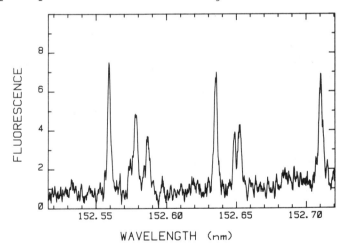

Fig. 1. Sample portion of CO spectrum. Lines are Q(44-46), P(40,41), R(49,50).

---

[*]This work performed at Sandia National Laboratories supported by the U.S. Department of Energy under contract number DE-AC04-76DP00789.

The observed spectrum from J = 25 to 62 is clean with no lines from other transitions. Limits on spectral range are set at the high J end by interference from the $e^3\Sigma^-$ (v=1) system and diminishing fluorescence due to declining population of high J states, and at the low J end by build-up of thermalized CO in the formaldehyde cell.

Numerous perturbations in the range $25 \leq J \leq 62$ are observed and assigned to known states of CO. Figure 2 is a plot of the perturbations for the Q branch of the X-A transition.

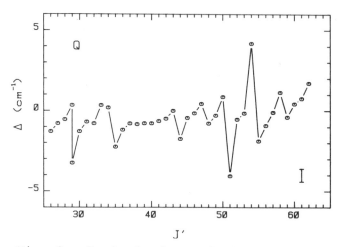

Fig. 2.   Perturbations of the Q branch from
J' = 26 to 62.

Assignment of the perturbations is unambiguous but some adjustment of rotational constants of perturbing levels is necessary.

REFERENCES

1.   J. H. Clark, C. B. Moore, and N. S. Nogar, J. Chem. Phys. 68, 1264-1271 (1978).
2.   S. C. Wallace and G. Zdasiuk, App. Phys. Lett. 28, 449-451 (1976).
3.   H. Junginger, H. B. Puell, H. Scheingraber, and C. R. Vidal, IEEE Journal of Quant. Elect. QE-16, 1132-1137, (1980).

Laser photoionization detection of impurities in solids and studies of hyperfine and isotope structures of radioactive atoms

G. I. Bekov, V. S. Letokhov, and V. I. Mishin*

Institute of Spectroscopy, USSR Academy of Sciences

142092 Troitzk, Moscow region

More than a decade ago the first experiments of atomic photoionization using lasers were made. With this method a number of problems in spectroscopy and its applications may be studied. These include high lying excited states, autoionization, isotope structure and isotope separation.

This report will describe the three beam laser installation for on- and off-line mass-separator experiments with isotopes far from stability. These experiments were carried out in collaboration with the Leningrad Nuclear Physics Institute at Gatchina. Also, in the second part of the report preliminary results for laser photoionization detection of aluminum and sodium traces in solid germanium will be presented.

Isotopes are prepared in the following way (Fig. 1a). The tantalum target is heated to a temperature of 2000-3000 °C and irradiated by a beam of accelerated protons. As a result of the interaction of the target with protons, a number of isotopes including unstable ones are produced. The spatial separation of isotopes is carried out by the mass-separator. Ions with energy of about 30 keV enter different regions of the tantalum foil and are implanted in the foil at the depth of about 100 Å (Fig. 1a).

---

*Guest Worker, National Bureau of Standards, Washington, D.C. 20234
ISSN:0094-243X/82/900101-11$3.00

There are three different ways for laser probing of the unstable isotopes. The first one (off-line) is when part of the foil with implant isotopes is cut off and placed into the hot crucible (Fig. 1b). At temperatures of more than 1200 °C isotopes are evaporated from the foil and enter the interaction region where they are photoionized by laser radiation. In the second method (on-line) a beam of separated radioacti ions enters the crucible from the top and are implanted at once in the hot crucible, neutralized, and enter the interaction region as above. The third way is by using the laser photoionization of isotopes in front of the mass-separator as an ion source instead of nonselective surface or plasma ion sources.

The lanthanide series of isotopes can be ionized very successfully by three step laser impulse excitation. In this technique the atom is raised to a highly excited, so-called Rydberg state, close to the ioniza limit or to an autoionization state above it. For high ion yields it is necessary to use high repetition frequency tuned pulse lasers because of the duty cycle. Fig. 2a shows schematically the laser systems which we are using at the mass-separator at Gatchina.

The pumping lasers are two manufactured copper vapor sealed-off selfheating lasers. The average power of each laser is 6W and the repetition rate is 10 kHz. We used transversally pumped dye lasers. An ethanol dye solution was pumped transversally with respect to both pumping and laser beams with a 1 liter/min flow rate through a rectangul 0.1 x 1.5 cm^2 slit in the cell made from four rectangular quartz plates. The output mirror of the laser cavity was a glass wedge 4 cm from the cell. A spherical lens was placed on the other side of the cell at the distance for improving the light beam divergence. After the lens

ed a six prism beam expander with 27 times magnification. A diffraction

ng with 1200 grooves/mm using Littrov's scheme in first order was

d after the prism. The dye lasers had output efficiencies from 10

% and bandwidths of 0.8 cm^{-1}. The power of these lasers as a

ion of wavelength is depicted in Fig. 2b. Using a dye amplifier

consisted of the same cell and a cylindrical lens which focused

umping beam, the dye laser power could be increased to the 1W level

an amplification efficiency equal to 30%. The frequency of the

ful light beam could be doubled by a KDP crystal with 5% efficiency.

d 20 mW average power of UV radiation from 400 mW input power of

laser beam. When a Fabry-Perot etalon with 0.5 cm air spacing and

sse equal to 15 was placed inside the laser cavity between the

nd and third prisms from the lens the bandwidth of the laser lines

eased to 0.015 cm^{-1}. The grating, etalon and prism expander were

osed in a pressure chamber for smooth wavelength tuning. First step

tation was produced by the narrow band dye laser and the following

s of excitation were produced by broadband ones.

Fig. 3 shows the resonance from the stable 176-ytterbium isotope

rded by three step ionization through the following chain of transitions

$$_0-6p^3P_1(\lambda_1 = 555.6 \text{ nm}),\ 6p^3P_1-7s^3S_1\ (\lambda_2 = 680.3 \text{ nm}),\ 7s^3S_1-17p^3P_1(\lambda_3 = 584.3 \text{ nm})$$

he 17p Rydberg state.[1] The Rydberg state was ionized by an applied

tric field of 15 kV/cm. The photoion yield, that is, the ratio of

recorded number of photoions to the number of isotope atoms in the foil is equal to 10^{-3}-5 x 10^{-4}. For estimating the sensitivity of the method it should be said that the noise was equal to one electrical pulse from the electron channel multiplier per second. With this syste a flow of unstable isotopes below 10^3 isotopes per second can be studie spectroscopically.

At this part of the report will be presented results of analytical applications of the laser photoionization method. For detecting impuri in a substance by an atomic approach it is necessary first, to atomize sample, second, to introduce the atomic cloud in the detection region and finally to detect impurities.

All analysis was produced by us in vacuum (Fig. 4a). Fig. 4b show the high temperature atomizer, which reaches a temperature of up to 3000 °C inside the 0.3 cm inner diameter, 4 cm length graphite tube. The thermal ion suppressor reduced the background to a 4 pulses per second level. The photoionization of impurities was produced in the atomic beam at 5 cm from the end of the graphite tube between two plate electrodes.

We detected sodium and aluminum in solid germanium using two step excitation of atoms to the Rydberg states and ionization of these atoms by an impulsive electric field.[2] Using two dye lasers of the type described above but with N_2-laser pumping, we excited sodium to the 13d state through the 3s-3p and 3p-13d transitions (λ_1 = 589.0 nm, λ_2 = 419.6 nm). Aluminum was excited through the 3p-4s and 4s-17p transitio (λ_1 = 396.15 nm, λ_2 = 447.4 nm). The energy of the laser pulses was about 10 μJ and the repetition rate was 10 pps.

The calibration of the registration part of the installation was realized by two methods; (a) using a totally controlled reference atomic beam of the same element as the impurity and (b) using the sample with the determined impurities.

The next two figures, 5 and 6, show the experimental results. The upper curves are the temperature variation with time. The middle curves are the ion current from the crucible without any sample. The bottom curves are the ion current with the samples. After each increase in temperature atoms desorbed from the surface of the crucible and the sample are detected. The impurities from the sample appear at later times. In these experiments it was possible to detect in germanium crystals aluminum impurities of magnitude 4×10^{-7} % with a s/n ratio of about 10 and sodium impurities of magnitude 2×10^{-8} %.

There are no problems in increasing the repetition rate of the lasers to 100 pps and the energy of one laser pulse to 1 mJ. This way the ion current can be increased by three orders and the sensitivity of the method should be the same. In our opinion, laser photoionization atomic spectral analysis has the following advantages: (1) sensitivity at the level of single atoms in the irradiation region during one laser pulse, (2) applicability for most elements of the periodic table, (3) operation with natural samples in vacuum without special preparation of samples, (4) separate analysis of surface and bulk impurities.

References

1. G. I. Bekov, V. S. Letokhov, O. I. Matveev, V. I. Mishin, Opt. Lett. 3, 159 (1978).

2. G. I. Bekov, V. S. Letokhov, and V. I. Mishin, Opt. Comm. 23, 85 (1977).

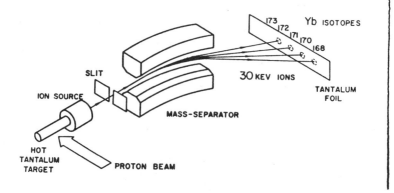

Fig. 1a   Schematic for the preparation of unstable isotopes, for example, ytterbium.

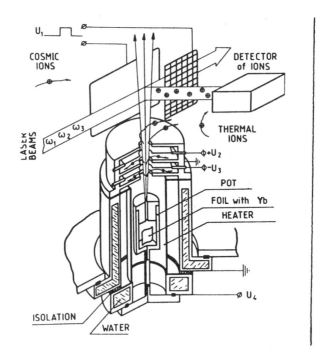

Fig. 1b   Experimental schematic for laser photoionization studies of on- and off-line implanted unstable isotopes.

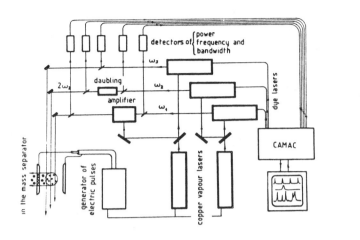

Fig. 2a   Scheme of the three color laser installation for isotope and hyperfine structure studies of unstable isotopes.

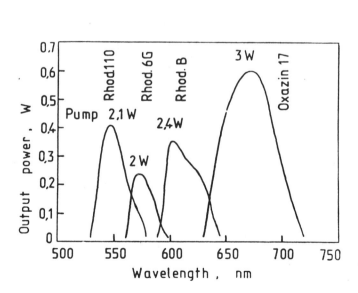

Fig. 2b   Output power of dye lasers pumped by the copper vapor laser.

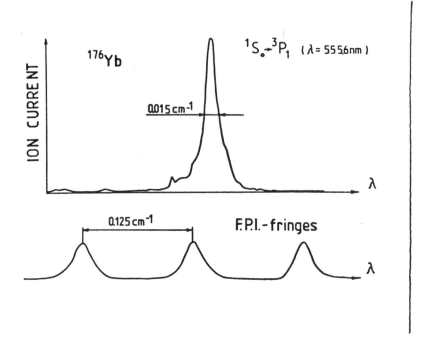

Fig. 3   Resonance of stable ytterbium isotope observed by
scanning the first isotope laser.

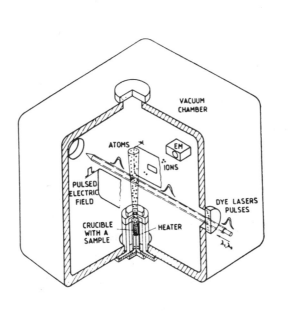

Fig. 4a   The vacuum chamber for laser photoionization atomic spectral analysis.

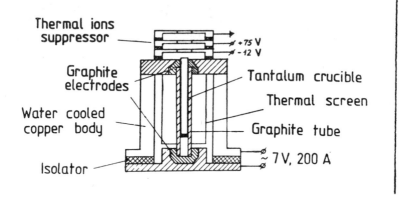

Fig. 4b   Scheme of the high temperature atomizer.

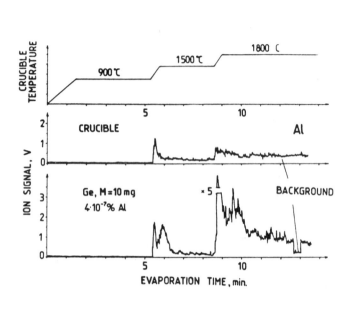

Fig. 5   The temperature of the crucible and ion current as
         a function of time. The center curve shows the
         ion current without sample and the bottom curve the
         ion current with the sample in place. The detection
         of aluminum.

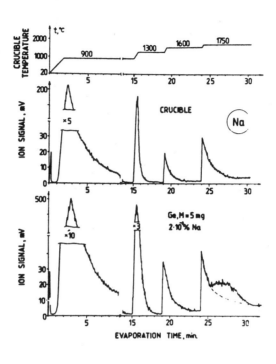

Fig. 6   The same as Fig. 5.   The detection of sodium.

GENERATION OF EXTREME ULTRAVIOLET RADIATION WITH EXCIMER LASERS

C. K. Rhodes
Department of Physics, University of Illinois at Chicago Circle
Post Office Box 4348, Chicago, Illinois 60680 USA

ABSTRACT

Coherent, tunable and intense radiation of pulse duration $\sim$ 10 ps at 193 nm has been used to generate coherent XUV radiation at $\sim$ 64 nm by frequency tripling. The present conversion efficiency (2 x 10^{-5}) may be significantly improved by appropriate phase matching. Tunable radiation at other frequencies can be produced by higher order harmonic conversion and frequency mixing.

SUMMARY

The development of excimer laser systems, particularly rare gas halogen (RGH) media, has marked a significant turning point in the development of coherent sources. Although standard commercially available RGH instruments readily generate radiation of high output power (> 10 MW) at several ultraviolet frequencies, they are far from ideal sources in regard to the frequency control and brightness of the output. Hence, experiments requiring high intensity radiation or narrow bandwidth and precise tuning of the frequency call for significant improvement in the spatial and spectral properties of these lasers. In the past few years, there have been concentrated efforts[1-6] in this direction. With transform limited[5] ArF* and KrF* sources[4], radiation which is $\sim 10^{10}$ better in spectral brightness in comparison with commercial lasers has been produced. In this configuration, it is possible to combine tunability, high output powers, (> 10 MW) narrow spectral width ($\sim$ 200 MHz) and low beam divergence ($\sim$ 30 μrad) in a single RGH system.

The peak power of these RGH systems can be significantly enhanced if the energy is extracted in a pulse duration significantly shorter than the lifetime of the radiating species. Fig. (1) shows the schematic of the apparatus used to obtain radiation with a pulse duration of $\sim$ 10 ps at $\sim$ 193 nm (ArF*). The output of a synchronously pumped, mode-locked dye laser (Coherent Radiation 599-04; $\lambda \sim$ 580 nm, pulse duration < 10 ps) is pulse amplified in a three stage XeF* excimer laser pumped dye amplifier chain. In order to suppress the amplified spontaneous emission, saturable absorber cells (250 μm thick), with DQOCI as the saturable absorber, are installed between two consecutive amplifier stages. A grating-pinhole arrangement suppresses the amplified spontaneous emission transmitted along with the 10 ps signal pulse. In this case, to make the relative pulse stretch caused by the tilt in the wavefront due to diffraction negligible, the grating must be chosen such that the number of lines illuminated on the grating $N \ll \frac{c\tau}{m\lambda}$ where τ is pulse duration, m is the order of diffraction, and λ is the wavelength of the pulse. Fig. (2) shows the autocorrelator trace of the pulse after the amplifier. The duration of this pulse, measured with a background free technique

ISSN:0094-243X/82/900112-05$3.00 Copyright 1982 American Institute of Phys

described by Wyatt and Marinero[7] is $\sim$ 8 ps. The 600 μJ output pulse is focussed into a Sr heat pipe with a 50 cm focal length lens where $\sim$ 2 nJ of the third harmonic of the dye laser signal is produced. This UV radiation at $\sim$ 193 nm is subsequently amplified in a double pass amplifier, spatially filtered and passed through a grating-pinhole arrangement with a 0.8 Å bandpass. Further amplification in two single pass amplifiers yields $\sim$ 10 mJ of $\sim$ 10 ps radiation.

XUV radiation at $\sim$ 64 nm has been obtained by third harmonic conversion of this short pulse at 193 nm. The UV radiation is focussed with a 1 m focal length lens into H_2 which acts as the nonlinear medium. Self absorption due to the nonlinear medium has been reduced by adopting a simple flow geometry for H_2 and sheathing it with a flow of Ne, which has negligible absorption at 64 nm, at a slightly higher pressure than the H_2. Coherent radiation of peak power $\sim$ 20 kW corresponding to a conversion efficiency $\sim$ 2 x 10^{-5} has been observed at 64 nm. No attempt has yet been made to optimize the conversion efficiency. With appropriate phase matching, a conversion efficiency of $\sim$ 10^{-3} may be obtainable. In addition, radiation at other frequencies can be produced by higher order harmonic conversion and frequency mixing.

Coherent, tunable, intense radiation in the XUV spectral region is useful for applications such as high resolution spectroscopy[8] and surface studies. Recently time resolved x-ray diffraction studies of lattice strain during pulsed laser annealing of implanted semiconductors have been made with nanosecond resolution using synchrotron radiation[9]. At the growth velocities (2 - 5 m/$_s$) believed to be characteristic of recrystallization[10] of annealed material, the solidification of the material within a few atomic diameters of the surface would be expected to occur in a time of $\sim$ 50 ps. With the availability of $\sim$ 10 ps pulses in the XUV, it may be possible to resolve the detailed dynamics of the recrystallization process itself. It must be noted here that even under the present suboptimal conditions, the 20 kW radiation at $\sim$ 64 nm provide us with $\sim$ 10^{11} photons which is very close to the number of photons required (10^{12}) in the XUV to conduct such photoemission studies.[10,11]

In conclusion, a high spectral brightness radiation source at $\sim$ 193 nm has been developed. With this source, tunable, coherent radiation of peak power $\sim$ 20 kW has been generated at $\sim$ 64 nm by frequency tripling. Generation of radiation at other frequencies in the XUV region appears possible through the use of higher harmonic processes and direct photoexcited systems. These radiation sources are expected to feature prominently in many important future scientific applications, particularly in the area of the solid state.

The author acknowledges the essential contributions of many colleagues including T. Srinivasan, H. Egger, H. Pummer, T. S. Luk, and K. Boyer in addition to the expert technical assistance of J. Wright, S. Vendetta, and M. Scaggs. This research was supported by the Air Force Office of Scientific Research, the Office of Naval Research, the National Science Foundation under grant no. PHY78-27610 and the Department of Energy under contract nos. DE-AC02-79ER10350, DE-AC02-80ET33065, and DE-AC08-81DP40142.

REFERENCES

1. J. C. White, J. Bokor, R. R. Freeman, and D. Henderson, Opt. Lett. 6, 293 (1981).

2. J. Goldhar, W. R. Rapoport and J. R. Murray, IEEE J. Quantum Electron. QE-16, 235 (1980).

3. T. R. Loree, K. B. Butterfield and D. L. Barker, Appl. Phys. Lett. 32, 171 (1978).

4. R. T. Hawkins, H. Egger, J. Bokor and C. K. Rhodes, Appl. Phys. Lett. 36, 391 (1980).

5. H. Egger, T. Srinivasan, K. Hohla, H. Scheingraber, C. R. Vidal, H. Pummer and C. K. Rhodes, Appl. Phys. Lett. 39, 37 (1981).

6. R. G. Caro and M. C. Gower, Opt. Lett. 6, 557 (1981).

7. R. Wyatt and E. E. Marinero, Appl. Phys. Lett. 25, 297 (1981).

8. M. Rothschild, H. Egger, R. T. Hawkins, J. Bokor, H. Pummer and C. K. Rhodes, Phys. Rev. A23, 206 (1981).

9. B. C. Larson, C. W. White and T. S. Noggle, Phys. Rev. Lett. 48, 337 (1982).

10. Photoemission in Solids I, edited by M. Cardona, and L. Ley (Springer - Verlag, Berlin, 1978) p. 1.

11. C. K. Rhodes, "Atomic, Molecular, and Condensed Matter Studies Using Ultraviolet Excimer Lasers," in Proceedings of the 29th Midwest Solid State Conference - Novel Materials and Techniques in Condensed Matter (Elsevier North Holland, New York, to be published).

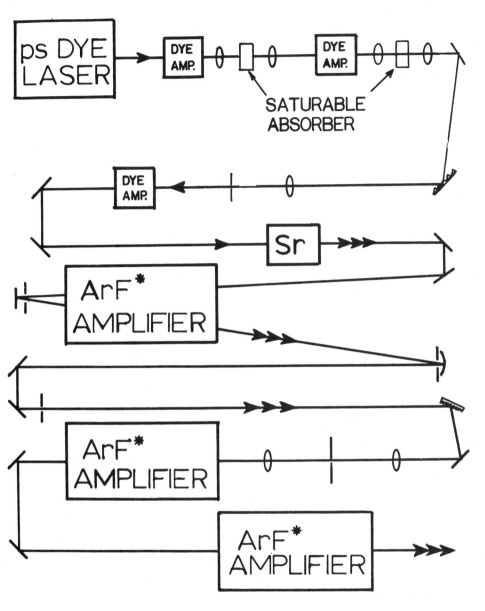

Fig. 1 Schematic of 10 ps, ArF* laser system with dye amplifier
chain, saturable absorber, Sr heat pipe and ArF* amplifiers.

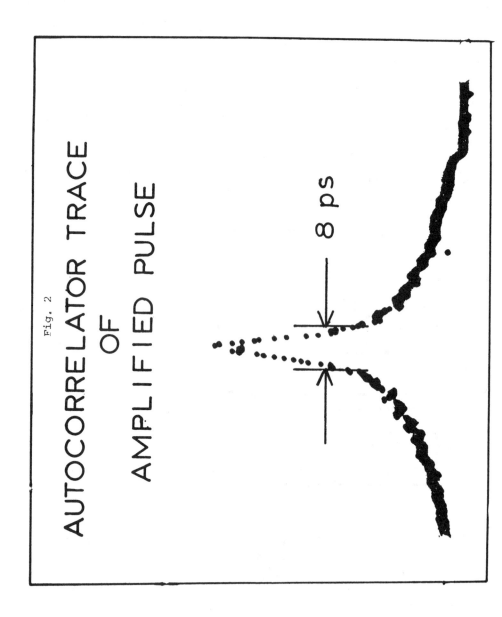

Fig. 2

AUTOCORRELATOR TRACE
OF
AMPLIFIED PULSE

8 ps

ANTI-STOKES RAMAN LASERS

by

J. C. White and D. Henderson
Bell Telephone Laboratories
Holmdel, New Jersey   07733

ABSTRACT

The first observations of UV and VUV anti-Stokes
Raman laser emission are reported. Anti-Stokes Raman
scattering of 532 nm and 355 nm pump lasers using
inverted metastable Tl atoms resulted in stimulated emission at 376 nm and 278 nm respectively. Also, a tunable
178 nm anti-Stokes laser using metastable I atoms has
been constructed.

INTRODUCTION

An anti-Stokes Raman laser may be defined as stimulated anti-Stokes Raman emission induced by a pump laser
between two levels of the same parity in which a population inversion exists between the upper and lower Raman
states. Such laser devices are particularly attractive
since they are tunable by tuning the pump laser; and
because the upper Raman state is often a metastable
level, large inversion densities and high anti-Stokes
output energies are possible. In this paper we report
the first UV and VUV anti-Stokes Raman lasers based upon
metastable $Tl(6p\ ^2P^o_{3/2})$ and $I(5p\ ^5\ ^2P^o_{1/2})$ inversions
created through selective photodissociation of diatomic
molecules.

Stimulated Stokes scattering (i.e., the coherent
Raman frequency down-shifting of a pump laser) has been
extensively investigated and is widely used in the
laboratory to frequency shift various laser sources. The
first observations of stimulated Stokes scattering were
made two decades ago by Woodbury and Ng[1] and Eckhardt
et al.[2] Since that time frequency conversion of high
power lasers has been achieved in a long list of liquids,
solids, and gases.[3] More recently Stokes scattering in
metal vapors has been used as an efficient means to frequency down-convert UV excimer lasers into the visible
spectral region.[4]

The construction and use of Raman lasers based upon stimulated anti-Stokes scattering (i.e., the coherent Raman frequency up-conversion of a pump laser) has received considerably less attention, primarily due to the difficulty in creating the metastable population inversions necessary for practical devices. The first observation of coherent anti-Stokes emission was made by Sorokin et al.[5] in K vapor, where a transient inversion between the $K(4p\ ^2P^o_{3/2})$ and $K(4p\ ^2P^o_{1/2})$ states was created by optical pumping near the $K(4s\ ^2S_{1/2} - 4p\ ^2P^o_{3/2})$ resonance transition. A 58 cm^{-1} Raman up-shift was observed in this case. Some years later by Carman and Lowdermilk[6] attempted to make an anti-Stokes Raman laser using large metastable population inversions. In that experiment measurable gain was observed in a single pass device using inverted I atoms, where the $I*(5p^5\ ^2P^o_{1/2})$ state was populated by flash photolysis of CF_3I. The anti-Stokes Raman gain was observed by pumping the inversion with the fundamental of a Nd:YAG laser at 1.06 μm and probing with a broad-band dye laser at 588 nm. Superfluorescent emission at the anti-Stokes wavelength was not observed, however.

We describe two experiments in which UV and VUV anti-Stokes Raman laser emission has been achieved for the first time. In the first experiment thallium vapor which was inverted by selective photodissociation of TlCl was used as the Raman medium. Using the second (532 nm) and third (355 nm) harmonics of a Nd:YAG laser as pump sources, stimulated anti-Stokes emission from the $Tl(6p\ ^2P^o_{3/2})$ to the $Tl(6p\ ^2P^o_{1/2})$ state was observed at 376 nm and 278 nm, respectively. In the second experiment an $I*(5p^5\ ^2P^o_{1/2})$ population inversion is created with respect to the $I(5p^5\ ^2P^o_{3/2})$ ground state. A 206 nm pump laser is used to drive the Raman process, resulting in stimulated emission at 178 nm with a pulse energy of 35 μJ (i.e., 7 kW).

THALLIUM ANTI-STOKES RAMAN LASER

The possibility of creating a population inversion between the $Tl(6p\ ^2P^o_{3/2})$ metastable state and the ground state by selective TlBr photodissociation at 266 nm was first considered by White and Zdasiuk.[7] Although a population inversion was found not to exist in this case, more recent work by VanVeen et al.[8] using a time of flight analysis of the photofragments of Tl salts produced by excimer laser irradiation has suggested that an inversion in the $Tl(6p\ ^2P^o_{3/2})$ state could be created by photodissociation of TlCl. In the present experiments absorption of an ArF* excimer laser photon at 193 nm results in photodissociation along the paths

$$TlCl + \gamma(193\ nm) \rightarrow Tl^*(6p\ ^2P^o_{3/2}) + Cl$$

or

$$\rightarrow Tl^*(6p\ ^2P^o_{3/2}) + Cl^*$$

where, since the splitting of the Cl ground state is small, both Cl and Cl* are probably produced in an undetermined ratio. Since no $Tl(6p\ ^2P^o_{1/2})$ ground state atoms are produced by this process, the $6p\ ^2P^o_{3/2}$ meta-stable state is inverted with respect to ground.

Absorption of a strong pumping field tuned near a dipole allowed state with opposite parity may connect the metastable and ground states via a two-photon, anti-Stokes Raman scattering process. If the input field is sufficiently strong, stimulated emission at the anti-Stokes wavelength will result. A schematic energy level diagram illustrating this concept is shown for the Tl atom in Fig. 1. In the present experiments anti-Stokes Raman emission was observed at 376 nm and 278 nm with 532 nm and 355 nm pumping lasers, as shown in Fig. 1a and 1b, respectively.

The anti-Stokes Raman gain cross section, σ_R, may be calculated assuming a near resonant, three level approximation. For the Tl case one has, in MKS units[4]

$$\sigma_R = \frac{e^4 f_1 f_2 \nu_R}{32\pi^3 \epsilon_0^2 m^2 hc^2 \nu_1 \nu_2 \Delta\nu^2 \Delta} \tag{1}$$

where f_1 and f_2 are the oscillator strengths connecting the initial and final states to the intermediate level, ν_1 and ν_2 are the respective frequencies, ν_R is the anti-Stokes frequency, $\Delta\nu$ is the virtual detuning, and Δ is the Raman linewidth. With 532 nm pumping as in Figure 1a, one has $f_1 = 0.135$, $f_2 = 0.125$, $\nu_R = 8.0 \times 10^{14}\ sec^{-1}$, $\nu_1 = 7.9 \times 10^{14}\ sec^{-1}$, $\nu_2 = 5.6 \times 10^{14}\ sec^{-1}$, $\Delta\nu = 3.1 \times 10^{12}\ sec$, and $\Delta = 3.0 \times 10^9\ sec^{-1}$, yielding an anti-Stokes gain cross section of

$\sigma_R(376 \text{ nm}) = 3 \times 10^{-23} \text{cm}^4/\text{W}$. Similarly for 355 nm pumping one has $f_1 = 0.036$, $f_2 = 0.145$, $V_R = 1.1 \times 10^{15} \text{ sec}^{-1}$, $V_1 = 1.1 \times 10^{15} \text{ sec}^{-1}$, $V_2 = 8.5 \times 10^{14} \text{ sec}^{-1}$, $\Delta V = 4.3 \times 10^{12} \text{ sec}^{-1}$, and $\Delta = 3.0 \times 10^9 \text{ sec}^{-1}$, which gives $\sigma_R(278 \text{ nm}) = 3 \times 10^{-24} \text{ cm}^4/\text{W}$. The anti-Stokes Raman gain is then calculated an $N\sigma_R I l$, where N is the population inversion density, I is the pump laser intensity, and l is the effective length of the medium.

The basic experimental apparatus is illustrated in Fig. 2. The 193 nm radiation used to dissociate the TlCl was generated using a commercial ArF* excimer laser. Approximately 80 mJ of 193 nm light was focussed into the salt heat pipe oven with a CaF_2 lens. The ArF* laser pulse was ~15 nsec in duration and was focussed to an area of about 3×10^{-2} cm^2 over 25 cm in the TlCl cell. The second and third harmonics of a Nd:YAG laser were used as pump sources and were generated using KDP crystals, yielding pulse energies of up to 300 mJ at 532 nm and 180 mJ at 355 nm. These beams had pulse lengths of approximately 7 nsec at 532 nm and 5-6 nsec at 355 nm. The pump beams were focused with a separate lens and combined with the ArF* laser using a dichroic mirror. The Nd:YAG harmonics were overlapped spatially with the ArF* beam and focussed to an area of 1×10^{-2} cm^2 over the 25 cm TlCl hot zone. In this configuration the pump laser beam area served to define the interaction region and the effective volume for the anti-Stokes Raman laser. Temporal overlap of the two laser beams was controlled using a precision delay generator stable to 1 nsec, thereby allowing the dissociation and pump lasers to be readily synchronized or delayed with respect to one another. The TlCl cell was a simple stainless steel oven with cold, unaligned CaF_2 windows. Argon buffer gas at 30 Torr was used to prevent TlCl vapor condensation on the windows. The cell was operated at 450°C providing a TlCl vapor density of about 6.9×10^{16} molecules/cm^3; however, observation of the stimulated Raman effect did not depend critically on this choice of operating conditions. The anti-Stokes lasers were studied using either a 0.5 meter scanning monochromator and photomultiplier tube combination or were dispersed using quartz prisms for easy energy measurements.

With only the 193 nm dissociating laser present, laser emission along the Tl(7s $^2S_{1/2}$ - 6p $^2P_{1/2}$) resonance line at 377 nm was observed. No laser emission along the Tl(7s $^2S_{1/2}$ - 6p $^2P_{3/2}$) transition was observed consistent with the data that the primary photofragment is the Tl(6p $^2P_{3/2}$) atom. At low (i.e., ≤15 mJ/pulse) 532 nm pump energies and anti-Stokes Raman rate was not

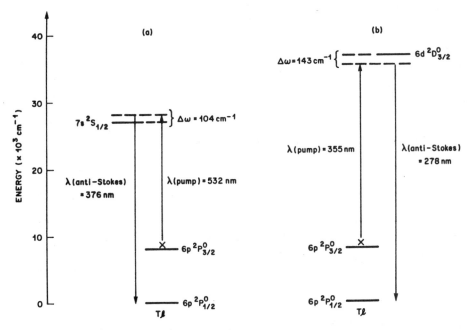

Fig. 1 Pertinent energy levels for anti-Stokes Raman lasing from inverted Tl(6p $^2P^o_{3/2}$) atoms. (a) Raman
emission at 376 nm using a 532 nm pump source.
(b) Raman emission at 278 using a 355 nm pump
source.

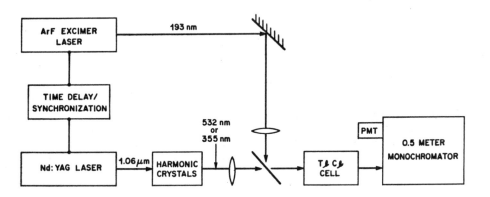

Fig. 2 Experimental apparatus used to observe stimulated
anti-Stokes Raman emission in thallium.

sufficient to compete with the alternate channels which resulted in premature filling of the Tl ground state. At a 532 nm energy of about 15 mJ the Raman process was stimulated and the 377 nm resonance emission disappeared. Above this threshold the Raman laser grew and saturated at a pump energy above approximately 25 mJ. Accounting for the stimulated emission from both ends of the cell, a pulse energy at 376 nm of 1.8 mJ was measured.

Since in principle all the stored energy can be extracted by increasing the pump laser intensity to saturation, the conversion efficiency is best defined in terms of the efficiency of the pumping process. In this manner one may define the efficiency as the ratio of the anti-Stokes output energy to the 193 nm energy absorbed in the active volume. Approximately 26 mJ of the 193nm pump was absorbed in the active volume, implying an efficiency of 7 percent.

The 532 nm laser could be delayed 30-40 nsec after the dissociating laser and still maintain Raman lasing. Under ideal conditions the $6p\ ^2P_{3/2}^o$ level is metastable for a considerably longer time; however, other competing processes, such as the 377 nm laser, tend to fill up the ground state and destroy the population inversion. The use of other dissociating wavelengths may give a more favorable distribution of photofragments and increase the effective storage time.[8]

Similar behavior for the 278 nm Raman laser using 355 nm as the pump wavelength (Fig. 1b) was observed. With an input energy of ~100 mJ/pulse at 355 nm, the output energy at 278 nm from both cell ends was approximately 2.5 mJ/pulse. The conversion efficiency in this case was 10 percent.

IODINE ANTI-STOKES RAMAN LASER

Anti-Stokes Raman laser techniques may also be applied to the construction of tunable VUV lasers. We have used a metastable $I*(5p^5\ ^2P_{1/2}^o)$ inversion to construct, for the first time, a tunable VUV laser based upon stimulated anti-Stokes scattering. A pumping laser at 206 nm was used to drive the Raman process resulting in stimulated emission at 178 nm with peak powers of 7 kW/pulse.

The $I*(5p^5\ ^2P_{1/2}^o)$ metastable population inversion was created by selective photodissociation of NaI at 248 nm using a KrF* excimer laser, via the reaction[9]

$$NaI + \gamma(248 \text{ nm}) \rightarrow Na + I^*(5p^5\ ^2P^o_{1/2})\quad .$$

Since no ground state $I(5p^5\ ^2P^o_{3/2})$ atoms are created by this process, the excited metastable state is inverted with respect to ground. NaI was chosen as the I donor molecule since the ground state Na atom liberated in the dissociation step does not strongly absorb either the 206 nm pump laser or the 178 nm Raman laser. Indeed, both of these wavelengths fall essentially within the single photon ionization Cooper minimum of Na, where the absorption cross section is less than 5×10^{-20} cm^2. Although various organic molecules that have high vapor pressures at room temperature (i.e, CF_3I) might make convenient halogen donors,[6] careful attention must be paid to the absorption characteristics of the residual photofragments.

Absorption of a strong pumping field tuned near 206 nm resonantly couples the I* metastable and ground states through the 6s $^2P_{3/2}$ level. If the input field is sufficiently strong, stimulated anti-Stokes emission at 178 nm will result. A schematic energy level diagram for I illustrating this process is shown in Fig. 3.

The anti-Stokes Raman gain cross section for this process may be calculated using Eq. (1) above. For the iodine case, one has $f_1 = 3.8\times10^{-3}$, $f_2 = 1.3\times10^{-1}$, $\nu = 1.45\times10^{15}$ sec^{-1}, $\nu_2 \equiv \nu_R = 1.68\times10^{15}$ sec^{-1}, and $\triangle = 1.5\times10^9$ sec^{-1}. The gain cross section is then $\sigma_R = 4.4\times10^{-21}\ (\triangle\nu)^{-2}$ cm^4/W, where $\triangle\nu$ is in units of cm^{-1}.

The experimental apparatus is illustrated in Fig. 4. The 248 nm radiation used to dissociated the NaI was generated using a KrF* excimer laser. Approximately 55 mJ of 248 nm light was focussed into the salt oven with a CaF_2 lens. The KrF* laser pulse was ~15 nsec in duration and was focused to an area of about 4×10^{-2} cm^2 over the 25 cm cell hot zone. The focus of the excimer laser was rectangular with a 2:1 aspect ratio. The tunable pump laser at 206 nm was generated as follows. The second harmonic of a Nd:YAG laser was used to pump a rhodamine 590 dye laser to yield 70 mJ/pulse at 555 nm with a measured linewidth of 0.2 cm^{-1}. This radiation was frequency doubled in a KDP crystal thereby generating ~15 mJ at 277 nm. Both the doubled light at 277 nm and the fundamental at 555 nm were focused into an H_2 Raman cell operated at 100 psi. The third anti-Stokes of this process at 206 nm was collected, focused to an area of 3×10^{-3} cm^2 over the 25 cm NaI hot zone, and spatially overlapped with the 248 nm laser. Approximately 75 μJ/pulse of tunable 206 nm radiation was delivered to

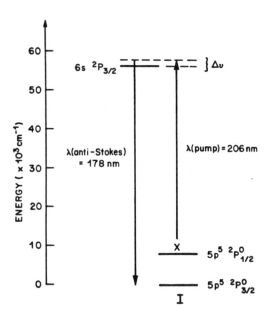

Fig. 3 Pertinent energy levels for anti-Stokes Ramar lasing from inverted I($5p^5\ ^2P^o_{1/2}$) atoms.

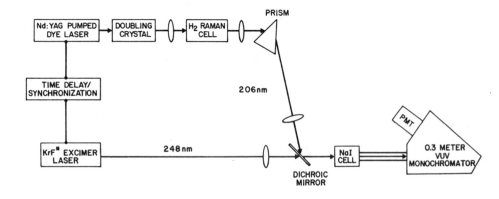

Fig. 4 Experimental apparatus used to observe stimulated
anti-Stokes Raman emission in iodine.

the cell; the pulse length was measured with a fast photodiode to be 2-3 nsec in duration. In this configuration the 206 nm pump beam spot size served to define the
interaction region. Temporal overlap of the two lasers
was controlled using a programmable, precision delay generator stable to better than 1 nsec, thereby allowing
accurate control over the arrival of the laser pulses in
the NaI cell. The NaI cell was a simple stainless steel
oven with cold CaF_2 windows. Argon buffer gas at 30 Torr
was used to prevent NaI vapor condensation on the windows. The cell was operated at 650°C providing a NaI
vapor density of approximately 3.0×10^{15} molecules/cm^3.
An evacuated tube sealed to the cell end served as a
vacuum light pipe coupling to a 0.3 meter VUV monochromator equipped with a photomultiplier tube.

With the 206 nm pump laser tuned 3 cm^{-1} to the blue
side of the $I(6s\ ^2P_{3/2})$ intermediate state, the 178 nm
anti-Stokes Raman laser was readily observed. The output
energy at 178 nm was estimated by comparing the PMT signal at 178 nm to the signal generated by the 206 nm laser
at constant monochromator slit width and photomultiplier
voltage. The 206 nm beam energy was then measured with a
calibrated thermopyle, and the response of the detection
system was normalized for the two different wavelengths
using manufacturers' data for the grating reflectivity
and the photomultiplier quantum efficiency. In this
manner, accounting for stimulated emission from both ends
of the cell, a pulse energy at 178 nm of 35 µJ was measured. The anti-Stokes laser followed the 206 nm pump and
had a pulse width of ~2 nsec, corresponding to an output
power of 7 kW. A total of 28 mJ of the 248 nm beam was
absorbed in the cell, implying an efficiency of 1.7 percent.

The effective storage time of the I* metastable
states was investigated by measuring the intensity of the
178 nm laser as a function of the time delay between the
dissociation and pump lasers. This data is shown in
Fig. 5, where the 206 nm wavelength was adjusted to be
resonant with the $6s\ ^2P_{3/2}$ level. A relative time delay
of zero corresponds to temporal overlap of the 248 nm and
206 nm beams, and a positive delay corresponds to the
206 nm beam arriving in the NaI cell after the 248 nm
dissociation laser. For negative delay the 206 nm laser
arrived prior to the dissociation pulse; and since no I*
metastables were created,[9] the 178 nm laser was not
observed. The 178 nm laser was observed up to a relative
delay of ~0.5 µsec. Past this point the 178 nm laser
emission abruptly terminated due to insufficient metastable population. This storage time is within an order
of magnitude of the diffusion time for I* atoms out of
the interaction region that one can simply estimate

assuming a cross section for I* - Ar collisions of 10^{-16} cm^2. The ability to delay the dissociation and pump lasers argues convincingly that a true I* population inversion was created and that parametric 4-wave processes were not responsible for the observed emission.

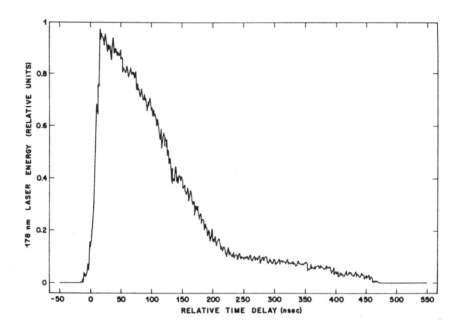

Fig. 5 Laser emission at 178 nm as a function of the relative delay between the 248 nm dissociation pulse and the 206 nm pump pulse. Positive relative delay corresponds to the 206 nm pulse arriving after the 248 nm beam in the NaI cell. The 206 nm wavelength was tuned into resonance with the I(6s $^2P_{3/2}$) level for this plot.

The 206 nm pump laser could be tuned 5-6 cm^{-1} to either side of the 6s $^2P_{3/2}$ intermediate state and still maintain lasing at 178 nm. From this we infer a tuning range of about 10 cm^{-1} for the 178 nm Raman laser. The resolution of the 0.3 meter monochromator used was not sufficient to directly verify that the 178 nm light tuned with the pump laser. However, since the maximum obtainable detuning was many laser linewidths from the 6s $^2P_{3/2}$ resonance, the assertion that the 178 nm emission tuned is reasonable.

SUMMARY

Anti-Stokes Raman lasers, due to their tunability and relatively high conversion efficiency could be useful for upconverting a variety of laser sources. Upconverters for CO_2 lasers could be constructed in alkali atoms to yield radiation in the 400 nm range. By utilizing the other halogen atoms or various metals as storage media, a wide variety of tunable, high power anti-Stokes lasers might be constructed in the 100 nm to 200 nm range. Since one is typically a few hundred to a thousand wavenumbers detuned from the intermediate state, dispersion should be minimum; and the anti-Stokes laser should prove a useful means of up-converting very short pulse length lasers.

A variety of other pumping methods might be utilized to create the necessary population inversions. Incoherent, flash photolysis has already been demonstrated to create very large metastable I populations, and such techniques could be readily applied to metal salts. Laser induced collisional pumping[10] has been proposed as a method of inverting metastable levels, and selective auto-ionization pumping should permit the creation of metastable inversions in ions.

REFERENCES

[1] E. J. Woodbury and W. K. Ng, Proc. IRE **50**, 2367 (1962).

[2] G. Eckhardt, R. W. Hellwarth, F. J. McClung, S. E. Schwartz, D. Weiner, and E. J. Woodbury, Phys. Rev. Lett. **9**, 455 (1962).

[3] Fred M. Johnson in: "Handbook of Lasers," Robert J. Pressley, ed., CRC Press (1971).

[4] N. Djeu and R. Burnham, Appl. Phys. Lett. **30**, 473 (1977).

[5] P. P. Sorokin, N. S. Shiren, J. R. Lankard, E. C. Hammond, and T. G. Kazyaka, Appl. Phys. Lett. **10**, 44 (1967).

[6] R. L. Carman and W. H. Lowdermilk, Phys. Rev. Lett. **33**, 190 (1974).

[7] J. C. White and G. A. Zdasiuk, J. Chem. Phys. **69**, 2256 (1978).

[8] N. J. A. VanVeen, M. S. DeVries, T. Batler, and A. E. DeVries, Chem. Phys. **55**, 371 (1981).

[9] P. Davidovits and D. C. Brodhead, J. Chem. Phys. **46**, 2968 (1967).

[10] R. W. Falcone, Appl. Phys. Lett. **34**, 150 (1979).

RECOMBINATION LASERS IN THE VACUUM ULTRAVIOLET

W. T. Silfvast and O. R. Wood, II
Bell Telephone Laboratories, Holmdel, New Jersey 07733

ABSTRACT

Recent observations of new recombination lasers in expanding plasmas include laser action on the 4f — 3d transition in Li I. Isoelectronic scaling suggests a possible 116.8 nm laser in C IV.

INTRODUCTION

Requirements for producing plasma recombination lasers are described. Theoretical output energy densities for such lasers are shown to exceed those of excimer lasers at wavelengths below 150 nm. Realistic gain scaling arguments are presented that predict useful gains in the VUV. As a specific example, a recombination laser in C IV is proposed which could produce up to 80 mJ of laser output energy at 116.8 nm.

A plasma recombination laser[1] operates by producing ions of an element E in charge state Z (see Fig. 1) and then allowing them to recombine into the next lower charge state (Z-1) in which inversions develop among certain characteristic excited states provided that the plasma density and temperature are in the range where collisional recombination occurs and the electron density does not exceed a specific value. Collisional recombination causes all of the electrons to move downward collisionally to the upper laser level thereby enhancing the quantum efficiency of the excitation process. Since E can be any element, and Z any charge state, a wide range of species is available from which to obtain lasers.

PLASMA FORMATION AND RECOMBINATION

A plasma recombination laser is generally produced in two steps. First, the plasma is formed by some energy input means such as an electrical discharge, an arc, or a laser-produced plasma. This accomplishes the production of state E(Z+) in Fig. 1. Second, the plasma is allowed to cool initiating electron-ion recombination, causing the bound electron to move downward through the excited states of E(Z-1)+ in Fig. 1 and establishing an inversion during the decay.

In order to efficiently produce a laser in ion stage E(Z-1)+, all of the ions must be in ion stage E(Z+) before recombination begins, i.e. as few ions as possible should be in either higher or lower charge

ISSN:0094-243X/82/900128-09$3.00 Copyright 1982 American Institute of Physics

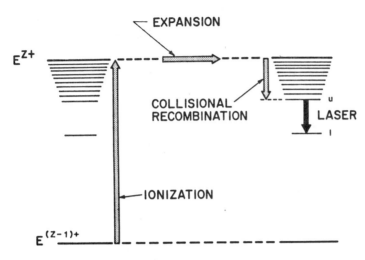

Fig. 1.   Population inversion mechanism for
recombination lasers in expanding plasmas.

states.   This would not be difficult for the case Z=1
since only enough energy to strip off one electron need
be supplied to the plasma.   In all cases it is necessary
to have a high electron density during plasma formation
because higher electron densities (typically 10^{15} to 10^{16}
cm^3 for low Z and higher for higher Z) help produce a
narrow initial distribution of charge states.
Furthermore, at high densities the electron temperature
is more easily adjusted to that required to produce a
specific ion stage because of the increased frequency of
inelastic collisions.   On the other hand, optimum laser
conditions require lower plasma electron densities (10^{13}
to 10^{14} cm^3 for Z=1 and higher for higher Z) than those
necessary for plasma formation.   This is because at the
higher densities electron collisional mixing[2] of upper
and lower laser levels during recombination would
destroy the inversions.

As the plasma cools, recombination moves the
population down to the upper laser level, u.   Inversions
then occur to a lower laser level, l, which has a rapid
decay path.   Since radiative decay rates from the upper
and lower laser levels are typically much faster than
recombination rates, a quasi-steady-state inversion is
formed which lasts for the duration of the recombination
process.   Because of this, energy input times are not
restricted to the lifetimes of the laser levels, and, in
fact, can be several orders of magnitude longer.   This
makes the collisional recombination laser scheme quite
different from other short-wavelength-laser pumping
schemes.   It can be argued that this quasi-steady-state
process will not produce the high gains that a transient

process could, but it will be shown in a later section that this scheme can produce ample gain at short wavelengths.

EXPERIMENTAL REQUIREMENTS FOR RECOMBINATION LASERS

While it is possible to make recombination lasers in a positive column discharge,[3] the requirement of a higher density for production than for inversion, together with the necessity for rapid cooling, can best be satisfied by plasma expansion to reduce the density after the energy input ceases. Rapid cooling of the electrons can be achieved either by adiabatic expansion into a vacuum[4] or by interaction with a background gas.[5] While inversions have been reported in laser-produced plasmas expanding into vacuum,[6] gains have not been high enough to detect significant stimulated emission or laser output. On the other hand, a large number of plasma recombination lasers have been produced[7] by allowing expansion to occur into a background gas that not only cools the electrons but also confines the inversion to a specific spatial region so that the output can be enhanced with a laser resonator.

Several other techniques have been used to optimize plasma recombination lasers. Segmentation of a long cylindrical plasma into a series of spherically expanding plasmas has been shown[8] to increase the output energy of a recombination laser 25-fold. The use of confinement channels[9] or plates[10] to guide the plasma expansion and delay the recombination until the plasma has expanded into the desired spatial region has been shown to enhance recombination and to increase the energy output of a plasma recombination laser more than a hundred-fold.[10] Other experimental techniques, which control the rate of decay of the input energy,[11] have also been shown to significantly enhance the laser output. All of these techniques should be applicable to the production of VUV recombination lasers.

Most theoretical treatments of plasma recombination lasers have considered only hydrogen-like species because the atomic properties of such systems are well known. Experimentally, however, most recombination lasers have been produced in non-H-like species that contain gaps in their energy level structure.[12,13] In such systems, collisional recombination moves essentially all of the population down to the top of these gaps and plays a similar role to that of rotational relaxation in molecular lasers and vibrational relaxation in dye lasers. More than 100 recombination laser transitions have been produced by expanding plasma SPER-type devices in 13 different metals.[7,12,13] These transitions arranged by ion stage and by species (with the number of laser transitions in each species given in parenthesis), are

as follows: Double ions, Al III (2), In III (5), Bi III
(2); Single ions, Mg II (1), Al II (7), Ca II (2), Cu II
(2), Zn II (3), Cd II (6), In II (4), Sn II (5) and Pb
II (4); Neutral atoms, Li I (4), C I (4), Mg I (2), Al I
(2), Ca I (5), Cu I (6), Zn I (7), Ag I (5), Cd I (5),
In I (4), Sn I (6), Pb I (7), Bi I (3). Although many
of these transitions may be potentially useful lasers at
wavelengths ranging from 298.3 to 5460 nm, the
possibility of scaling to higher ion stages and thus
shorter wavelengths is equally important. An example of
the use of scaling to design a short wavelength laser
will be given later in the paper.

ELECTRON DENSITY AND ENERGY OUTPUT LIMITATIONS

An upper limit on the plasma electron density
during recombination must be maintained to prevent
electron collisional mixing from thermalizing the upper
and lower laser levels. By equating the electron
collisional deexcitation rate to the radiative rate for
a hydrogen like level it has been shown[14] that the
electron density limit can be written as a function of
wavelength λ(cm) and electron temperature Te (^{O}K) as:

$$n_e = 0.13 \sqrt{T_e} / \lambda^3 \qquad (1)$$

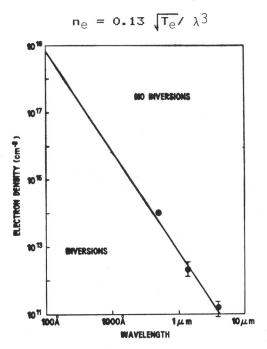

Fig. 2. Maximum electron density in a plasma
recombination laser as a function of wavelength.

The wavelength dependence of the limiting electron

density (given by Eqn. 1) at a temperature of 3000 °K is shown in Fig. 2 as the solid line. Also shown in Fig. 2 (as solid points) are limiting electron densities determined in experiments on recombination lasers operating at three different wavelengths. While this expression was derived for hydrogen-like levels, it agrees with observations in non-H-like systems operating over a decade variation in wavelength. Using this limiting electron density, assuming that all of the ions in stage $E(Z+)$ pass through the laser levels during recombination and contribute to the laser energy, and that the concentration of ions is primarily in $E(Z+)$, the maximum specific output energy, E_{max} (J/cm^3), for a plasma recombination laser as a function of T_e and λ is approximately given by

$$E_{max} = 2.5 \times 10^{-24} \sqrt{T_e} / \lambda^4. \tag{2}$$

This wavelength dependence for a temperature of 3000 °K is shown in Fig. 3 as a solid line. Also shown in Fig.

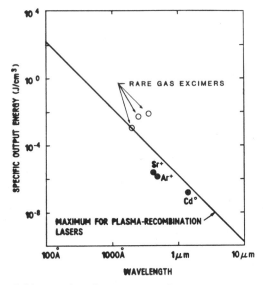

Fig. 3. Specific output energy for plasma-recombination lasers as a function of wavelength.

3 (as solid points) are experimental values for the maximum observed specific output energy for plasma-recombination lasers in Sr II at 373.7 nm,[3] Ar II at 488 nm[15] and Cd I at 1.433 μm.[12] The experimental point for the Cd SPER laser ($E_{max} = 1.5 \times 10^{-7}$ J/cm^3, $\lambda = 1.433$ μm) indicates that this device provided nearly 50 % of the maximum available specific output energy at this wavelength. Even so, the limiting value, 3.2×10^{-7} J/cm^3, is well below that achieved by many existing

infrared lasers. For example, the specific output energy achieved by Nd lasers can be as high as 0.6 J/cm^3 at 1.06 μm[16] and by high-pressure CO2 lasers can be as high as 0.1 J/cm^3 at 10.6 μm.[17] In the near ultraviolet, rare gas halogen excimer lasers (open circles in Fig. 3)[18] are also expected to outperform plasma-recombination lasers. At still shorter wavelengths (150 nm and shorter), however, the performance of plasma-recombination lasers is expected to become equal to or exceed lasers based on all other population inversion mechanisms.

GAIN SCALING WITH Z

The dramatic increase in potential output energy/unit volume for plasma recombination lasers at shorter wavelengths (shown in Fig. 3) is due primarily to the higher electron densities that can be tolerated in higher ion stages at shorter wavelengths before electron collisional mixing thermalizes the upper and lower lasers levels. This increase in electron density leads to an increase in gain at shorter wavelengths. By rewriting the mathematical expression for the small signal gain, G=1.74 x 10^{-6} λ^3 A_{ij} M/T($N_2 - \frac{g_2}{g_1} N_1$) to show the Z dependence of each parameter explicitly, several authors[19,20] have shown that the gain should scale as $Z^{7.5}$. Recently, however, a semi-empirical approach[21] to the gain scaling problem has been taken that involves the introduction of an experimentally measurable parameter, r=N_u/n_e (where N_u is the density of the upper laser level) into the gain expression. Using this approach, it can be shown that G should scale as $rZ^{4.5}$. For the results of the semi-empirical approach to agree with the results of the earlier approach, r would have to scale as Z^3 . Values for r from experiments at a number of wavelengths, however, are inconsistent with a Z^3 scaling but still suggest an attractive enhancement in G with Z.

Following this semi-empirical approach, it is possible to write an approximate gain scaling formula for H-like transitions as

$$G = 3.2 \times 10^{-7} A_{ij}^H r Z^{4.5} \qquad (3)$$

where A_{ij}^H is the transition probability for atomic hydrogen between levels i and j. This approximate expression can be used when the laser wavelength scales as λ_{ij}^H /Z^2 , where λ_{ij}^H is the wavelength of the i to j transition in atomic hydrogen. A similar formula for non-H-like transitions (using the likely 4 - 3 transition as a model) can be written

$$G = 2.0 \ r\sqrt{M} \ Zeff^4 \qquad (4)$$

where M is the atomic mass number and Zeff = $(I_P^E(Z-1)/I_P^H)$ where $I_P^E(Z-1)$ is the ionization potential of the ion stage in which laser action is expected to occur. Eqn. (4) would be most accurate for non-H-like transitions in those cases when $\lambda_{ij}^H/I_P^E(Z-1) = 5 \times 10^{-2}$ (the ratio of the 4-3 transition in H to the ionization potential of H) and would tend to underestimate the gain on transitions characterized by larger ratios.

Experimental measurements of r suggest that an inverse scaling with Z might be more realistic than the Z^3 scaling implied earlier. In SPER-type devices r has been found to be $\simeq 10^{-3}$ for neutral lasers (i.e., Z = 1) and $\simeq 5 \times 10^{-5}$ for single-ion lasers (i.e., recombining from double ions).[22] It is known, however, that the initial double ion species was not being populated efficiently, hence, the value of r in the single-ion SPER device may be artifically low. In the future, laser-produced plasmas are expected to provide a more efficient means of producing higher ion stages. In fact, an r of 4×10^{-5} has already been achieved for the C VI transition at 18.2 nm by Key, et al.[6] in a Nd laser-produced carbon plasma. Hence,values for r between 5×10^{-5} and 10^{-3} can already be achieved in the laboratory and with this in mind, it is possible to examine some potential laser systems.

A RECOMBINATION LASER AT 116.8 nm

One particularly attractive laser transition is the 4f-3d isoelectronic sequence consisting of Li I (1870 nm), Be II (467.5 nm), B III (207.8 nm) and C IV (116.8 nm) shown in Fig. 4. Laser action has already been achieved on the 1870 nm transition in Li I in a single plasma SPER-type device[23] where the gain was measured to be $\simeq 0.01$ cm^{-1} and r was calculated to be 0.5×10^{-3}. Recombination laser action has also been seen in Be II at 467.5 nm in the afterglow of a positive column discharge.[24] This particualr sequence is attractive for recombination lasers for two additional reasons. First, the ratio $\lambda/I_P^E(Z-1)$ is 0.165. This is considerably larger than the ratio for 4-3 transitions in H-like atoms and ions, and means that significantly higher values of ne can be used than could be used in an H-like 4-3 transition in the same species. Consequently, higher values of r and thus higher gain can be achieved. Second, the ionization potential of the next higher ion stage (C 4+) is greater by a factor of 6 than the stage of interest (C 3+) and, as a consequence, during the heating process it should be fairly simple to make C 4+ ions the dominant ion species. This should increase the efficiency of the excitation process and increase the value of r.

The gain for the C IV 116.8 nm transition can be

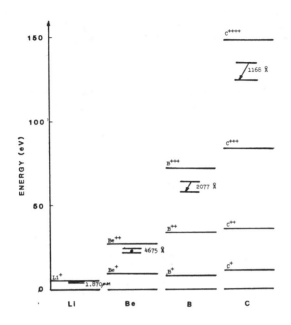

Fig. 4. Partial energy level diagram illustrating isoelectronic scaling sequence in Li, Be, B and C.

written as $G = 7.6 \times 10^2$ r (cm^{-1}) using known A-values, assuming Doppler broadening, assuming Te $\simeq$ Ti during recombination, and assuming that recombination will take place at the optimum value of ne (4×10^{15} cm^3). Using experimental values for r in the range 5×10^{-5} to 10^{-5} yields gains in the range 0.04 to 0.76 cm^{-1}. With state of the art mirrors at 116.8 nm (73% reflectivity) laser action should be possible with a gain length of 1 cm (for the highest r) and with a length of 7 cm (for the lowest r).

A 1 Joule pulse from a TEA CO$_2$ laser, focused to an area of 1 mm^2, should provide sufficient intensity to produce C 4+ at a density approaching 10^{19} cm^3. Subsequent expansion of this plasma to 1 cm would reduce this density to the optimum value required for maximum inversion in C IV at 116.8 nm. Although as many as 8 plasmas in a row may be necessary to produce enough gain to overcome the mirror losses, when laser action occurs, an output energy as high as 80 mJ should be available from this device.

REFERENCES

1. F. V. Bunkin, V. I. Derzhiev, S. I. Yakovlenko, Sov. J. Quantum Electron. 11, 981 (1981).
2. W. L. Bohn, Appl. Phys. Lett. 24, 15 (1974).

3. V. V. Zhukov, V. S. Kucherov, E. L. Latush and M. F. Sem, Sov. J. Quantum Electron. 7, 708 (1977).

4. P. T. Rumsby and J. W. M. Paul, Plasma Phys. 16, 247 (1974).

5. W. T. Silfvast, L. H. Szeto and O. R. Wood, II, Opt. Lett. 4, 271 (1979).

6. F. E. Irons and N. J. Peacock. J. Phys. B: Atom. Molec. Phys. 7, 1109 (1974); R. J. Dewhurst, D. Jacoby, G. J. Pert and S. A. Ramsden, Phys. Rev. Lett. 37, 1265 (1976); V. A. Bhagavatula and B. Yaakobi, Opt. Comm. 24, 331 (1978); M. H. Key, C. L. S. Lewis and M. J. Lamb, Opt. Comm. 28, 331 (1979).

7. W. T. Silfvast and O. R. Wood, II, Opt. Lett. 7, 34 (1982) and references therein.

8. W. T. Silfvast, L. H. Szeto and O. R. Wood, II, Appl. Phys. Lett. 34, 213 (1979).

9. J. F. Reintjes, R. H. Dixon and R. C. Elton, Opt. Lett. 3, 40 (1978).

10. W. T. Silfvast, L. H. Szeto and O. R. Wood, II, Appl. Phys. Lett. 36, 500 (1980).

11. O. R. Wood, II, L. H. Szeto and O. R. Wood, II, Unpublished.

12. W. T. Silfvast, L. H. Szeto and O. R. Wood, II, Appl. Phys. Lett. 36, 615 (1980).

13. W. T. Silfvast, L. H. Szeto and O. R. Wood, II, Appl. Phys. Lett. 39, 212 (1981).

14. O. R. Wood, II and W. T. Silfvast, Submitted for publication.

15. E. M. Campbell, R. G. Jahn, W. F. von Jaskowsky, and K. E. Clark, Appl. Phys. Lett. 30, 575 (1977).

16. W. E. Martin, J. B. Trenholm, G. J. Lunford, S. M. Yarena and C. A. Hurley, IEEE J. Quantum Electron. QE-17, 744 (1981).

17. O. R. Wood, II, Proc. IEEE. 62, 355 (1974).

18. C. A. Brau, in Excimer Lasers, Vol 30, edited by C. K. Rhodes (Springer-Verlag, New York, 1979), pp. 87-133.

19. G. J. Pert, J. Phys. B: Atom. Molec. Phys. 9, 3301 (1976).

20. R. C. Elton, J. Opt. Engr. 21, March-April (1982).

21. W. T. Silfvast and O. R. Wood, II, Submitted for publication.

22. W. T. Silfvast and O. R. Wood, II, Unpublished result.

23. J. J. Macklin, O. R. Wood, II and W. T. Silfvast, submitted for publication.

24. V. V. Shukov, V. G. Il'yushko, E. I. Latush and F. M. Sem, Sov. J. Quantum Electron. 5, 757 (1975).

ANTI-STOKES SCATTERING AS AN XUV RADIATION SOURCE AND FLASHLAMP[*]

S. E. Harris, J. F. Young, R. W. Falcone, Joshua E. Rothenberg,
J. R. Willison, and J. C. Wang
Stanford University, Stanford, Ca. 94305

ABSTRACT

Anti-Stokes scattering from atoms in an electrically excited discharge provides a means of translating many properties of visible lasers into the XUV. These properties include tunability, narrow linewidth, picosecond time scale, and prescribed polarization. This paper describes the use of anti-Stokes scattering for absorption and emission spectroscopy of core-excited levels and as a flashlamp for XUV lasers.

INTRODUCTION

By using spontaneous anti-Stokes scattering from atoms in an electrically excited discharge, many of the properties of visible lasers may be translated into the extreme ultraviolet.[1] The result is a radiation source which is tunable, narrow band, has prescribed polarization and, of special interest, may be of picosecond time scale. As demonstrated by Zych, et al.,[2] the maximum intensity of this radiation source is determined by the effective temperature of the storage level; for example, the $(1s2s)^1S$ level of He or Li^+. Since this level does not spontaneously decay, its accumulated population and, therefore, the peak source intensity, may exceed that of a radiating level by a factor of several thousand. To reach this peak intensity the applied visible laser must cause the media to become two-photon opaque to the generated XUV radiation. In this sense the radiation source functions as a two-photon blackbody.

---

[*]The work described here was supported by the Office of Naval Research, the Air Force Office of Scientific Research, and the Army Research Office.

ISSN:0094-243X/82/900137-16$3.00 Copyright 1982 American Institute of Physics

Our interest in this radiation source is two-fold: first, we are interested in using it as a flashlamp to pump a 200 Å laser; second, and closely related, we are designing instrumentation which uses the tunability and, before long, will use the picosecond nature of the source, to study the properties of autoionizing levels of core-excited alkali atoms. In the following sections we summarize progress in these areas.

LITHIUM SPECTROSCOPY

An early absorption spectra of core-excited Li was taken on the NBS synchrotron by Ederer and co-workers.[3] One immediately notes that all of the observed transitions have broad linewidths, implying autoionizing times of 10^{-12} to 10^{-14} seconds, and low radiative yields.

Early workers[4,5] noted the existence of core-excited levels of atomic Li which are metastable against autoionization and, in some cases, also against radiation. The longest lived of these is the quartet level $1s2s2p \; ^4P_{5/2}$. Since the spins of the three electrons are aligned, in the absence of a spin-spin interaction there is no lower configuration into which the atom may decay. The level lies 57.4 eV above ground and has a measured lifetime of 5.8 μs. Of even more interest to us are those levels in the doublet series which are stable against autoionization but are strongly radiatively allowed. An example of such a level is $1s2p^2 \; ^2P$. This level decays primarily by emission at 207 Å with a lifetime of about 35 ps. The metastability of this level is dependent on the extent to which L and S are good quantum numbers. In Li those levels with even parity and odd angular momentum or of odd parity and even angular momentum, and which also lie below the $(1s2p)^3P$ Li^+ limit, are of the required type. These have been studied extensively by Bunge,[6,7] and much of our work is based on his theoretical results. An energy level diagram of the even parity levels is shown in Fig. 1.

About two years ago we constructed a microwave heated Li plasma[8] and used it to take an emission spectra of Li over the region from

160 Å to 250 Å. Results are shown in Fig. 2 and tabulated in Table I.

Many of the transitions noted in Table I are candidates for lasers in the 200 Å region of the spectrum.[9] Besides having good radiative yields, the lower levels of these transitions may be readily emptied using an incident laser beam. Also, since the core-excited levels lie far above the first ionization potential, the photoionization cross section of the valence electron is less than 10^{-19} cm^2.

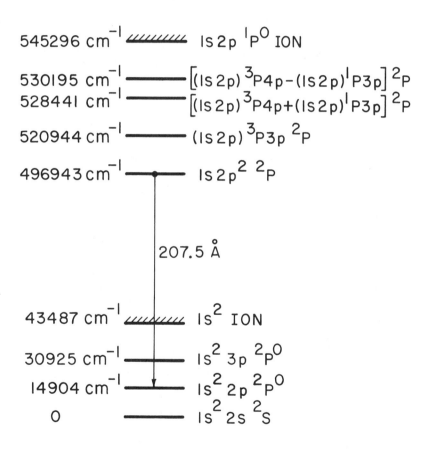

Fig. 1. Even parity core-excited levels in neutral Li.

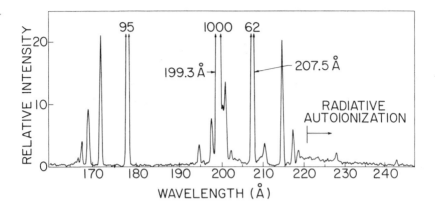

Fig. 2. Emission spectrum of Li.

Table I. Transitions from core-excited ^2P
states in neutral lithium.

Observed Wavelength (± 0.1 Å)	Relative Intensity	Transition	Calculated [a] Wavelength
207.5	62	$1s2p^2\ ^2P \rightarrow 1s^22p\ ^2P^o$	207.45
214.6	12	$\rightarrow 1s^23p\ ^2P^o$	214.58
217.1	6	$\rightarrow 1s^24p\ ^2P^o$	217.16
218.4	2	$\rightarrow 1s^25p\ ^2P^o$	218.37
220.5		$\rightarrow 1s^2\ ^1S + e^-$	
197.8	8	$(1s2p)\ ^3P\ 3p\ ^2P \rightarrow 1s^22p\ ^2P^o$	197.61
194.8	3	$[(1s2p)\ ^3P\ 4p + (1s2p)\ ^1P\ 3p]\ ^2P \rightarrow 1s^22p\ ^2P^o$	194.72
201.0	10	$\rightarrow 1s^23p\ ^2P^o$	200.99
203.3	0.3	$\rightarrow 1s^24p\ ^2P^o$	203.26
204.3	0.2	$\rightarrow 1s^25p\ ^2P^o$	204.32
200.3	8	$[(1s2p)\ ^3P\ 4p - (1s2p)\ ^1P\ 3p]\ ^2P \rightarrow 1s^23p\ ^2P^o$	200.29
202.5	1.5	$\rightarrow 1s^24p\ ^2P^o$	202.54

[a]Calculation by Bunge (Ref. 6).

We have recently reported an experiment[10] in which an intense laser is used to cause the intercombination transfer of metastable $1s2s2p$ ^4P atoms to the $1s2p^2$ ^2P level with subsequent radiation at 207 Å, thereby providing the first direct tie between the quartet and doublet manifolds in neutral Li (Fig. 3).[9] A schematic of the apparatus is shown in Fig. 4. The cell consisted of a ridged X-band waveguide and was driven by a 50 kW, 9 GHz pulsed magnetron. The incident laser beam had an energy of about 10 mJ and was focused to an area of about 2×10^{-3} cm^2 beneath the ridge.

The same apparatus was used to demonstrate anti-Stokes scattering from the singlet and also triplet-singlet transfer in the Li^+ ion. Experimental results for the three cases are shown in Figs. 5, 6, and 7. It is of interest to note that, although the intercombination oscillator strength of the triplet transfer is about six orders of magnitude smaller than that for the singlet transfer, the laser still saturates the transition and the peak triplet transfer signal was only about five times smaller than that of the singlet transfer. The peak quartet signal was about 1/100 times smaller than the singlet signal and the data of Fig. 7 required a total scanning time of about 2 h.

LITHIUM LASER

Harris has proposed three methods of accessing the non-autoionizing (^2P and $^2D^o$) core-excited levels of Li. These are: quartet to doublet transfer,[9] two-photon pumping,[8] and laser designation.[13]

Quartet to Doublet Transfer

The quartet to doublet transfer technique (Fig. 3) has been studied by Rothenberg and Harris.[11] A principal result of that paper is that as one moves to alkali elements that are heavier than Li, the breakdown of LS coupling increases the intercombination oscillator strength, and reduces the autoionizing time of the radiating level. The interplay of these effects reduces the necessary transfer laser power density from 1.3×10^{11} W/cm^2 in Li to 6×10^6 W/cm^2 in K. We have also recently realized that, in the case of Li,

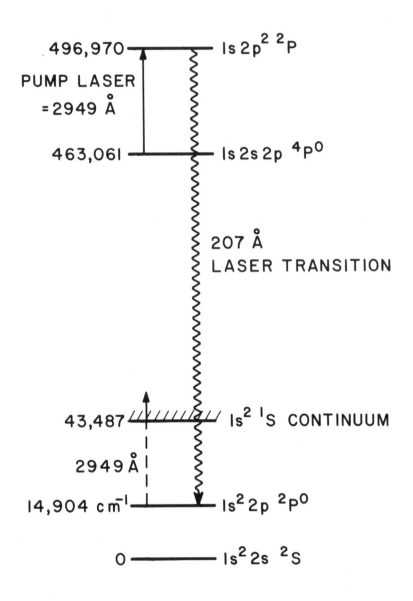

545,303 //////// 1s2p ¹P⁰ CONTINUUM

519,522 //////// 1s2s ³S CONTINUUM

496,970 ——————— 1s2p² ²P

PUMP LASER
=2949 Å

463,061 ——————— 1s2s2p ⁴P⁰

207 Å
LASER TRANSITION

43,487 //////// 1s² ¹S CONTINUUM

2949 Å

14,904 cm⁻¹ ——————— 1s²2p ²P⁰

0 ——————— 1s²2s ²S

Fig. 3--Energy level diagram for a quartet-doublet
transfer in neutral Li.

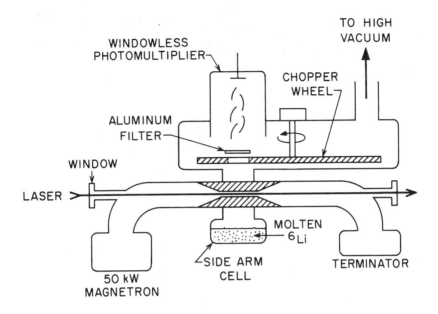

Fig. 4. Experimental apparatus for intercombination transfer.

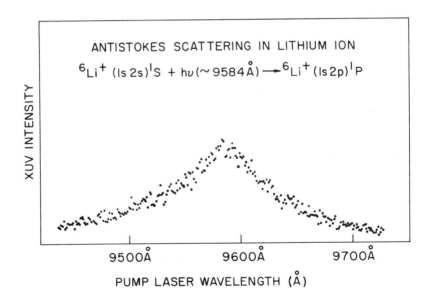

Fig. 5. XUV intensity vs. laser wavelength for
$1s2s \ ^1S \rightarrow 1s2p \ ^1P$ transfer in $^6Li^+$.

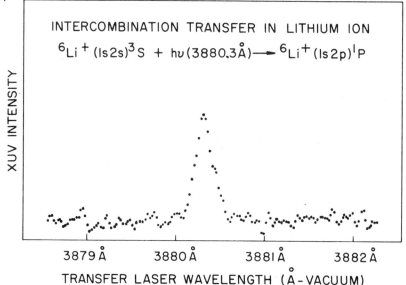

Fig. 6. XUV intensity vs. laser wavelength for
$1s2s$ $^3S \rightarrow 1s2p$ ^1P transfer in $^6Li^+$.

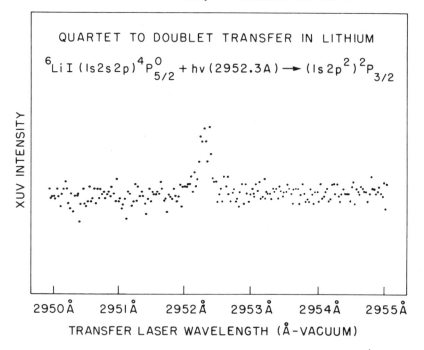

Fig. 7. XUV intensity vs. laser wavelength for $1s2s2p$ $^4P_{5/2} \rightarrow$
$1s2p^2$ $^2P_{3/2}$ transfer in neutral $^6Li^+$.

radiative trapping of the terminal laser level $1s^2 2p\ ^2P$ may allow a reduction of the necessary transfer power by about a factor of six.

Two-Photon Pumping

Energy level diagrams for "anti-Stokes flashlamp" two-photon pumping of the Li $(1s2p)^1P\ 3d\ ^2D^o$ level are shown in Figs. 8 and 9. In Fig. 8 the flashlamp is fired by a 9536 Å laser (second Stokes in H_2 of 5320 Å) and the absorption of the generated photons is caused by a 6537 Å laser. In Fig. 9 the flashlamp is fired by a 1.06 μ laser and the two-photon absorption is caused by a near resonant 6104 Å laser. In both cases the initial level $(1s^2 2p\ ^2P^o)$

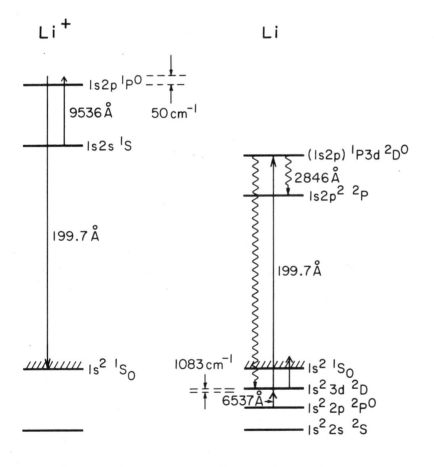

Fig. 8. Two-photon pumping of neutral Li using a near resonant anti-Stokes flashlamp.

146

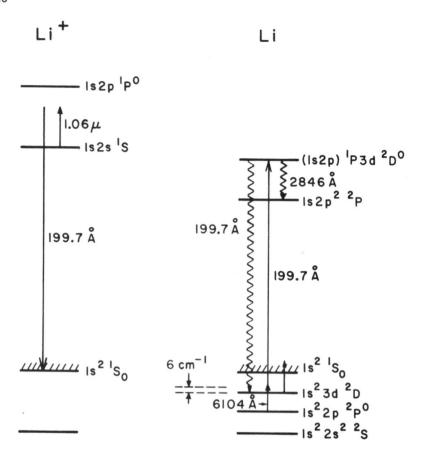

Fig. 9.   1.06 μ flashlamp pumping of neutral Li.

is populated by the discharge and the terminal level ($1s^2\,3d\ ^2D$) is assumed to be emptied by an additional laser in the wavelength window 8194 Å − 7554 Å, which ionizes the lower but not the upper laser level.

At this time it is not clear whether the technique of Fig. 8 or that of Fig. 9 is preferable. The technique of Fig. 8 has the advantage of readily reaching the two-photon blackbody limit of the flashlamp [4×10^4 W/cm^2 at a ratio of Li$^+$ ($1s2s\ ^1S$) to Li$^+$ ($1s^2\ ^1S$) ions if 1/100], but has the disadvantage that if the 6537 Å laser is applied at a sufficient intensity to cause complete absorption of the flashlamp photons it will ionize the upper laser level and reduce the effective pumping time. Using a pulsed hollow cathode discharge,[12] Falcone and co-workers have attained a $Li^+ (1s2s)^1S$ storage density of 1.5×10^{12} atoms/cm^3, a ground ion density of $Li^+ (1s^2) < 4 \times 10^{14}$ atoms/cm^3, and an initial Li $(1s^2 2p)^2 P^0$ population of 4×10^{16} atoms/cm^3. For these conditions we estimate a gain, depending on the value of the ionization cross section of the upper laser level, of between 1% and 3% per cm at 199.7 Å.

The 1.06 μ technique of Fig. 8 is simpler and leads to predictions of somewhat larger gain. The principal uncertainty is beam blow-up of the near resonant 6104 Å laser.

We also note that there is a "magic wavelength" equal to 7757 Å which fires the flashlamp, causes its absorption, and ionizes the lower laser level while not ionizing the upper laser level.

Laser Designation

It is also possible that a laser can be constructed by a simple technique which we term as laser designation.[13] In this case, one tunes a laser to the $1s^2 3d$ ^2D level, thereby "designating" the orbit of the outer electron. Hot electrons then cause 1s-2p core excitation to populate the $(1s2p)^1P$ $3d$ $^2D^0$ level. We estimate a trapping time and, therefore, an allowable excitation time, of several ns. Following this excitation a second laser is used to rapidly ionize the $(1s2p)^1P$ $3d$ $^2D^0$ level, allowing lasing at 199.7 Å.

Using laser designation we have observed fluorescence on the transition $(1s2p)^1P$ $3d$ $^2D^0 \rightarrow 1s2p$ ^2P at 2846 Å.

We also note that, due to its narrower Doppler width, the 2846 Å transition has a gain 18.7 times larger than the 199.7 Å transition. Gain measurmeents at 2846 Å should be useful for determining the relative merits of the different transfer techniques.

Excitation Technology

Our work to date has emphasized pulsed hollow cathode technology which has been developed by Falcone and co-workers. This work is the subject of a separate paper in the volume.[12]

Another possible method of excitation (Fig. 10) would make use of an intense 1.06 μ laser focused onto a tantalum target so as to

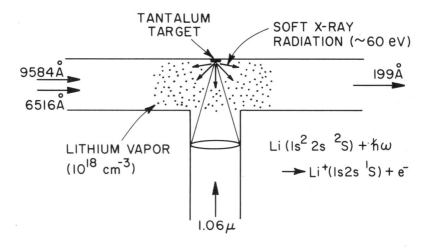

Fig. 10. Soft x-ray excitation of the Li[+] flashlamp.

produce a burst of incoherent, broadband XUV radiation with a temperature of perhaps 50 eV. This radiation would be absorbed by Li
vapor with an absorption cross section of about 10^{-18} cm^2 to produce
Li[+] $(1s2s)^1$S ions, thereby integrating the broadband radiation over
frequency and time. A picosecond pulse near 9584 Å would then cause
an intense burst of flashlamp radiation at 199 Å, which, as described
earlier, would then be absorbed by the neutral Li target.

ANTI-STOKES ABSORPTION SPECTROSCOPY

In recent months we have used the anti-Stokes process as an external source of radiation for high resolution absorption spectroscopy of the 3p[6] shell of K.[14] A schematic of the apparatus is shown
in Fig. 11. Table II shows the spectral regions studied and Fig. 12
shows the results of a typical scan.

For the operating conditions of the hollow cathode glow discharge (2 torr − 300 mA), we estimate that the metastable singlet He
population is about 5×10^{11} atoms/cm^3. For photons at 6000 Å the
anti-Stokes scattering cross section is 4×10^{-23} cm^2. An incident
laser pulse of 50 mJ produces about 2×10^8 XUV photons in the 60 cm

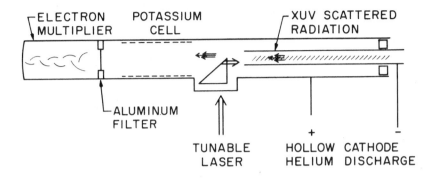

Fig. 11.   Schematic of apparatus used for absorption
spectroscopy of K.

Table II.   Spectral regions studied.

Laser Dye	Range of Laser Wavelength (Å)	Range of XUV Frequency (cm^{-1})
Kiton Red 620	5990 — 5740	182975 — 183695
Rhodamine 590	5760 — 5470	183635 — 184549
Coumarin 500	5220 — 4880	185435 — 186769

cell length. The solid angle subtended by the detector reduces the
measured signal (through the aluminum filter) to about 20 photons/
pulse or 200 photons/sec. This signal is about three times larger
than that produced by the background light from the plasma (Fig. 11)
and, therefore, no monochromator need be used with the device. We
note that the narrower features in Fig. 11 have not been previously
observed.

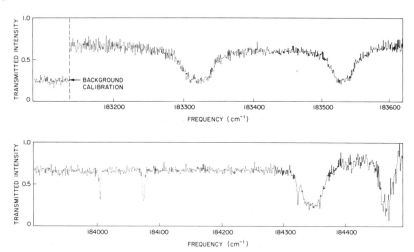

Fig. 12.   Absorption scans of K. Vapor pressure is 10^{15} atoms/cm^3, and cell length is 5 cm.

In order to increase the intensity of the radiation source we have constructed a high power microwave pumped He discharge (Fig. 13) which operates in the recombination mode. At a few torr pressure and 500 kW of pulsed X-band power we estimate a singlet storage density of about 10^{13} atoms/cm^3, with an increase in output signal (compared

MICROWAVE PUMPED ANTI-STOKES SOURCE

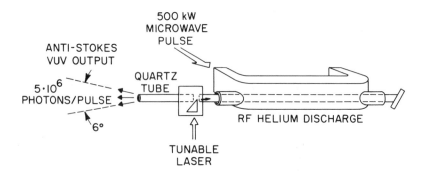

Fig. 13.   Microwave pumped anti-Stokes radiation source.

to the cw hollow cathode device) of about a factor of 1000. Part of this increase results from better use of grazing incidence in this geometry. Unfortunately, the higher power discharge also produces much more background radiation than did the cw device and, at this time, it is necessary to use a monochromator as a bandpass filter in order to take spectra.

The microwave pumped device may be particularly useful for taking spectra by detecting subsequent fluorescence or ejected electrons. In such cases the narrowband absorption of the target species acts as a filter against the background radiation.

ACKNOWLEDGEMENTS

The authors thank C. F. Bunge for preprints of his papers. Many helpful discussions with R. Caro, T. Lucatorto, T. McIlrath, S. Manson, and A. Weiss are gratefully acknowledged.

REFERENCES

1. S. E. Harris, Appl. Phys. Lett. 31, 498 (1977).

2. L. J. Zych, J. Lukasik, J. F. Young, and S. E. Harris, Phys. Rev. Lett. 40, 1493 (1978).

3. D. L. Ederer, T. Lucatorto, and R. P. Madden, Phys. Rev. Lett. 25, 1537 (1970); J. Phys. Paris C4, 85 (1971).

4. P. Feldman and R. Novick, Phys. Rev. Lett. 11, 278 (1963); Phys. Rev. 160, 160 (1967).

5. H. G. Berry, Phys. Scr. 12, 5 (1975); H. G. Berry, E. H. Pinnington, and J. L. Subtil, J. Opt. Soc. Am. 62, 767 (1972).

6. C. F. Bunge, Phys. Rev. A 19, 936 (1979).

7. Rocio Jauregui and Carlos F. Bunge, Phys. Rev. A 23, 1618 (1981).

8. J. R. Willison, R. W. Falcone, J. C. Wang, J. F. Young, and S. E. Harris, Phys. Rev. Lett. 44, 1125 (1980).

9. S. E. Harris, Optics Lett. 5, 1 (1980).

10. J. R. Willison, R. W. Falcone, J. F. Young, and S. E. Harris, Phys. Rev. Lett. 47, 1827 (1981).

11. Joshua E. Rothenberg and Stephen E. Harris, IEEE J. Quant. Elect. QE-17, 418 (1981).

12. R. W. Falcone and K. D. Pedrotti, Optics Lett. 7, 74 (1982); R. W. Falcone, D. E. Holmgren, and K. D. Pedrotti, "Pulsed Hollow Cathode Discharge for XUV Lasers and Radiation Sources," in Laser Techniques for Extreme Ultraviolet Spectroscopy, R. R. Freeman and T. J. McIlrath, eds. (AIP, New York, 1982) (to be published).

13. S. E. Harris, R. W. Falcone, M. Gross, R. Normandin, K. D. Pedrotti, J. E. Rothenberg, J. C. Wang, J. R. Willison, and J. F. Young, in Laser Spectroscopy V, A. R. W. McKellar, T. Oka, and B. P. Stoicheff, eds. (Springer-Verlag, New York, 1981).

14. Joshua E. Rothenberg, J. F. Young, and S. E. Harris, Optics Lett. 6, 363 (1981).

PHOTO-AUTOIONIZATION PUMPED BA ION LASER

by

J. Bokor
Bell Laboratories
Holmdel, NJ   07733

R. R. Freeman
Bell Laboratories
Murray Hill, NJ   07974

W. E. Cooke
Department of Physics
University of Southern California
Los Angeles, CA   90007

ABSTRACT

Doubly excited autoionizing states in barium of the configuration $6p_{3/2}np$ (n $\geq$ 12) are shown to decay preferentially to yield the $6p_{1/2}$ excited state of Ba^+. The resulting population inversion is used to produce amplified spontaneous emission at 493 nm and 650 nm corresponding to the $6p_{1/2} - 6s_{1/2}$ and $6p_{1/2} - 5d_{3/2}$ transitions in Ba^+. Under appropriate conditions, strong coherent output at 167 nm was simultaneously produced and could be attributed to a four-wave parametric process in the excited Ba^+. This principle of selective autoionization is expected to be applicable to the construction of a new class of ion lasers extending into the vacuum-ultraviolet wavelength region.

Introduction

In general, when an atom is excited to an autoionizing state whose energy is greater than several of the atomic ionization limits all of the energetically possible final ion states are produced. The branching ratios to different final ion states are determined by the degree of configuration interaction coupling the ion-plus-bound-electron configuration to the various ion-plus-continuum electron configurations. In certain cases, preferential decay to specific excited ionic states occurs. Such behavior has been identified in Ba,[1-4] Yb,[3] and Pb.[5] Gallagher, et al.[1,2] first pointed out that this effect might be useful in creating ionic population inversions resulting in laser action. We refer to this phenomenon as selective autoionization.

ISSN:0094-243X/82/900153-10$3.00 Copyright 1982 American Instiute of Physics

We report the first demonstration of laser action in an atomic ion pumped by selective autoionization. Amplified spontaneous emission (ASE) was observed on two visible transitions in Ba^+ following step-wise two-photon excitation of selected autoionizing states in neutral Ba. A schematic energy level diagram showing the relevant barium states and optical transitions is shown in Figure 1. The autoionizing states of interest are of the nominal configuration 6pnp and were excited via the 6snp resonant intermediate state. Ultraviolet laser radiation tunable near 247 nm was used for the first transition, $6s^2$ - 6snp, while visible laser radiation tunable near 455 nm was used to make the second transition, 6snp - 6pnp. Two Rydberg series of 6pnp autoionizing states exist, each converging to one of the two spin-orbit split 6p $^2P_{1/2,3/2}$ levels in the ion (see Figure 1). These two series are labeled as $6p_{1/2}np$ and $6p_{3/2}np$. States belonging to the $6p_{3/2}np$ series with $n > 12$ lie above the $6p_{1/2}$ ion limit and were found to decay preferentially to $6p_{1/2}$ ions. Population inversion of the $6p_{1/2}$ ion level with respect to both lower lying states in the ion (6s and 5d) was demonstrated in a Ba vapor cell by the observation of ASE at 493 nm and 650 nm corresponding to the $6p_{1/2}$ - $6s_{1/2}$ and $6p_{1/2}$ - $5d_{3/2}$ transitions respectively.

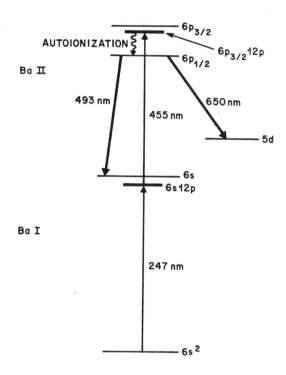

Fig. 1

Schematic energy level diagram showing states of both neutral and singly ionized barium. The selective decay of the $6p_{3/2}12p$ autoionizing state to yield the $6p_{1/2}$ excited ionic state is shown as well as the optical transitions in Ba^+ which exhibited amplified spontaneous emission (ASE).

In addition, spectroscopic investigations of the 6pnp states were performed in an atomic beam apparatus. Good correlation was found between the excitation spectrum for ASE found in the cell experiments and the ionization spectrum measured in the beam experiments. Linewidths for the $6p_{3/2}np$ ($n \geq 12$) states were found to be a factor of up to 17 times wider than the corresponding $6p_{1/2}np$ states. This behavior is qualitatively similar to that found for the 6pns states in barium[1,2] and arises because these $6p_{3/2}nl$ states can decay into the newly opened $6p_{1/2}$ channel.

The selective production of Ba$^+$ 6p$_{1/2}$ was further exploited to generate additional output wavelengths.^{2} It was found that when the input excitation beams were mildly focused, the excitation spectrum for ASE (taken by tuning the visible input laser) broadened dramatically by the depletion mechanism.^{6} Strong ASE at 650 nm persisted even as the visible input was tuned over 6 nm to the blue. This fact made it possible to maintain selective production of Ba$^+$ 6p$_{1/2}$ and then resonantly excite these ions to higher lying excited states using the same two input lasers. Two examples are shown in Figure 2. Tuning to 453 nm resulted in excitation from 6p$_{1/2}$ - 7s$_{1/2}$. ASE output was then observed on the 7s$_{1/2}$ - 6p$_{3/2}$ transition at 490 nm. When the visible input was tuned near the 6p$_{1/2}$ - 6f$_{5/2}$ two-photon transition at 451 nm, doubly resonant four-wave parametric mixing was observed, whereby two of the input 451 nm photons mixed with the 650 nm photons to produce an intense coherent output at 167 nm.

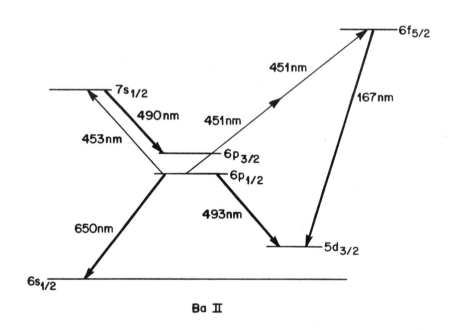

Ba II

Fig. 2

Schematic energy level diagram for Ba$^+$ showing two higher order excitation processes.

Apparatus

The ultraviolet radiation tunable near 247 nm used for the first excitation step was produced using a pulsed dye laser pumped by the second harmonic output of a Nd:YAG laser and standard nonlinear optical techniques (second harmonic generation and sum frequency generation in KDP crystals). Up to 1 mJ pulses of 5 nsec duration were produced at 247 nm in a linewidth of 2 cm^{-1} at a repetition rate of 10 pps. Visible radiation, tunable near 450 nm was derived from a second dye laser. Up to 10 mJ pulses of 5 nsec duration were produced in a linewidth of 0.5 cm^{-1} at 10 pps. The visible and ultraviolet beams were spatially and temporally overlapped and directed either into a Ba vapor cell or a Ba effusive atomic beam apparatus.

The atomic beam apparatus is similar in construction to that used by Gallagher, et al.[1] Total ionization yield spectra were taken by fixing the ultraviolet wavelength to the $6s^2\ ^1S_0$ - $6snp\ ^1P_1$ transition, and then scanning the visible wavelength laser in the vicinity of the $6s\ ^2S_{1/2}$ - $6p\ ^2P_{3/2}$ transition in Ba^+. Doubly excited autoionizing states of the nominal configuration $6p_{3/2}np$ were thus excited. (However, as discussed below, the data indicate significant configuration mixing.) Both of the excitation lasers were well polarized and experiments were performed with the relative polarizations either parallel or crossed. The selection rules are easily derived and show that with parallel polarization, only $J = 0$ and $J = 2$ states are excited, while for crossed polarization, only $J = 1$ and $J = 2$ states are excited. In this way, it was possible to assign a value of total J for each state observed.

Atomic Beam Data

In Figure 3 the ionization spectra for the $6p_{3/2}12p$ states are shown for both parallel and crossed polarizations. Our assignments of total J are also shown. Similar spectra were taken for $12 \leq n \leq 16$. As mentioned in the previous section, the shape of the spectrum was strongly dependent on the input energy, showing saturation and broadening behavior. In all linewidth measurements, the visible laser was heavily attenuated in order to avoid saturation. For the same total J, the $6p_{3/2}np$ states were generally much broader than what was found for $6p_{1/2}np$ states. For example, we measured the width of the $6p_{1/2}12p$ $J = 0$ state to be 2 cm^{-1} while the $6p_{3/2}12p$ state has a linewidth of 34 cm^{-1}. More typically, $6p_{3/2}np$ linewidths were 2 to 3 times the corresponding $6p_{1/2}np$ linewidths, while in certain cases the $6p_{3/2}np$ linewidths were comparable with the corresponding $6p_{1/2}np$ linewidths.

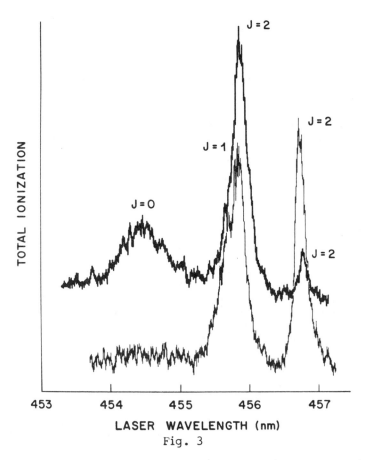

LASER WAVELENGTH (nm)

Fig. 3

Total ionization spectra for barium taken in an atomic beam apparatus. These data correspond to the nominal configuration $6p_{3/2}12p$. The upper, bold trace is the spectrum obtained using parallel laser polarizations. The lower, light trace was taken with orthogonal laser polarizations. The total angular momentum quantum number J is identified for each peak. The J = 1 peak is a shoulder on the blue side of the central J = 2 peak in the lower curve.

We attempted to predict these spectra using JK, jj, and intermediate coupling schemes assuming a pure single configuration. No satisfactory predictions were obtained in this way, indicating the influence of configuration interaction. A multichannel quantum defect theory (MQDT)[8] analysis of the spectrum would therefore be desired. However, such an analysis requires additional spectroscopic data. For the purpose of simplicity, in the remainder of the discussion we will continue to use the single configuration labels $6p_jnp$, bearing in mind that such a designation is an oversimplification.

Cell Experiments

The apparatus used in the cell experiments was 70 cm in length, sealed at both ends with LiF windows and had a central hot zone of approximately 25 cm in length. Typical running conditions were at temperatures of 800-1000 C, with 20 to 50 Torr of argon used as a buffer gas. In most of the experiment, the ultraviolet radiation was set to excite the $6s^2 \, ^1S_0$ - $6s12p \, ^1P_1$ transition[7] at 40428.7 cm^{-1}. The visible radiation was then scanned, just as in the beam experiments. The visible ASE outputs were isolated from the input lasers using either narrowband interference filters or a 0.35m monochromator. Both the 650 nm and 493 nm outputs were confirmed as due to ASE by three criteria. First, they exhibited threshold behavior as a function of either input pumping power with fixed vapor density or as a function of vapor density with fixed pumping power. Second, the outputs were well defined, well collimated beams. Third, approximately equal outputs were observed emerging from both ends of the cell. This last test was performed in order to eliminate coherent parametric interactions which would be phasematched in the forward direction only. The strongest output was observed at 650 nm, corresponding to the $6p \, ^2P_{1/2}$ - $5d \, ^2D_{3/2}$ transition in Ba^+. Using collimated input beams of approximately 0.05 cm^2 area, with 0.5 mJ per pulse at 247 nm and 2.0 mJ per pulse at 455 nm, the 650 nm output which was copropagating with the input beams was measured to be 1 μJ per pulse. Both forward and backward outputs were also detected at 493 nm, corresponding to the $6p \, ^2P_{1/2}$ - $6s \, ^2S_{1/2}$ transition in Ba^+. The output energy at this wavelength was approximately one order of magnitude smaller than the output at 650 nm.

Figure 4 shows the excitation spectra for the 650 nm ASE output. At low input pumping power, the excitation spectrum is narrowed due to the highly nonlinear nature of the gain process. Threshold behavior is quite clearly shown in Figure 4. The dependence of the 493 nm output on input power was qualitatively similar. The excitation spectra for both the 650 nm and 493 nm ASE outputs were also measured as a function of the relative polarization of the two pump lasers. The relative heights of the three main peaks were qualitatively similar to those found in the ionization spectra as shown in Figure 3.

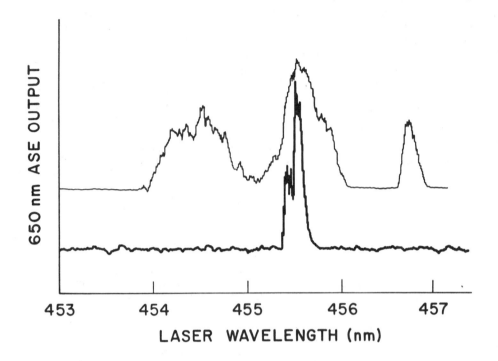

Fig. 4

Excitation spectra for the 650 nm ASE output taken in the cell experiments. The input laser polarizations were parallel for both curves.
The lower, bold trace was taken with an input visible laser pump
power which was a factor of 10 lower than that used to obtain the
upper, light trace.

Higher Order Processes and VUV Generation

For the study of higher pumping processes the input beams were
focused into the cell using a 250 nm quartz lens. Care was taken to
ensure that the foci overlapped properly. When the visible beam was
tuned to exact resonance with the $6p_{1/2} - 7s_{1/2}$ transition at 452.6
nm, efficient transfer of population to the $7s_{1/2}$ level was observed.
The $6p_{1/2}$ level was sufficiently depleted that the 650 nm and 493 nm
ASE outputs were fully quenched. Instead, strong ASE output was then
observed at 490 nm, corresponding to the $7s_{1/2} - 6p_{3/2}$ transition.
This output satisfied all three criteria for ASE.

In an effort to obtain stimulated emission in the VUV, the visible beam was tuned to resonance with the $6p_{1/2} - 6f_{5/2}$ two-photon transition at 451 nm. The output window of the cell was sealed to the input of a 1 meter vacuum ultraviolet monochromator. In this instance, no quenching of the 650 nm ASE was detectable. Nevertheless, strong collimated emission was observed at 167 nm which corresponds to the $6f_{5/2} - 5d_{3/2}$ transition in Ba^+. Between 10^{11} and 10^{12} photons per pulse were obtained at this wavelength. A special evacuated chamber was constructed around the input end of the cell and a dichroic VUV reflector was used to direct any backward propagating VUV emission into a solar-blind photomultiplier tube. No such emission was detected. Thus, the 167 nm output could not be attributed to ASE. Instead, this output is attributed to a doubly resonant four-wave parametric interaction involving the excited Ba^+ $6p_{1/2}$ ions as the nonlinear medium. In this process, two 451 nm input photons mix with the internally generated 650 nm photons to generate the 167 nm output photons. This process bears some similarity to four-wave parametric oscillation.[9] Further confirmation of the assignment of the VUV output to a parametric process was obtained by a careful examination of its excitation spectrum. A single, narrow line was observed in excitation with a linewidth of approximately 3 cm,$^{-1}$ much larger than the laser linewidth of 0.5 cm^{-1} or the transition Doppler width of 0.04 cm.$^{-1}$ The peak was actually found to occur approximately 7 cm^{-1} on the red side of the position of the two-photon resonance $6p_{1/2} - 6f_{5/2}$ as given in Moore's table.[10] Finally, the linewidth and precise position of the peak were found to be temperature dependant. As the cell temperature was raised from 700°C to 900°C, the peak shifted by approximately 2 cm^{-1} to the red and broadened by approximately 20%. The behavior of this excitation spectrum is typical of phasematching behavior of parametric mixing processes in atomic vapors.

Discussion

Selective autoionization behavior has been previously observed in several atomic systems via ejected electron spectroscopy.[2-5] These systems all share the common feature that multiply excited states of the neutral atom which lie energetically between two spin-orbit split excited ion limits are found to decay preferentially to the nearby ion level corresponding to the lower member of the doublet.[2-5] Rosenberg, et al.[3] have given a qualitative explanation for such preferential autoionization decay. A more rigorous analysis, involving an MQDT fit to the spectrum would directly reveal the channel interactions and give the autoionization branching ratios. To our knowledge, MQDT has not yet been applied to the problem of predicting autoionization branching ratios.

Selective production of ion excited states by photoexcitation may also occur in regions of the continuum devoid of autoionizing states. Green and Falcone[11] produced an inversion of the resonance line of Sr^+ by photoionizing a bound two-electron excited state of neutral Sr. In their experiment, the selectivity is achieved by radiatively coupling the bound excited state to the desired continuum. The well known proposal of Duguay and Rentzepis[12] for pumping an X-ray laser by photoionization to selectively populate certain inner-shell vacancy states is based on similar principles. In selective autoionization, it is the dynamics of the two-electron interaction which leads to the selective decay of a doubly excited autoionizing state to the desired ion continuum. Selective autoionization thus represents a qualitatively new physical mechanism for the production of population inversions in ions.

In summary, selective autoionization has been demonstrated as a new physical principle for the production of population inversion and laser action in atomic ions. Since the phenomenon of selective autoionization has been observed in several inner-shell excited atoms[2-5] it is believed to be quite general. We therefore anticipate the construction of a new class of optically excited ion lasers operating in the deep ultraviolet region of the spectrum.

REFERENCES

[1] T. F. Gallagher, K. A. Safinya, and W. E. Cooke, Phys. Rev. A21, 148 (1980).

[2] W. E. Cooke and T. F. Gallagher, Phys. Rev. Lett. 41, 1648 (1978).

[3] R. A. Rosenberg, S. T. Lee, and D. A. Shirley, Phys. Rev. A21, 132 (1980).

[4] J. P. Connerade and M. A. P. Martin, J. Phys. B13, L373 (1980); H. Hotop and D. Mahr, J. Phys. B8, L293 (1980).

[5] J. P. Connerade, J. Phys. B14, L141 (1981).

[6] W. E. Cooke, S. A. Bhatti, and C. L. Cromer, paper MA3, this conference.

[7] J. A. Armstrong, J. J. Wynne, and P. Esherick, J. Opt. Soc. Am. 69, 211 (1979).

[8] See e.g., U. Fano, J. Opt. Soc. Am. 65, 979 (1975).

[9] J. Bokor, R. R. Freeman, R. L. Panock, and J. C. White, Opt. Lett. 6, 182 (1981).

[10] C. E. Moore, Atomic Energy Levels, NSRDS–NBS35 (1971).

[11] W. R. Green and R. W. Falcone, Opt. Lett. 2, 115 (1978).

[12] M. A. Duguay and P. M. Rentzepis, Appl. Phys. Lett. 10, 350 (1967).

STIMULATED VACUUM ULTRAVIOLET EMISSION FROM RARE GAS CRYSTALS

N. Schwentner[*], O. Dössel and H. Nahme
Institut für Experimentalphysik, Olshausen-
straße 40 - 60, D-2300 Kiel, West-Germany

ABSTRACT

Investigations of Xe, Kr and Ar crystals with
respect on applications as laser media yield stimulated
emission from Ar crystals at 126 nm.

INTRODUCTION

The broad bandwidth emission continua from noble
gas excimers, extending from 60 nm in the case of He to
172 nm for Xe, favour these gases for short wavelength
lasers. The noble gas excimers are bound with dissociation energies of about 1 eV in the excited state. They
decay radiatively to the strongly repulsive part of the
ground state potential curve. The repulsive ground state
ensures an extremely fast depopulation by dissociation
$(10^{-13}s)$ thus facilitating inversion. It is responsible
for the large width of the emission bands which should
allow a broad tunability region. The particular problems
encountered in the construction of short wave length
laser follow from an λ^{-5} scaling law for the pumping
power per unit volume which is required to produce a
gain coefficient of unity per unit length. In 1965 it
has been recognized that noble gases emit essentially
the same excimer bands in the solid, liquid and gaseous
phase[1]. The idea to exploite the higher density in the
condensed phases for obtaining a sufficient high
density of excimers lead to the first successful
operation of an excimer laser by the Basov group[2] in
1970. Amplified stimulated emission from electron beam
excited liquid Xe at 178 nm has been observed. This
project has not been continued because of the low
efficiency which has been attributed mainly to
reabsorption of the light. Besides a proposal for an
ablation laser[3] no further attempts to reach stimulated
emission from condensed noble gases have been reported.
The gas phase excimer lasers made their way[4] as will be
demonstrated by Ch. K. Rhodes in this meeting. Despite
of the progress in gas phase excimer lasers and in
frequency multiplication it is still difficult to cover
the region of wavelengths shorter than about 180 nm by

[*] now at: Institut für Atom- und Festkörperphysik, Freie
Universität Berlin, D-1000 Berlin 33, West-Germany.

lasers. Any additional way should be pursued. Since two years we try to obtain stimulated emission in the region below 180 nm from Ar, Kr and Xe crystals. Most of our emphasis has been concentrated on Ar crystals which radiate around 126 nm. Several experimental observations indicate that we succeeded in obtaining stimulated emission from Ar crystals. The expenses in pumping power and in equipment for achieving a comparable radiation power seem to be lower than for the Ar gas phase laser. Considering this success and the relatively little efforts devoted up to now to solid phase lasers it seems to be promising to continue also this way.

EMISSION BANDS AND LIFETIMES IN SOLID NOBLE GASES

A comparison of the emission bands of noble gases demonstrates the close correspondence in the gaseous (at high pressure), liquid and solid phase (Fig. 1).

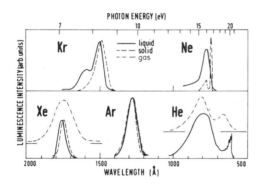

With Ne crystals the wavelength range for laser crystals can be extended to 70 nm. Helium causes problems because external pressure is required for solidification. We restrict ourself to Ar, Kr and Xe crystals because different parts of the optical path can be separated by LiF windows. Time resolved emission spectra of the crystals

Fig. 1. Emission bands of gaseous, liquid and solid Xe[1,5], Kr[6], Ar[7,10], Ne[8], He[9].

contain a shortliving component ($\tau \sim 3 - 10$ ns) and one or more long living components ($\tau \sim 20 - 20000$ ns) for each material. By detecting the emission intensity with a time window of 100 ns during or after the time of excitation it is possible to resolve two bands within each band shown in Fig. 1. These bands which are separated by 40 - 110 meV are reproduced in the upper row of Fig. 2. In the case of Xe two pairs of bands with a temperature dependent intensity ratio are observed[5,12,13]. All short lifetimes do not depend on sample temperature (Fig. 2) whereas the long lifetimes are rather sensitive to the temperature except of Ar

(Fig. 2). All the bands presented in Fig. 2 are attributed to excimer like centers in the crystals and constitute the potential for laser emission. There exist additional higher energetic emission bands of different origin which are by about two orders of magnitude lower in intensity[11,14,15,16,17]. These bands are not immediately essential for laser applications but they play an important role for an understanding of the formation of the excimer centers.

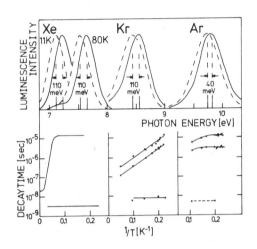

Fig. 2. Results of time-resolved luminescence experiments for solid Xe, Kr and Ar. Full lines in the intensity curves (upper panel) correspond to the components with short decay times, dashed lines, to long decay times (this work). The decay times (lower panel) have been compiled by Zimmerer[11].

Besides those established emission bands we observed a flat continuum emission in Kr and Xe crystals starting at 1.5 eV and extending to beyond 3 eV (Fig.3).

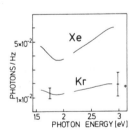

In Ar this continuum is absent or at least one order of magnitude smaller than in Kr. This emission is only present during the excitation time of 3 ns. The appearance of this emission seems to be connected to our high excitation densities, therefore it is discussed with the loss processes.

Fig. 3. Continuum emission of Xe (60 K) and Kr (11 K) crystals in 4π sterad.

NATURE OF EMITTING STATES

The emission bands shown in Fig. 2 are attributed to excimer like centers because of the close relationsship of the bands in all three phases[1,10]. This assign ment is supported by the appearance of the short living component which is also observed in the gas phase and which is explained by a transition from the $^1\Sigma$ excimer state to the ground state. This strongly allowed transition is the only laser active band that will be discussed in the following. A counterpart to the long living component in the solid phase is also known in the gas phase due to the $^3\Sigma$ state. The transitions from $^1\Sigma$ (0_u^+) and $^3\Sigma(0_u^-, 1_u)$ are indicated in the potential curves for Xe in part I of Fig. 4. The excimer formation in the crystal is interpreted by a movement of one of the 12 next nearest neighbour ground state atoms towards the excited atom. The distance between the two atoms is reduced by about 30 % compared to the equilibrium separation of the crystal in the ground state. The excimer in the crystal has a lower symmetry (D_{2h}) as the free excimer ($D_{\infty h}$). The $^3\Sigma(0_u^-, 1_u$) state splits into three states Γ_1', Γ_2', Γ_4' whereas the $^1\Sigma(0_u^+)$ state is nondegenerate and changes its symmetry to Γ_3'[19]. The relations between these states are depicted in part II and III of Fig. 4. The transition from Γ_3' to the ground state remains strongly allowed and theconditions for stimulated emission remain essentially unchanged in the crystal. The cross sections σ_s,given in table I for stimulated emission follow from

$$\sigma_s = \frac{1}{8\pi\,n_r^2} \cdot \frac{\lambda^2}{\tau_s\,\Delta\nu} \tag{1}$$

with the index of refraction n_r, the lifetime τ_s and the linewidth $\Delta\nu$.

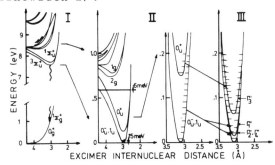

Fig. 4. Xe excimer potential curves in gaseous[18] (part I, II) and solid phase[16,20,21] (part III). Nearest neighbour separation in the crystal: 4.3 Å.

The Γ_1', Γ_2' and Γ_4' states are responsible for the long living components. The temperature dependence and the appearance of several components in the long living parts have been explained by the splittings of these states and a temperature dependent redistribution of the population of these states[13,16,17,20,21,22]. The doubling of the pairs in Xe is attributed to the formation of excimers associated with a vacancy in the crystal (low energetic pair) and of excimers with a complete surrounding (high energetic pair)[23]. The cross sections for stimulated emission from the Γ_1',

Table I. Emission wavelength λ , bandwidth Δv , radiative lifetime τ_s and cross section σ_s of stimulated emission from the $^1\Sigma$ state in noble gas crystals.

	λ (nm)	Δv $(10^{14}Hz)$	τ_s $(10^{-9}s)$	σ_s $(10^{-18}cm^2)$
Ar	126	1.21	3-4	7
Kr	145	1.06	8	4
Xe	162	0.82		
	171	0.87	3	13

Γ_2^- and Γ_4^- states are some orders of magnitude lower than those of the Γ_3^- states due to the longer radiative lifetime.

The formation of excimers has been confirmed further by transient absorption experiments carried out by the Kanzaki group[24,25], by V. Saile[26] and by us. The time dependence of the transmitted intensity and the spectral dependence of the transient absorption have been measured on crystals which we used also for stimulated emission experiments (Fig. 5, 6). The time dependence of the transmission T (t) after exciting the crystal at t = 0 reflects the cross section for transient absorption σ_a, the lifetime τ of the initial excimer state ($^3\Sigma$), the initially excited density n_o of excimer states and the length of the optical path L in the crystal due to

$$T(t) = \exp \left(-\sigma_a L n_o \exp \left(-t/\tau \right) \right) \qquad (2)$$

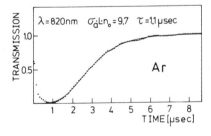

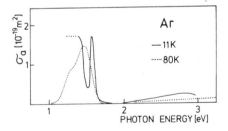

Fig. 5. Measured and fitted time course of the transmission.

Fig. 6. Transient absorption cross section at 11 K (this work) and 80 K[24] (arb.units).

The first absorption bands in the spectrum around 1.5 eV can be correlated to transitions from the $^3\Sigma$ states to higher excimer states. The broad band above 2 eV indicates a severe modification of very high excimer states in the crystal. The temperature dependence gives a feeling for the changes in transition energies of the excimers in the crystal due to thermal expansion of the crystal and due to different phonon populations.

The additional weak VUV emissions in the crystals mentioned above contain emissions due to vibrationally excited excimers, due to the formation of so called bubbles around excited atoms and due to resonant emission of free excitons which are created in the undistorted crystal[11,14,15,16,17].

CRYSTAL GROWTH AND EXPERIMENTAL SET UP

The growth even of single crystals from noble gases is well known[27,28]. The demands for laser crystals are less severe since only optically clear but free standing crystals with a volume of about 1 cm^3 and a front surface of 2 x 1 cm^2 are necessary. We simplified a device described by Schuberth et al.[29]. A pyrex box is mounted below a copper block which can be cooled down to 10 K by a closed cycle refrigerator (Fig. 7).

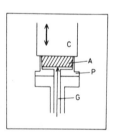

Fig. 7. Crystal growth device with refrigerator C, pyrex box P, crystal A and gas inlet G.

With the copperblock pressed versus the box and gas flowing through the pipe at the bottom we grow a crystal within about 30 minutes. By shifting the refrigerator and the crystal upwards with a bellow sealed translation stage we bring the free standing crystal into the optical path of the ultrahigh vacuum sample chamber. The pure gases are contained in an ultrahigh vacuum handling system and are sometimes additionally purified with a hot Ti getter.

The crystals are excited by a Febetron 706 delivering electron pulses of up to 10 J (600 keV, 5 kA, 3 ns) through a 0.025 mm thick Ti foil into the sample chamber. The distribution of the deposited energy across the front surface and the penetration depth into the crystal have been measured with a thermopile and a Faraday cup (Fig. 8). The deposited energies, the

depth d of homogenous excitation and the deposited
energy densities are listed in table II. The crystals
withstand several hundred shots without damage for a
deposited energy of 0.06 J/cm^2. For 0.14 J/cm^2 cracks
appear after several shots which lead to light
scattering. No other effects like evaporation or
temperature raise have been observed.

The VUV light is dispersed by a Sey.a Namioka
monochromator. A complete spectrum per shot is recorded
with a resolution of 0.1 nm by a spatially resolving SIT
vidicon. The vidicon has been gated (Δt = 100 ns) to
analyse the time course of the emission spectra. The
absolute VUV intensity emitted into 10^{-2} sterad of the
amplifying direction has been displayed on a fast
storage oscilloscope by a calibrated vacuum diode
(ITT F4115). In the next future it will be possible
to measure directly the angular distribution of the
light and to mount mirrors for a laser cavity.

Transient absorption spectra have been measured
simultanously to control the optical quality during
excitation, to derive information about the excited
centers and to get a feeling for reabsorption of light

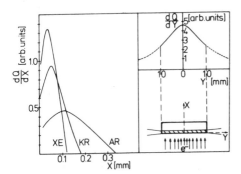

Fig. 8. Top view of the crystal
and distribution of deposited
energy across the surface (top)
and in the depth (left) for
400 keV equivalent electron
energy.

by excimers. For this
purpose light from a
high pressure Xe lamp
is focused into the
excited area of the
crystal at a glancing
angle of 7°. The
totally reflected light
is dispersed and then
recorded by a second
gateable SIT vidicon.
Transmission spectra
at different delay
times are taken in this
way. Alternatively the
time course of transient
absorption is monitored
at a fixed wavelength
by a fast photomulti-
plier.

AMPLIFIED SPONTANEOUS EMISSION

In the present configuration without laser mirrors
it is necessary to create a density of excimers which is
high enough for a significant amplification of the light
during a single path through the excited area. The
intensity I (t) due to spontaneous emission and due to
amplified spontaneous emission within the cone dΩ into

the amplifying direction is given by

$$I(t) = \frac{d\ h}{\sigma_s\ \tau_s}\ (\exp(n(t)\ \sigma_s\ L)-1)\ d\Omega \qquad (3)$$

with the excimer density $n(t)$, the penetration depth d, the length L and the height h of the crystal. The gain α due to amplification is given by

$$\alpha = n(t)\ \sigma_s\ L \qquad (4)$$

Table II. Deposited energy Q, mean energy M for excimer creation, calculated total number N^c_{tot} and measured total number N_{tot} of $^1\Sigma$ excimers, conversion efficiency q into $^1\Sigma$ photons, penetration depth d (50 % decrease), mean excimer separation r_{e-e} (from N^c_{tot}) and measured maximal $^1\Sigma$ excimer density n_m for an electron equivalent energy of 500 keV.

	Q (J)	M (eV)	N^c_{tot} (10^{16})	N_{tot} (10^{15})	q (%)	d (mm)	r_{e-e} (nm)	n_m (10^{16}cm^{-3})
Ar	0.22	19.3	7.1	4	6	0.31	8.6	5
Kr	0.22	17.7	7.8			0.16	6.6	
Xe	0.22	14.5	9.5			0.10	5.3	

As criteria for amplification within the $^1\Sigma$ component we discuss a) the dependence of the width of the $^1\Sigma$ emission band on excitation power, b) the time course of the $^1\Sigma$ emission, c) the degree of collimation and d) the absolute intensity of the $^1\Sigma$ component. Amplification has been observed only for Ar crystals. Therefore we restrict us in the following to Ar crystals.

a) A first indication for amplification is a decrease by 10 % of the fwhm of the $^1\Sigma$ band for an increase of the deposited energy from 0.09 J to 0.22 J yielding $\alpha = 0.7$ for the highest excitation power. This interpretation is confirmed by the fact that the $^3\Sigma$ line shape remains unchanged.

b) The time course of the emission within $d\Omega = 10^{-2}$ sterad is shown in Fig. 9. The long living part in the insert corresponds to the $^3\Sigma$ contribution

with a lifetime of about 10^{-6} s. An intermediate part
from 5 ns to 15 ns can be correlated with ordinary
spontaneous emission of $^1\Sigma$ excimers and a lifetime
τ_s = 4 ns. The dashed line shows for comparison the
time duration of the exciting Febetron pulse. Evidently
most of the $^1\Sigma$ contribution is emitted synchronously
with the exciting Febetron pulse and much faster than
a convolution of the exciting pulse with τ_s would
predict. This part contains an essential part of
amplified emission on a background of spontaneous
emission. The amplified contribution vanishes when
the excimer density falls below the amplification
threshold which happens around t = 5 ns. From the
enhancement of the intensity a gain coefficient α = 1
is estimated.

c) The intensities of the $^1\Sigma$ and $^3\Sigma$ components
from gated vidicon spectra versus the deposited energy
are given in Fig. 10 for two typical crystals. They are
corrected for the time windows. Thus the total fluxes
from the $^1\Sigma$ and $^3\Sigma$ components into the amplifying
direction are compared. The $^1\Sigma$ contribution increases
progressively in contrast to the $^3\Sigma$ contribution which
increases linear or even slightly sublinear with
deposited energy. The sum of both parts increases
progressively too. Similar results are obtained from
the photon fluxes in the short and long living parts
of the diode outputs (Fig. 9). The enhancement of the
$^1\Sigma$ component is caused by an increasing collimation
of the $^1\Sigma$ radiation into the amplifying direction. The
$^3\Sigma$ radiation is emitted isotropically because of the
much smaller cross section for stimulated emission.
This experiment yields a gain which increases from
α = 0.4 to α = 1.3 for enlarging the deposited energy
from 0.09 J to 0.22 J.

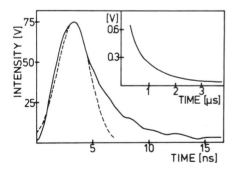

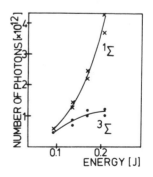

Fig. 9. Emission intensity
versus time. Dashed line:
time course of Febetron
pulse.

Fig. 10. Time integrated
intensity of the $^1\Sigma$ and
$^3\Sigma$ components versus
deposited energy.

d) The measured diode outputs of up to 4 A for a deposited power of 0.22 J correspond to maximal intensities of 8 x 10^{20} photons/s into dΩ = 10^{-2} sterad. From equation (4) with the known value for σ_s we can derive a maximal density n_m = 5 x 10^{16}cm^{-3} of $^1\Sigma$ excimers and a gain α = 0.3. The total number of $^1\Sigma$ excimers N_{tot} = 4 x 10^{15} in the crystal follows from the time course and the excited volume. These absolute intensity measurements indicate also that we reached amplification. A peak power of 1300 watts due to amplified spontaneous emission has been detected in 10^{-2} sterad. Furthermore they show that a total of 4 x 10^{15} photons due to $^1\Sigma$ excimers is emitted yielding a peak power into all directions of 1 x 10^6 watts. Finally the long component yields a density of 8 x 10^{16}cm^3 $^3\Sigma$ excimers or a total number of 4 x 10^{15}.

Summarizing point a) to d) we can state that stimulated emission with a gain between 0.3 $\leq \alpha \leq$ 1.4 has been observed in Ar crystals.

LOSS PROCESSES

The number N^c_{tot} of excimers which could be created if no loss process would be present follows from the deposited energy divided by a mean energy M which includes the mean energy for electron hole pair creation[30] and the percentage of directly created excitons[30]. A comparison of these values in table II indicates an efficiency q = 6 % for a conversion of N^c_{tot} into $^1\Sigma$ photons. For the calculated values N^c_{tot} the mean separation r_{e-e} between two excimer centers in the crystal is only about 10 lattice constants (table II).Quantitative models for loss processes at such high densities of excimers have not been developed for noble gas crystals in contrast to the gas phase[4].

Reabsorption by ground state atoms and ionisation of excimers by the emitted radiation are of similar nature in the gas and solid phase but the cross sections are not known in the solid phase.

The selfquenching processes of excited states differ in the gaseous and solid phases because in the crystal two types of excited states namely highly movable free excitons and strongly localized excimer centers exist[1,10,11,14,15,16,17,31]. Creation of an electron hole pair in the crystal can lead to a selftrapping of the hole, capture of an electron[32] and relaxation within the localized excimer states. Free excitons can be created directly or via capture of an electron by a free hole or by dissociation of an electronically excited excimer state. There is no model available to sort out the contributions due to

excimer-excimer, excimer-exciton, excimer-electron,
exciton-exciton and exciton-electron processes but
there is a qualitative difference in the quenching
efficiencies expected for excimer centers on the one
hand and free excitons on the other hand. Since the
energies of the localized $^1\Sigma$ and $^3\Sigma$ excimer states
are lower by their dissociation energies than that of
the free excitons it is impossible for the excimers
to move resonantly in the crystal. Thus self quenching
of excimers can take place only by a Förster-Dexter
type[33] energy transfer process between two excimers
which has an effective range of the order of only some
nm. The free excitons on the other hand move resonantly
either by a diffusive or coherent motion with a high
group velocity through the crystal and can quench each
other by collisions. This transport process would also
enhance nonradiative decay at defects. Thus a high
radiative efficiency is expected when the free excitons
are localized very fast. The lowest free excitons and
excimers are separated by a potential barrier of some
meV which has to be crossed for localisation [11,14] and
which determines the lifetime of free excitons ·
concerning localisation. The weak intensity of free
exciton emission indicates that the free exciton lifetime
is only of the order of 10^{-10} to 10^{-12} s. Nevertheless
this lifetime is sufficient that the free excitons
travel distances up to several hundred Å[31]. When the
balance lies on the side of localized excimers then the
efficiency q in the crystal can be higher than in the
gas phase since excimer-excimer quenching by collisions
is a very severe loss process at high gas densities. Ar
crystals seem to be a favourite example in this sense.

The continuum emissions (Fig. 3) offer an explanation
for the failure to reach amplification in Kr and Xe
crystals. The continuum emission is attributed to an
electron plasma emission since only in this way the
spectral[34] and temporal behaviour can be explained.
Transition and Cerenkov radiation[35] has to be excluded
because of the unpolarized nature and the too high
intensity of the observed emission. Ordinary bremsstrahlung[36] depends only weakly on the atomic number,
not on temperature and should be also of lower intensity.
From the intensity approximate temperatures of the
electron plasma of 7000 - 8000 K for Xe and Kr are
derived. The intensity of the plasma radiation decreases
with increasing sample temperature whereas the excimer
emission shows the opposite trend in our case of high
excitation densities. Our observations suggest that
the plasma is cooled down in Ar so effectively that it
does not radiate and consequently the excimers are
localized so fast that those excimers surviving the
quenching processes are still enough for amplification.
In Kr and Xe the free charge carriers live long enough

to form a radiating plasma in crystals at low temperatures. Thus the quenching rate will be very high until the Febetron pulse terminates. At elevated crystal temperatures cooling of the plasma seems to be more efficient which shows perhaps a way for obtaining stimulated emission also from Kr and Xe crystals.

COMPARISON GAS-SOLID

A typical Ar gas excimer laser[4,37,38] requires gas pressures between 10 - 100 bars in a stainless steel tube, excitation by a Pulserad electron source (2 MeV, 15 KA, 20 ns, 500 J) and a gas purification system. The handling of the gas, its purification and the combination of high pressure and VUV radiation require an equipment which is at least as expensive as the crystal growing device. The lifetime of the crystals, which might be increased by a better understanding of the degradation mechanismus is comparable with the gas laser lifetimes determined by the tube lifetime and by the gas purity[38]. The deposition of the energy in the crystal is easy and very efficient. In the gas phase a delicate compromise is necessary between the gas pressure, the thickness of the tube wall and the penetration depth given by the energy of the electrons.

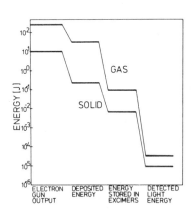

Fig. 11 shows a comparison of the efficiencies for an Ar gas laser[38,37] and a solid state laser starting with the electron gun. Evidently the crystal can be more efficient concerning the conversion of deposited energy into photons. The higher conversion efficiency is based on the lower selfquenching rate of the localized excimers in the crystal compared to the collisional excimer-excimer quenching in the gas phase.

Fig. 11. Approximate conversion efficiencies for Ar gas[37,38] and our Ar crystals.

Excimer centers in liquids transport energy over quite large distances[39] therefore the efficiency of liquid rare gases is an open question.

CONCLUSIONS

This paper aimed to demonstrate that stimulated emission from noble gas crystals can be obtained. Even at this early stage of collecting experiences with noble gas crystal laser it seems that the crystals offer advantages in handling and efficiency compared to gas lasers. Therefore they are promising candidates for VUV solid state lasers. At the moment experiments showing laser action in combination with mirrors have to be done. More elaborate models for the energy flow in noble gas crystals could perhaps guide the experimentalist in choosing optimal conditions for amplification.

ACKNOWLEDGEMENT

It is a pleasure to acknowledge the strong support by R. Haensel. The experiments gained essentially by the contributions of A. Quirin, H. Wilcke, G. Balssen and T. Schröder. We thank for financial support of the Deutsche Forschungsgemeinschaft DFG.

REFERENCES

1. J. Jortner, L. Meyer, S. A. Rice and E. G. Wilson, J. Chem. Phys. 42, 4250 (1965).
2. N. G. Basov, V. A. Danilychev, Yu M. Propov and D. D. Khodkevich, JETP Lett. 12, 329 (1970)
3. R. W. Dreyfus, S. C. Wallace, Optics Comm. 13, 218 (1975).
4. M. H. R. Hutchinson, Appl. Phys. 21., 95 (1980).
5. N. G. Basov, E.M. Balashov, O. V. Bogdankevich, V. A. Danilychev, G. N. Kashnikov, N. D. Lantzov and D. D. Khodevitch, J. Luminescence 1/2, 834 (1970).
6. O. Chesnovsky, B. Raz and J. Jortner, J. Chem. Phys. 59, 5554 (1973).
7. O. Chesnovsky, B. Raz and J. Jortner, J. Chem. Phys. 57, 4628 (1972).
8. R. E. Packard, F. Reif and C. M. Surko, Phys. Rev. Lett. 25, 1435 (1970).
9. M. Stockton, J. W. Keto and W. A. Fitzsimmons, Phys. Rev. A5, 372 (1972).
10. J. Jortner in "Vacuum Ultraviolet Radiation Physics" eds. E. E. Koch, R. Haensel and C. Kunz, Vieweg, Pergamon, New York, p. 263 (1974).
11. G. Zimmerer, J. Luminescence 18/19, 875 (1979).
12. O. Chesnovsky, A. Gedanken, B. Raz and J. Jortner, Sol. State. Comm. 13, 639 (1973).
13. G. Zimmerer in "Proc. of Intern Summer School on Synchrotron Radiation Research" ed. N. Mancini and I. F. Quercia, Alghero Vol. I, p. 453 (1976).

14. I. Ya Fugol', Adv. Phys. 27, 1 (1978).

15. F. Coletti and A. M. Bonnot, Chem. Phys. Lett. 55. 92 (1978).

16. N. Schwentner, E. E. Koch and J. Jortner in "Rare Gas Solids", Vol. 3, ed. M. K. Klein and J.A. Venables, Academic New York (1982).

17. N. Schwentner, Appl. Opt. 19, 4104 (1980).

18. W. C. Ermler, Y. S. Lee, K. S. Pitzer and N.W. Winter, J. Chem. Phys. 69, 976 (1978).

19. A. G. Molchanov, Fiz. Tverd. Tela 4, 9 (1972).

20. U. Hahn, N. Schwentner and G. Zimmerer, Opt. Comm. 21, 237 (1977).

21. R. Kink, A. Lohmus, M. Selg and I. Soovik, phys. stat. sol. (b) 84, K61 (1977).

22. R. Heumüller and M. Creuzburg, Opt. Comm. 26, 363 (1978).

23. R. Kink and M. Selg, phys. stat. sol. (b) 96, 101 (1979).

24. T. Suemoto and H. Kanzaki, J. Phys. Soc. Japan 46, 1554 (1979).

25. ibid. 49, 1039 (1980).

26. V. Saile, Appl. Optics 19, 4115 (1980).

27. J. Hingsammer and E. Lüscher, Helvetia Physica Acta 41, 914 (1968).

28. J. A. Venables and B. L. Smith, in "Rare Gas Solids" Vol. II, eds. M. L. Klein and J.A. Venables, Academic, New York, 609 (1977).

29. E. Schuberth and M. Creuzburg, phys. stat. sol. (b) 71, 797 (1975).

30. T. Doke, A. Hitachi, S. Kubota, A. Nakamoto and T. Takahashi, Nucl. Instr. and Methods 134, 353 (1976).

31. N. Schwentner, G. Martens and H.W. Rudolf, phys. stat. sol. (b) 106, 183 (1981).

32. J. W. Keto, R. E. Gleason and F. K. Soley, J. Chem. Phys. 71, 2676 (1979).

33. T. Förster, Ann. Phys. 2, 55 (1948).

34. A. Unsöld, Ann. Phys. 33, 607 (1938).

35. G.M. Garibian, JETP 6, 1079 (1958).

36. E. A. Edelsack, W. E. Kreger, W. Mallet and N. E. Scofield, Health Phys. 4, 1 (1960).

37. W.-G. Wrobel, H. Röhr, K.-H. Steuer, Appl. Phys. Lett. 36, 113 (1980).

38. W.-G. Wrobel, Max Planck Institut München, IPP1/186 (1981).

39. J. W. Keto, F. J. Soley, M. Stockton and W. A. Fitzsimmons, Phys. Rev. A 10, 872 and 887 (1974).

HIGH AVERAGE BRIGHTNESS RARE-GAS
HALIDE LASER TECHNOLOGY[*]

Stephen E. Moody, George J. Mullaney,

William Grossman, Philip E. Cassady,

and Stanley Byron

Mathematical Sciences Northwest Inc.

2755 Northup Way, Bellevue, Wa. 98004

ABSTRACT

We describe several high repetition rate (100-1000Hz) discharge pumped rare-gas halide laser systems having near diffraction-limited divergence, resulting in unusually high average brightness.

Discharge pumped rare-gas halide lasers are attractive pump sources for non-linear far UV generation techniques such as multiwave mixing and Raman conversion because of their high peak power at UV wavelengths. To extend the performance of these lasers to much higher average powers, we have designed and built several high velocity closed loop gas flow systems to allow both 100 and 1000 Hz operation. Using this approach, devices which produce average powers ranging from 1 to greater than 100 Watts at 308 nm have been successfully operated.

At the highest powers, active heat exchangers and sidewall acoustic wave dampers are required. Careful design of upstream flow components and the laser cavity-flow interface is also required to

[*] Work supported by U. S. Department of Energy.

prevent flow separation. A laser head incorporating these features and designed for 100 W average output power is shown schematically in Figure 1. A flow duct system designed using windtunnel techniques is used and includes a centrifugal circulator and heat exchanger. Flow velocity is 45 m/sec, and in steady state operation, the heat exchanger must dissipate about 10 kW while providing 2°C temperature uniformity at its output. Additional flow nonuniformity can be introduced by pressure waves originating in the laser discharge which can interact with loop components and propagate back into the active laser volume. To reduce these disturbances to acceptable levels, acoustic dampers have been included in the system both upstream and downstream of the laser head. These dampers consist of perforated side walls in the flow channel backed by acoustically lossy material.

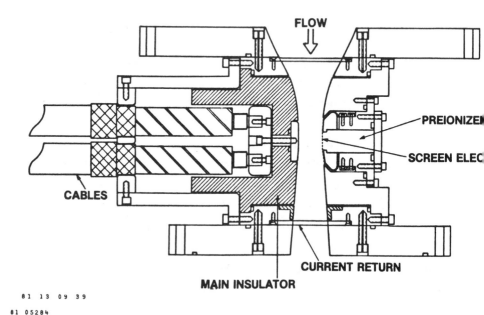

81 13 09 39
81 05284

Figure 1. Laser Head Design for 1 kHz Operation

The level of transient pressure disturbances has been measured at various locations around the loop, and the effectiveness of the acoustic dampers in lowering the level of these disturbances has been clearly shown. Direct measurements of the gas density uniformity in the laser head have also been made with pulsed laser interferometric techniques during high repetition rate operation. It is essential that the gas density recover during the 1 millisecond interpulse interval needed for 1 kHz operation. Figure 2 shows the measured recovery, indicating successful achievement of useful levels of flow uniformity.

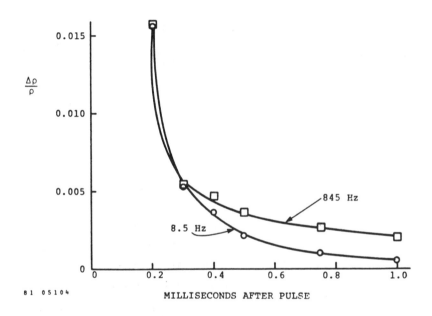

Figure 2. Measured Medium Homogeneity

The best measure of the success of such a laser system is its ability to provide individual optical pulses whose characteristics are independent of repetition rate. Figure 3 shows the energy output of this device as a function of repetition rate, with nearly flat performance up to more than 1000 Hz. This data was taken with

a stable optical resonator. The measured pulse length is
approximately 35 nsec fwhm.

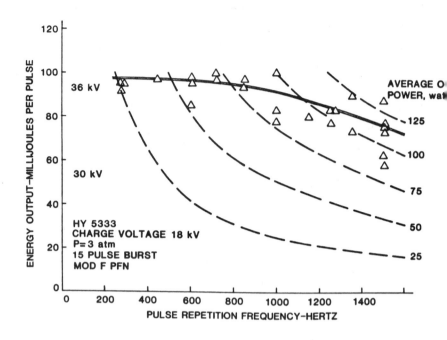

Figure 3. Measured High PRF Laser Performance

For applications using nonlinear conversion, brightness is
more relevant than power for describing the usefulness of a given
laser. Rare-gas halide lasers with simple two mirror resonators
typically demonstrate very poor beam quality because of their high
Fresnel numbers and extremely high gain, with beam divergences of
order 100-1000 times diffraction limited being not atypical. As has
been shown by many workers, the beam quality of these devices can be
improved substantially by the use of positive branch unstable
resonators.

There are two important mechanisms which can limit the
ultimate beam quality achieved in these systems. For flowing
systems, it is necessary to ensure that the internal medium
homogeneity is sufficient to support a high quality intraresonator

field. This requirement has been met in our systems using the type
of designs already described. There is also a finite time for mode
formation of a high quality beam in a given unstable resonator. If
this time is not short compared with the laser gain time, then the
output will not demonstrate good beam quality. This time is given
approximately by

$$\tau_r = \frac{21}{c} \frac{\ln[N(M-1)/\eta M]}{\ln[M]}$$ [1]

where M is the resonator magnification, N the Fresnel number, l the
resonator mirror spacing, and η is a geometry dependent number of
order 0.5.[1] For Fresnel numbers of several hundred, and
magnifications of order 4, which are typical, τ_r is roughly 5
resonator round-trips, and does not vary strongly with either N or
M. Since the gain times of these devices do not tend to scale
strongly with device size, it is generally true that small
oscillators must be used when good beam quality is required.

Figure 4 shows a measurement of beam quality as achieved in a
very small scale device (which is commercially available as the MSNW
Excilite[TM] 401). The active volume of this device is
$0.7x0.7x15$ cm^3. Repetition rate is 100 Hz, and typical output
energy with a flat-flat resonator in XeCl is 5-7 millijoules. A
positive branch unstable resonator was built using external optics,
with output coupling from a dielectric coated scraper mirror having
an uncoated elliptical hole at the center. Magnification of 5 was
used, and antireflection coated windows were mounted on the laser
head to prevent parasitics. Because of the very small size of the
laser head, total resonator length could be kept to 40 cm.

The beam quality measurement was made by scanning a focal spot
with a straight knife-edge. The resulting extinction curve is
compared with the extinction function for idealized Gaussian and
Airy focal spots of various sizes relative to the diffraction
limit. For the miniature oscillator, these scans give results

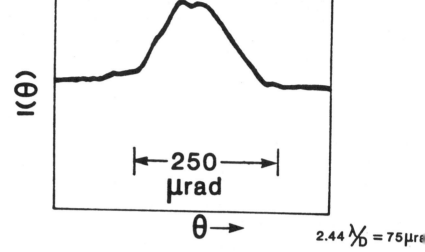

Figure 4. Measured Far-Field Beam Shape at 1 kHz

consistent with beam quality of better than two times an ideal flat
phase annular beam. Average power at the time of these measurements
was 0.25 Watt, resulting in an average brightness of approximately
2×10^7 W/(cm^2-steradian). Although all of the results described here
were measured during operation in XeCl, there is no fundamental
reason why comparable performance could not be achieved using the
other rare-gas halides.

If excellent beam quality is required from a large scale
rare-gas halide laser, either very long gain times must be achieved,
or the beam quality must be determined in a small scale oscillator,
and transferred to the larger device by operating either as an
amplifier or as an injection locked oscillator. The small scale
oscillator just described has been used to injection lock an
oscillator with discharge dimensions of $3\times3\times75$ cm^3, and successful
improvement of the resultant high energy beam quality has been
demonstrated. Results of this experiment will be described elsewhere.

A positive branch unstable resonator (M=4) was also installed on the 1 kHz device described above. Far field beam quality was measured by focussing a sample of the beam onto the target of a TV camera vidicon. The shape of the resulting spot can be directly measured by observing one raster line of the TV camera video output, which gives a direct read out of intensity versus position along the raster line, as shown in Figure 4.

An ideal beam having a flat intensity and phase distribution over the annular output beam would have intensity nodes at approximately 75 μrad (full width). The resultant beam is therefore between roughly 3 and 4 times diffraction limited. At the time of this measurement the average power was greater than 30 Watts, corresponding to an average brightness of 5×10^8 W/(cm^2-steradian). The average power was limited by poor filling of the active laser volume, which resulted from a poor choice of hole size in the output scraper mirror. We believe that a more optimum choice of resonator parameters would result in nearly the full 100 Watt output achieved with a stable resonator, while maintaining the 3-4 times diffraction limited beam quality. Average brightness in this case would be greater than 10^9 W(cm^2-steradian).

In conclusion, by combining careful flow design to achieve high repetition rate and good optical quality, and careful discharge design to result in long excimer gain times, with positive branch unstable resonators, it has been possible to achieve unprecedented levels of average brightness in a UV beam. The application of such lasers as drivers for nonlinear far UV upconversion sources could result in an extraordinary tool for extending UV spectroscopy.

REFERENCE

1. R. S. Hargrove, E. Grove, and T. Kan, IEEE QE-15, 1228 (1979).

ANALYSIS OF VUV AND SOFT X-RAY LASING BASED ON CHARGE-EXCHANGE MECHANISMS

J. C. Bellum, W. W. Chow, K. Drühl and M. O. Scully
University of New Mexico, Albuquerque, NM 87131

ABSTRACT

A clashing beam charge exchange laser is analyzed where a dense helium plasma pulse is incident on a cesium vapor target. The coupled rate equations for the charge exchange process and spontaneous emission from the excited He* are solved in closed form, and the resulting population inversion is calculated. Several competing processes are discussed, the most important being probably penning ionization by metastable He*.

BASIC PRINCIPLES

In the helium-cesium clashing beam laser population inversion is sought in the 2p state of helium by means of a charge transfer reaction.[1] A dense helium plasma pulse is generated by a plasma gun and swept into the interaction region where it collides with a cesium vapor target traveling in the opposite direction. The cesium vapor is obtained by flash heating a thin layer of cesium metal to evaporation and allowing the vapor to expand freely. With this vapor source densities as high as $N = 10^{17}/cm^3$ have been achieved. A typical plasma density is $n = 10^{16}/cm^3$, and drift velocities range from $v = 3 \times 10^6 cm/sec$ to 10^7 cm/sec.

In the interaction region Cs^+ ions and excited He* atoms are formed by the charge exchange reaction:

$$He^+ + Cs \rightarrow He^* + Cs^+ + \Delta E \tag{1}$$

The theoretical work of Olson and Smith[2] predicts a cross section of $\sigma \simeq 2 \times 10^{15}$ cm^2 for the excited 2^1P state of He, which is of interest to us as upper state in the lasing transition. The total cross section including the metastable 2^3S, 2^1S and 2^3P states was estimated to be $\sigma_{tot} \simeq 8 \times 10^{15} cm^3$. Experimental studies of relative cross sections[3] however are at variance with these predictions and indicate that the metastable He states are populated much more strongly than was predicted. The present uncertainties in absolute cross sections however need not concern us at this point, since the results of our gain calculations scale in a simple way with parameters like cross sections and drift velocity. Various other processes compete with the charge exchange reaction, and will be briefly discussed below.

COMPETING PROCESSES

The following processes poase a threat to propulation inversion by depleting either the Cs target or the excited He population.

1. Photoionization of Cs
2. Electron-impact excitation and ionization of Cs
3. Penning ionization of Cs by metastable He*
4. Photoionizations of He*
5. Electron-impact ionization of He*
6. Spontaneous emission

1. Photoionization of Cs by light from the plasma and the He(2^1P) - (1^1S) transition is negligible. UV radiation from the evaporation flash lamp can have a strong effect and must be eliminated through filtering.
2. Electron impact excitation will populate preferentially the 6P levels, which will also participate in the charge exchange reaction. Electron impact ionization on the otherhand is an important depleting mechanism, with cross section only by a factor 2 to 3 smaller than charge exchange.
3. Both ionization cross-sections and absolute concentrations of metastable He* species are not sufficiently well known at present. The latter may be substantially reduced by 4. and 5.
4. and 5. Cross sections for both processes are significantly smaller than the stimulated emission cross section of about $\sigma \simeq 4 \times 10^{-16} \text{cm}^2$.
6. The spontaneous decay rate is $\tau^{-1} \simeq 2 \times 10^9 \text{sec}^{-1}$. Excited He* is produced by the charge exchange reaction at a rate $\tau_0^{-1} \simeq \sigma N \cdot v$. Hence for population inversion to occur the Cs density N has to exceed a critical value N_{cr} given by

$$\sigma N_{cr} \cdot v > \tau^{-1} \tag{2}$$

$$N_{cr} > 10^{17}/\text{cm}^3 \text{ for } \sigma = 2 \times 10^{-15} \text{cm}^2, \ v = 10^7 \text{cm/sec}$$

A more precise calculation of N_{cr} will be given below.

These brief remarks are intended to justify our following analysis to the extent permitted by the limits of this presentation. A more detailed discussion of competing processes in our charge exchange scheme will be published elsewhere. In following we shall study the set of rate equations for the charge exchange reaction and for spontaneous emission, neglecting all other competing mechanisms.

RATE EQUATIONS AND SOLUTIONS

The rate equations for the He-Cs charge exchange reaction, including spontaneous emission from the He*(2 P) state are:

$$\frac{\partial}{\partial z} n_+ = - \sigma n_+ N_0 \tag{3}$$

$$\frac{\partial}{\partial z} n_* = \sigma n_+ N_0 - \gamma n_* \tag{4}$$

$$\frac{\partial}{\partial z} n_0 = \gamma n_*$$ (5)

$$\frac{\partial}{\partial z} N_0 = - \sigma n_+ N_0$$ (6)

The reacting species, symbols for their densities and other parameters are as follows

species	Cs	Cs^+	He^+	He^* (2 P)	He
density	N_0	N_+	n_+	n_*	n_0

σ = charge exchange cross section

$\gamma = (v \cdot \tau)^{-1}$, v = drift velocity of He^+ plasma
               τ = spontaneous decay time of He^*.

$\lambda = vt - z$.

    The Cs target is assumed to be at rest and situated in the
region $z > 0$, while the He^+ plasma is moving into the positive z-
direction with velocity v, being confined to the region $z < 0$ at
$t = 0$. Hence $(-\lambda)$ is equal to the z-position in the He^+ rest frame.
The initial pulse shapes are given by

$$n_+(\lambda, 0) = n_+(\lambda)$$

$$N_0(0, z) = N_0(z)$$ (7)

    The solutions of equations (3) and (6) subject to the initial
conditions (7) are

$$n_+(\lambda, z) = n_+(\lambda) \frac{e^{B(\lambda) - A(z)}}{1 - e^{-A(z)} + e^{B(\lambda) - A(z)}}$$ (8)

$$N_0(\lambda, z) = N_0(z) \frac{1}{1 - e^{-A(z)} + e^{B(\lambda) - A(z)}}$$ (9)

where

$$A(z) = \sigma \int_0^z dz' N_0(z')$$

$$B(\lambda) = \sigma \int_0^\lambda d\lambda' n_+(\lambda')$$ (10)

    It is instructive to consider the case where both the He^+ and
the Cs initial densities are constant

$$N_0(z) = N_0 \cdot \Theta(z) \quad : \quad A(z) = \sigma N_0 z = \alpha \cdot z$$

$$n_+(\lambda) = n_+ \Theta(\lambda) \quad : \quad B(\lambda) = \sigma n_+ \lambda = \beta \cdot \lambda \tag{11}$$

For $\sigma = 2 \times 10^{-15} cm^2$ and $N_0 = 10^{17}/cm^3$ we have $\alpha = 2 \times 10^2/cm$, showing that the transient effects at the leading edge of the Cs target are dying out after a distance $z \gg 10^{-2} cm$. Neglecting these effects a steady state dynamics is obtained

$$n_+(\lambda,z) = n_+(\lambda) \frac{e^{\beta\lambda - \alpha z}}{1 + e^{\beta\lambda - \alpha z}}$$

$$\beta\lambda - \alpha z = \sigma n_+ v(t - \frac{1}{v_{eff}} \cdot z); \quad v_{eff} = \frac{v}{1 + N_0/n_+}$$

At any given time t the charge exchange reaction is confined to a limited interaction region of width $\delta z < \alpha^{-1}$. In a situation where population inversion will occur $N_0 > 10^{17} cm^{-3}$ (Eq. (2)) and the interaction region is quite narrow: $\delta z < 10^{-2} cm$. The interaction region moves into the Cs target at an effective velocity $v_{eff} < v$.

The excited He^* density and the population inversion Δ are obtained from (4) and (8)

$$n_*(\lambda,z) = \int_0^z dz' e^{\gamma(z'-z)} \left(-\frac{\partial}{\partial z'} n_+(z') \right)$$

$$\Delta(\lambda,z) = \frac{1}{3} n_*(\lambda,z) - n_0(\lambda,z) = \frac{4}{3} n_*(\lambda,z) + n_+(\lambda,z) - n_+(\lambda)$$

$$= \int_0^z dz' \left[\frac{4}{3} e^{\gamma(z'-z)} - 1 \right] \left(-\frac{\partial}{\partial z'} n_+(\lambda,z') \right) \tag{12}$$

In the case of constant initial densities (11) Δ can be expressed as a function of dimensionless variables:

$$\Delta(\lambda,z) = n_+ \cdot \tilde{\Delta}(t_0(\lambda,z),r) \tag{13}$$

$$r = \sigma N_0/\gamma$$

$$t_0 = \gamma(z - z_0(\lambda)) \quad ; \quad z_0(\lambda) \text{ is solution of } A(z_0) = B(\lambda)$$

For $r \to \infty$ the function $\tilde{\Delta}(t_0, r)$ is simply

$$\tilde{\Delta}(t_0) = n_+ \left[\frac{4}{3} e^{-t_0} - 1 \right] \quad t_0 > 0$$

$$= 0 \qquad t_0 < 0 . \tag{14}$$

These results remain valid for arbitrary pulse shapes provided the rise distance ΔZ of the Cs pulse is much larger than the width

δz of the interaction region (quasistatic approximation): $\Delta z \gg \alpha$. Since $\Delta z \simeq 1$ cm this condition is well satisfied. In this case the parameters n_+ and r have to be replaced by thier local values:

$$n_+ \to n_+(\lambda) \quad ; \quad r \to \delta N_0(z_0)/\gamma.$$

The function $\Delta(t_0,r)$ is plotted below for various values of the parameter r. It can be seen from this that positive inversion occur for $r > 3$. Furthermore maximal inversion occurs at $z \simeq z_0$, although the exact position of the maximum depends on r.

Figures 1, 2, and 3: Normalized population inversion $\tilde{\Delta}(t_0,r)$

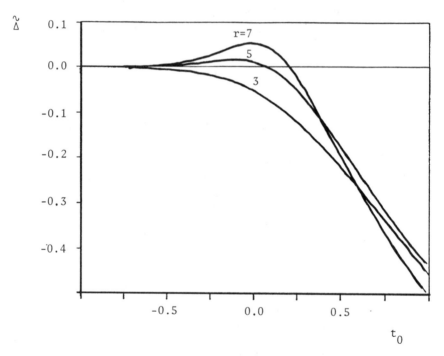

FIGURE 1

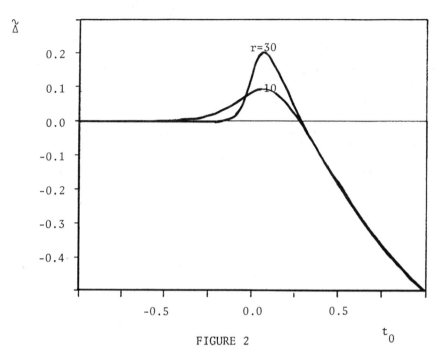

FIGURE 2

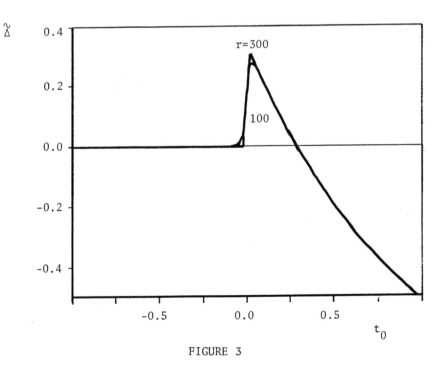

FIGURE 3

SUMMARY AND CONCLUSION

The rate equations for the clashing beam charge exchange laser can be solved in closed form, if only charge exchange and spontaneous emission are taken into account. The population inversion is calculated in the quasistatic approximation as a function of time and position. The charge exchange cross section σ enters in terms of a dimensionless parameter, which is easily rescaled as more reliable values for σ become available. The most relevant competing process appears to be Penning ionization of Cs and He* by metastable He* species, since due to their long lifetimes appreciable concentrations of these species can be expected to build up. Before the inportance of this process can be assessed quantitatively, more reliable absolute values for the charge exchange cross sections into the various He* states concerned are needed.

REFERENCES

1. D. Anderson, J. McCullen, M. O. Scully, J. F. Seely, Optics Commun. 17, 226 (1976) and references given there.
2. R. E. Olson, F. T. Smith, Phys. Rev. A7, 1529 (1973).
3. C. Reynaud, J. Pommier, Vu Ngoc Tuan, M. Barat, Phys. Rev. Lett. 43, 579 (1979).

TRANSFORM-LIMITED-BANDWIDTH INJECTION LOCKING OF AN XeF LASER

Irving J. Bigio and Michael Slatkine
Los Alamos National Laboratory
Los Alamos, NM 87545

ABSTRACT

A pulsed Ar-ion laser, operating on the 3511 Å line of the doubly ionized species, has been used to injection lock an unstable resonator XeF laser. A single longitudinal mode of the Ar-ion laser was selected with a Fabry-Perot étalon, and the resulting XeF bandwidth was measured to be $\sim$ 50 MHz which is the Fourier-transform limit corresponding to the XeF laser pulse duration of $\sim$ 20 ns. Since the XeF B → X emission originates from more than one upper vibrational level, for this pulse duration it was determined that $\sim$ 60% of the total output was available to be locked to the reference oscillator line. The 2 MW output beam was also found to be near (< 1.5 x) diffraction limited.

INTRODUCTION

The channeling of the energy of high-power excimer lasers into very narrow spectral bandwidths in low divergence beams is a prerequisite for many potential applications of these lasers. High spectral brightness is required for spectroscopic applications and optical harmonic generation,[1,2] and the long coherence length associated with a narrow bandwidth is required for nonlinear phase conjugation, backward stimulated Brillouin and Raman scattering[3] among other applications. Ultranarrow bandwidth is virtually essential for most nonlinear vuv generation schemes.

It is by now well recognized that the simplest way to control the spectral bandwidth and beam divergence of high-gain excimer lasers is by the combined implementation of an unstable resonator and injection locking. This conclusion is not new, nor is it without precedent. Injection locking has long been recognized as a way to lock the frequency of a high-gain broadband oscillator to the frequency of a low-power narrow-bandwidth reference oscillator.[4] This technique has been applied successfully to unstable resonators for CO_2 lasers[5] as well as for dye lasers.[6] It has also been used previously to narrow the bandwidth and reduce the divergence of high-gain unstable-resonator excimer lasers.[7,8]

The technique is attractive because near-diffraction-limited output and ultranarrow linewidth can be achieved with no loss of laser efficiency. Moreover the attractiveness of the technique is enhanced by the fact that very little reference oscillator power is required to achieve those parameters. The contribution of our work has been to determine (and obtain) the theoretical minimum required injected power for reliable locking,[9] and to achieve transform-limited bandwidth in an injection-locked excimer system[10] (also with ultralow reference oscillator power).

ISSN:0094-243X/82/900191-05$3.00 Copyright 1982 American Institute of Physics

In this paper we review the simple technique employed for achieving transform-limited bandwidth in an XeF laser.

EXPERIMENTAL PROCEDURE

The experimental setup used for injection locking the XeF laser is schematically depicted in Fig. 1. As a reference oscillator we used a small pulsed Ar III-ion laser followed by a Fabry-Perot étalon, which passed a single longitudinal mode of the laser. The Ar-ion laser is inexpensive and simple to construct, and its design closely follows that of Ref. 11. It emits a polarized 10-W, 0.5 -μs-duration pulse whose divergence is near the diffraction limit. Because of the selectively reflective optics, most (∿ 80%) of the output energy was at the 3511-Å transition of the doubly ionized Argon atom. It was composed of several longtitudinal modes within the inhomogeneous Doppler linewidth (∿ 1 GHz).

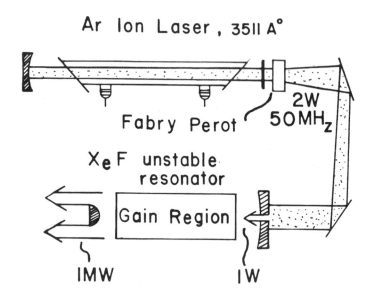

Fig. 1

The single longitudinal mode transmitted by the étalon ($\sim$ 2 W) had a bandwidth of < 50 MHz. About 50% (1 W)of the Ar-ion beam was injected through a 1-mm hole in the rear mirror of the positive-branch confocal unstable-resonator cavity of an XeF laser. The unstable resonator is $\sim$ 100-cm long (variable ± 5 cm), and the magnification is 3. The discharge dimensions are 2 cm x 2 cm x 60 cm, and the device is similar to the one in Ref. 9. The free-running XeF laser's total output energy was $\sim$ 40 mJ, and the beam was unpolarized. The timing of the XeF (25-ns pulse duration) and the Ar-ion laser are easily adjusted for injection locking since the Ar-ion laser-pulse duration is as long as 0.5 μs. The single-pass gain of the XeF amplifier at 3511 Å was measured to be $\gtrsim$ 10.

RESULTS

In contrast to the KrF, XeCl, and ArF lasers, there are inherent limitations to the injection locking of a XeF laser on the B → X transition. The XeF emission is composed of many distinct lines with several different upper vibrational levels,[11-13] and there is no necessarily fast energy transfer among these upper levels. Consequently, only part of the total energy is available to be pulled into a specific locked line. The effect is similar to that of an inhomogeneously broadened laser when one attempts to force it to run on a single longitudinal mode.

Since our 1-m spectrometer's resolution was $\sim$ 0.1 Å ($\sim$ 50 GHz), the spectrometer was not useful for measuring the locking efficiency into a 50-MHz bandwidth. Thus we determined the energy transfer by measuring the amount of XeF output energy pulled by the injection process into the polarization state of the Ar III laser. This measurement was accomplished by a deceptively simple yet accurate technique. First the laser energy was measured in both locked and unlocked (i.e. with the reference oscillator blocked) conditions. Then the measurements were repeated with a linear polarizer placed in front of the detector and oriented so as to block the polarization state of the reference oscillator. Some simple algebraic manipulation of the four numbers yields a locking fraction of $\sim$ 60%. (This can be compared with $\sim$ 100% for KrF or ArF.) The fraction locked can be expected to improve with higher temperature, higher pressure, and longer pulse duration.

The beam divergence (for the locked fraction) was found to be very near (< 1.5 x) the diffraction limit when compared with the theoretical optimum for an Airy disc of the same aperture. When unlocked, the divergence was 5 - 10 times greater.

In Fig. 2 we present a high resolution Fabry-Perot interferogram of the locked XeF output, demonstrating the 50-MHz bandwidth.

Fig. 2

CONCLUSIONS

We have demonstrated the channeling of ∿ 60% of the output energy of an unstable-resonator XeF laser into a 50 MHz Fourier-transform-limited bandwidth at 3511 Å. This has been achieved by injection locking the XeF laser with ∿ 1 W of power from a Fabry-Perot-filtered Ar-ion laser. If one wants to remove most of the remaining 40% broadband XeF output, it is possible to do so in a simple manner by extracavity filtration with an étalon after the unstable resonator or by spatial filtering since the unlocked portion has much larger divergence. Finally, we have been able to observe backward-stimulated Brillouin and Raman scattering from several liquids by use of the injection-locked XeF laser. These stimulated processes were observed only when the laser was locked, thus providing an additional indication of the reduction of the XeF laser bandwidth.[14,15]

These injection locking techniques demonstrate that ultahigh spectral brightness and excellent beam quality are available from excimer lasers with relatively uncomplicated and moderate-cost systems. This should enable a much larger number of researchers to probe nonlinear optics in the uv and vuv.

REFERENCES

1. R. T. Hawkins et al., Appl. Phys. Lett. 36, 391 (1980),
    H. Egger et al., Opt. Lett. 5, 282, (1980).
2. J. Reintjes, Appl. Opt, 19, 3889 (1980).
3. I. J. Bigio et al., CLEO '81 postdeadline paper THS3, IEEE J.
    Quantum Electron. QE-17, No. 12-II, 220 (1981).
4. R. Adler, Proc. IRE 34, 351 (1946).
5. C. J. Buczek, R. J. Frieberg, and M. L. Skolnik, Proc. IEEE 61,
    1411 (1973), C. Cason et al., J. Appl. Phys. 48, 2531 (1977).
6. Irving J. Bigio, in High-Power Lasers and Applications,
    K. L. Kompa and H. Walther, eds. (Springer-Verlag, Berlin
    1978), p. 116.
7. J. Goldhar and J. R. Murray, Opt. Lett. 1, 199 (1977);
    J. Goldhar et al., IEEE J. Quantum Electron. QE-16, 235 (1980).
8. T. J. Pacala and C. P. Christensen, Appl. Phys. Lett, 36, 336
    (1981).
9. Irving J. Bigio and M. Slatkine, Opt. Lett. 6, 336 (1981).
10. Irving J. Bigio and M. Slatkine, Opt. Lett. 7, 19 (1982).
11. J. C. Hsia et al., Appl. Phys. Lett. 34, 208 (1979); M. Rokni
    et al., Appl. Phys. Lett. 36, 243 (1980).
12. J. Tellinghuisen et al., J. Chem. Phys. 64, 4796 (1976).
13. J. Tellinghuisen et al., J. Chem. Phys. 68, 5177 (1978).
14. I. J. Bigio, B. J. Feldman, R. A. Fisher, and M. Slatkine, CLEO
    '81 postdeadline paper FK3, IEEE J. Quantum Electron. QE-17,
    No. 12-II, 228 (1981).
15. Michael Slatkine, Irving J. Bigio, B. J. Feldman, and
    R. A. Fisher, Opt. Lett. 7, 108 (1982).

REVIEW OF UPCONVERTED ND-GLASS LASER
PLASMA EXPERIMENTS AT THE
LAWRENCE LIVERMORE NATIONAL LABORATORY*

K. R. Manes
Lawrence Livermore National Laboratory
Livermore, California 94550

INTRODUCTION

Systematic scaling experiments aimed at deducing the
dependence of laser-plasma interaction phenomena on target plasma
material and target irradiation history have been underway in
laboratories all over the world in recent years.[1-5] During
1980 and 1981 the Livermore program undertook to measure the
laser light absorption of high and low Z plasmas and the
partition of the absorbed energy amongst the thermal and
suprathermal electron populations as a function of both laser
intensity and wavelength.[3-5] Simulations suggested that short
wavelength laser light would couple more efficiently than longer
wavelengths to target plasmas.[6,7] Shorter wavelength heating
of higher electron plasma densities would, it was felt, lead to
laser-plasma interactions freer of "annomalous" absorption
processes.[8,9] The following sections review LLNL experiments
designed to test these hypotheses.

ABSORPTION STUDIES

The experiments discussed here were carried out using the
Argus and Shiva lasers.[10,11] Figure 1 shows the Argus target
area where upconverted laser light was typically directed onto
600 μm diameter (Au, Be, Ni, Ti, CH) disk targets by f/2.2
lenses.[3] Many of the instruments mounted on the target chamber
are x-ray spectrometers. The first step in measuring the
disposition of laser energy is to determine the optical
absorption of the target plasma. Absorption, in this context, is
taken to mean the fraction of the incident laser energy not
accounted for by reflected beam diagnostics. A schematic diagram
of the Argus experiments, Figure 2, shows one of the 28 cm
diameter Argus 1.06 μm laser beams down collimated to 10 cm,
frequency doubled and tripled by KDP crystals and focused onto a
target located within an enclosing "box calorimeter."
Target irradiation conditions for absorption studies and the
x-ray measurements which followed are summarized in Table I.
Laser intensities noted are spatial average values at the peak of
the temporal pulse. Although the near field Argus laser beam

* Work performed under the auspices of the U. S. Department of
Energy by the Lawrence Livermore National Laboratory under
Contract No. W-7405-ENG-48.

Fig. 1   Argus laser facility with the target area in the foreground. The target chamber with its optical and x-ray diagnostic instruments is located in the center of the room.

presented to the focusing lens was very uniform (typically less than 10% intensity modulation on the beam), passive phase errors from the many optical components accumulated to produce highly structured beam profiles at the targets. Figure 3 quantifies this situation with intensity distribution functions measured on typical shots at each wavelength. Beam quality degraded with upconversion so that the $3\omega_0$ beams contained the most fine structure and showed very strong modulation. It should be noted here that each optical element employed met or exceeded the $\lambda/4$ optical flatness criterion before it was used in the laser system. Because of their large output apertures, future large Nd-glass lasers, such as Nova, will probably use arrays of conveniently sized nonlinear crystals in order to upconvert their pulses. Still more distorted target plane intensity distributions can be expected in these devices due to slightly mismatched crystals and diffraction from crystal interstices. Designers of directly illuminated capsules will encounter great difficulty in meeting their beam uniformity requirements. In the following, "intensities" will always refer to spatial mean intensity and error bars will denote plus or minus one standard deviation about this mean value.

Enclosing or "box calorimeters" have often been used at LLNL.[22] Unabsorbed laser light passes through transparent ion

198

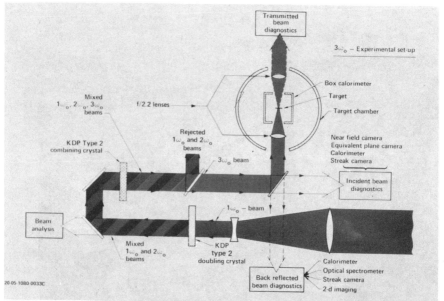

Fig. 2 Experimental layout for wavelength scaling experiments in the Argus facility.

shields (Schott WG-280) indicated in Figure 4 and is absorbed in the walls of the calorimeter. Each laser wavelength imposes a different set of criteria on the calorimeter components. For example, the infrared fundamental and the second harmonic green light can be absorbed on calorimeter walls clad with Schott NG-1 while ultraviolet third harmonic radiation was measured using Schott GG-19 absorber. At intervals the WG-280 ion shields which intercept target debris were replaced. On several occasions during the experiments quartz shields were substituted, but no difference in box calorimeter performance was noted. The calorimeter was recalibrated before and after each experiment and its integrity checked during the shot sequence.

Figure 5a groups absorption measurements made on Au plasmas according to incident laser wavelength while Figure 5b shows similar measurements carried out on low Z Be plasmas. The general trends characteristic of classical collisional absorption, inverse bremsstrahlung, are apparent in these data. At a given intensity, shorter wavelength light was absorbed more efficiently by higher Z plasmas. The scattering of incident third harmonic light by the plasmas was often too low to register above background in the calorimeter.

In preparation for x-ray conversion efficiency measurements and to check for any residual resonance absorption contribution, absorption versus target tilt for p-polarized irradiation was measured.[22] As expected, the f/2.2 focusing system introduced a wide enough range of incident ray angles so that the 600 to

TABLE I   Irradiation Conditions at $1\omega_0$, $2\omega_0$ and $3\omega_0$

- **Incident laser energy:**

  $3\omega_0$ $4J \leqslant E_{inc} \leqslant 40J$

  $2\omega_0$ $4J \leqslant E_{inc} \leqslant 200J$

  $1\omega_0$ $4J \leqslant E_{inc} \leqslant 10kJ$

  - **Minimum energy: diagnostics limited**

  - **Maximum energy: damage limited**

- **Laser pulse duration: 600 to 700 ps FWHM**

- **Focus:**

  Argus: f/2.2, P polarization

  Shiva: 20 f/6 beams, radial polarization

- **Spot diameters:**

  $\sim 150~\mu m$ for $I \sim 3 \times 10^{13}$ W/cm^2

  $\sim 150~\mu m$ and $\sim 100~\mu m$ for $I \sim 3 \times 10^{14}$ W/cm^2

  $\sim 100~\mu m$ and less for $I \geqslant 10^{15}$ W/cm^2

700 ps Gaussian laser pulses were absorbed with only a weak dependence on the angle between the beam axis and the target normal. Figure 6 shows that no detectable difference in absorption was noted for tilts as large as plus or minus 30° from the converging laser beam axis. This leads to the conclusion that resonance absorption plays a relatively minor role in these measurements.

Laser light scattered from the target into the focusing lens cone was analysed by time resolved backscatter spectrometers. A decided red shift characteristic of scattering from laser driven ion waves, stimulated Brillouin scattering, can be seen in all of this data; typical examples of which are reproduced in Figure 7c.[5] If the observed spectral shifts are viewed as the result of competition between stimulated Brillouin scattering and the Doppler shift imposed by an expanding plasma, the measured backscattered spectra as the targets are tilted provide a crude estimate of coronal electron temperature.[12] Figure 7a and b are reproduced from Rosen et al (Ref. 12) wherein it is observed that the Brillouin red shift, f_B, should be reduced by the Doppler blue shift, f_D, of outward moving matter. Since that

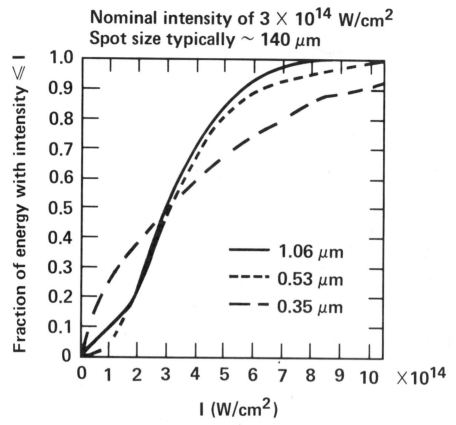

Nominal intensity of 3×10^{14} W/cm²
Spot size typically $\sim$ 140 μm

Fraction of energy with intensity $\leq$ I

I (W/cm²)

——— 1.06 μm

- - - - 0.53 μm

— — 0.35 μm

Fig. 3   Target plane intensity distribution function at 1 μm, 0.5 μm and 0.35 μm.

matter blows off roughly normal to the face of the disk, tilted disk spectra should show a smaller Doppler blue shift. In making an estimate of matter velocity from spectral shifts and in the soft x-ray conversion measurements which will be described next, the assumption is implicit that laser heated plasmas of like material which absorb like amounts of energy of a given wavelength, intensity and pulse duration are essentially identical even though they are irradiated at different incidence angles. With this in mind, the observed average spectral shifts may be written

$$\Delta f(\theta) \cong f_B + f_D \cos \theta \qquad (1)$$

Where θ is the angle between the target normal and the laser beam/observation axis. Table II summarizes the velocities inferred from the spectra in Figure 7 using Lasnex estimates for the average ionic charge state, $\bar{Z}$, at Ne $\approx$ 1/3Nc.[5,12]

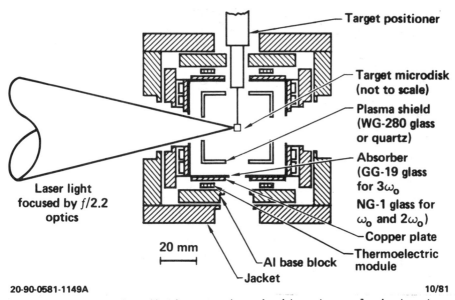

20-90-0581-1149A 10/81

Fig. 4   Target irradiation was done inside a box calorimeter to
measure scattered laser light.

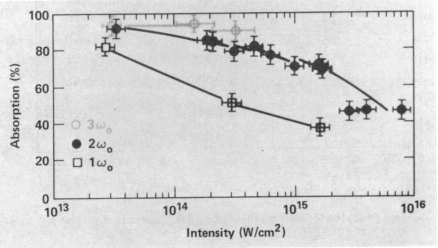

Fig. 5a $1\omega_0$, $2\omega_0$ and $3\omega_0$ absorption - Au disk targets.

The picture which emerges from simulation calculations is of
a progressively more "classical", cooler corona plasma as I or λ
is decreased, particularly for high Z targets.[4] Particle-in-
cell calculations by Eastabrook et al were carried out in order
to arrive at the predictions reproduced in Figure 8.[12,13]

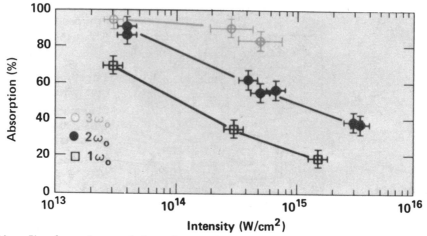

Fig. 5b $1\omega_0$, $2\omega_0$ and $3\omega_0$ absorption - Be disk targets.

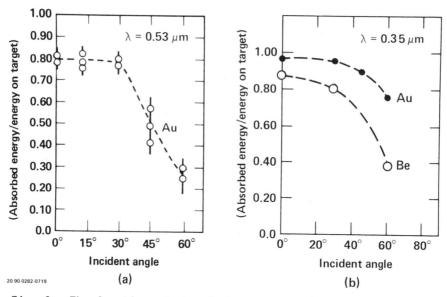

20-90-0282-0719

(a)   (b)

Fig. 6 The fraction of absorbed energy was investigated for
        angle of incidence dependence for both high and low
        Z-targets.

While stimulated Brillouin scattering plays a limited role in
these calculations, backscattered spectral predictions have thus
far failed to account for the large red shifts observed in, for
example, Figure 7. The gross features of absorption are
satisfactorily modeled now, but the present generation of
simulations fails to reproduce reliable low density corona
information and is thus limited in predicting optical scattering

and x-ray production adequately.

Table II

Wavelength	Intensity	$\bar{Z}$	$\langle T_e \rangle + \dfrac{3\langle T_i \rangle}{Z}$	Matter Velocity
1.06 μm	3×10^{14} W/cm^2	51	(5 ± 3) keV	$\sim 3 \times 10^7$ cm/sec
1.06 μm	3×10^{15} W/cm^2	58	(20 ± 10) keV	$\sim 7 \times 10^7$ cm/sec
0.53 μm	3×10^{14} W/cm^2	50	(2.8 ± 1) keV	$\sim 1.8 \times 10^7$ cm/sec

SOFT X-RAY CONVERSION STUDIES

With absorption established, measurements of x-ray conversion efficiency began. Laser heated targets, particularly those irradiated for several hundred picoseconds or more, expand in an axially symmetric two dimensional flow which naturally causes the x-ray emitting region to move with time.[12,14] Shiva experiments using a streaked x-ray microscope, observed $\sim 10^{14}$ W/cm^2 irradiations of Au disk targets.[15] Striking photographs such as that in Figure 9 show a thin emission zone moving steadily away from the target's initial surface at a

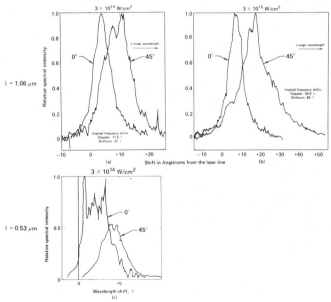

Fig. 7   From the angular dependence of the mean red shift, the Doppler and Brillouin shifts can be separated.

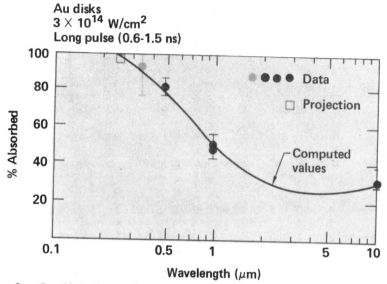

Fig. 8   Predicted wavelength scaling of absorption by high Z
materials match experimental results.

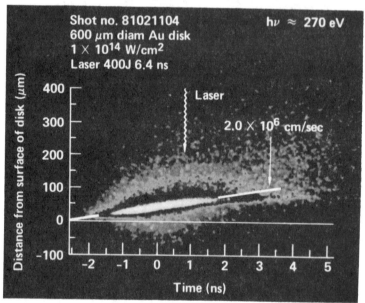

Fig. 9   Streaked x-ray image of a laser irradiated gold disk.

velocity which scales approximately as the 0.44 power of the peak
incident laser intensity. Such observations make it clear that
x-ray emitting target surfaces evolve rapidly in time and space.

The assumption of either a spherical or a planar emitting surface can not be safely made and observations at many angles are needed to derive an angular integral of x-ray emission. By the same token, time resolved spectral measurements made over the past several years have consistently reported that laser heated target plasmas are not in local thermodynamic equilibrium.[12] High Z plasmas, such as Au, often emit copious N and M line radiation. Without five or more time resolved channels between approximately 0.1 and 1.5 keV, an accurate x-ray energy accounting is not possible. LLNL x-ray diagnosticians have faced the challenge of performing absolutely calibrated measurements on very bright pulsed sources in hostile environments for over thirty years. The "Dante" instruments used in these studies properly belong to that technology in which spectral resolution is given up for absolute calibration.[5,25] Figure 10 contains a sampling of x-ray spectra recorded in 0.53 μm target irradiations.[4]

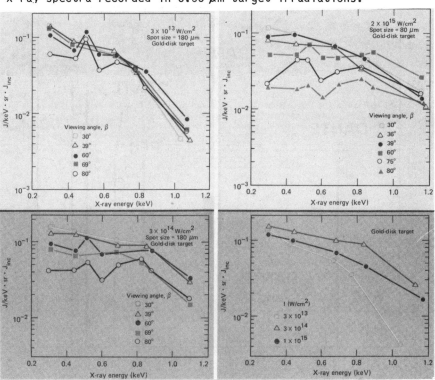

Fig. 10 Typical x-ray spectra from 0.53 μm heated gold disk
targets.

Clearly, both an angular and a spectral integration must be carried out in order to measure the x-ray conversion efficiency.
At both the Argus and the Shiva target irradiation facilities, ten channel "Dante" x-ray K&L edge filtered x-ray

diode systems measured spectra such as those in Figure 10 on each target shot. Three such instruments labled T, L and W and independently absolutely calibrated viewed the targets in the Argus chamber from widely separated angles. As Figure 6 suggests, target tilts in the range of plus or minus thirty degrees have no detectable effect on absorption. It is assumed that soft x-ray emission will be likewise unaffected by target tilts within this range (as no other observable was seen to change). The lines of sight relative to the target normal obtained by +30, +10 and -30 degree tilts, for example, adequately span the important viewing angles as indicated in Figure 11. Note, for example, that Dante L and W share the same polar angle but widely separated azimuthal angles at a target tilt of -30°. This provides a check of the assumed azimuthal symmetry of the x-ray flux.

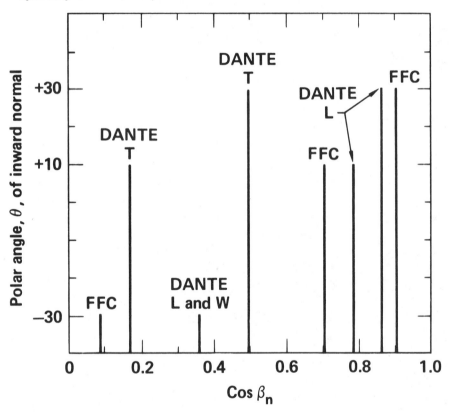

Fig. 11   Soft x-ray diagnostic viewing angles are plotted versus target tilt. Three Dante spectrometers and two broadband "fast flat calorimeters", FFC, were used.[4,5]

The angular emission of soft x-ray flux from Au disks heated by 25J, 600 ps, 0.53 μm pulses is displayed in Figure 12. On

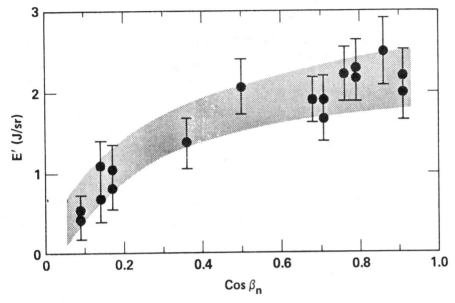

Fig. 12 Angular distribution of x-ray emission from Au at
$I \sim 3 \times 10^{14} W/cm^2$ (25 Joules/600 psec).

such a plot, an ideal planar source, a Lambertian emitter, would be represented by a straight line passing through the origin. An ideal spherical source would deliver the same amount of energy into any angle and so would be represented by a horizontal line on Figure 12. The time varying source measured, however, can not be represented by either of these simple shapes without substantial error. Many similar measurement sets provide the conversion efficiency data summarized by Figure 13.[3,4,16] Significantly, the shorter wavelength laser light is converted to soft x-rays with an efficiency of $\sim$60%.

The roll-off in conversion efficiency observed in the 0.53 and 0.35 μm data was not anticipated and is not yet completely understood.[16] Rather, simulations contended that the conversion efficiency should decline monotonically with $I\lambda^2$ at approximately the rate exhibited by the 1.06 μm data.[4]

The soft x-ray conversion efficiency also varies with target material atomic number, Z. Although angular x-ray distributions were not measured when all of these data were taken, Figure 14 shows that higher Z target plasmas heated with 1.06 μm laser pulses emitted soft x-rays more efficiently than lower Z plasmas.[25,26] Since sub keV x-ray production far from spectral regions dominated by line emission are probably generated by bremsstrahlung of low energy electrons colliding with screened nuclear potentials, it may be more meaningful to plot normalized

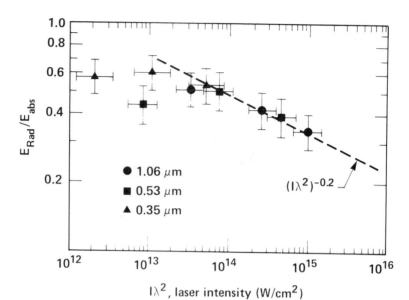

Fig. 13 Thermal x-ray conversion efficiencies from Au disk
targets.

x-ray energy versus $\bar{Z}$, where $\bar{Z}$ is the LASNEX calculated charge
state. As Figure 14 shows, the soft x-ray conversion efficiency
appears to increase almost linearly with $\bar{Z}$.[26]

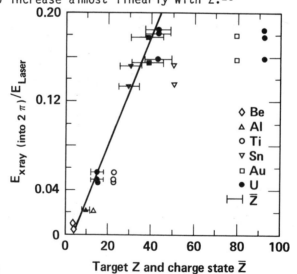

Fig. 14 X-ray energy for various targets as a function of target
Z and charge state $\bar{Z}$, at incident intensity of
5×10^{14} W/cm^2. Error bars for values of
E_{x-ray}/E_{laser} are all $\sim \pm 25\%$.

X-RAY LINE PRODUCTION

High resolution crystal spectroscopy has revealed very complex line structures radiating from non-LTE laser heated targets.[25] This review can not even brush the surface of this fascinating field and must be restricted to a small part of the data accumulated during the past year. Fundamental to this sort of study is an absolute measurement of the amount of laser energy required to produce a given amount of "line" emission. The laser intensity, pulse duration and wavelength play a significant role. Qualitatively, long low intensity short wavelength laser pulses create enough hot plasma to allow copious production of highly ionized atoms, which radiate He- and Li-like lines. Short high intensity long wavelength laser pulses produce suprathermal electrons which induce fluorescence lines in cold matter; these lines can dominate the radiated x-ray spectrum.

Motivated by a search for x-ray backlighter sources suitable for radiography of ICF targets, Rupert, Matthews and Koppel conducted a survey of lines in laser heated plasmas; concentrating on the K lines of Ti, Ni and Zn.[4] They defined a conversion efficiency, ξx, for an x-ray line emitter as the number of photons per line or line cluster divided by the incident laser energy. For all materials and laser irradiation conditions tested, $\xi x \cong 10^{11}$ to 10^{12} photons/J. The maximum measured conversion efficiencies are summarized in Figure 15.[4]

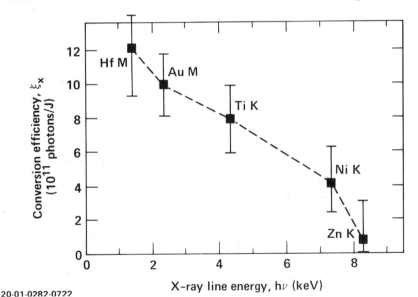

20-01-0282-0722

Fig. 15 Maximum measured conversion efficiencies for various backlighters.

Figure 16, drawn from Ref. 4 and 17, shows the variation in

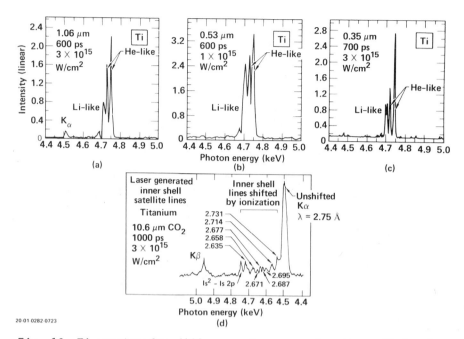

Fig. 16 Ti spectra for difference laser wavelengths. Plots a),
         b) and c) are taken from Reference 4 while d) comes from
         Ref. 17.

spectral emission with laser wavelength. At the shorter
wavelengths, all emissions are "thermal" (He- and Li-like); the
spectra show no detectable level of the cold K_a line that is seen
at $\lambda = 1.06 \mu$m and dominates the spectrum at $\lambda = 10.6 \mu$m.
Variations in ξx with incident laser intensity are illustrated by
titanium data in Figure 17.[4] Below a fairly sharp threshold
which depends on laser wavelength and pulse duration, x-ray lines
were not observed. Above this threshold, the conversion
efficiency increased with decreasing laser intensity as was the
case with soft x-ray conversion efficiency. Although somewhat
higher conversion efficiencies were observed for shorter
wavelengths, at a given laser intensity there were no significant
differences in ξx seen when the laser pulse duration was varied
from 100 ps to 2 ns.

Much effort has gone into measuring the temporal and spatial
extent of the x-ray emission from laser heated
plasmas.[4,17,18,19] Typically, the x-ray flashes were the same
or longer in duration than the laser pulses with K_a emission
slightly (50 to 100 ps) delayed as though the suprathermal
electrons exciting these lines were produced late, traveled
outwards and reflected from coronal potentials before interacting
with the cold matter. Spatially, He- and Li-like line emission
was confined to the most vigorously heated region of the target

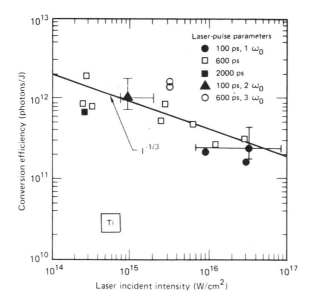

Fig. 17 Line-conversion efficiency as a function of laser
         intensity for Ti.

while K_α emission was reported from somewhat larger areas.

SUPRATHERMAL ELECTRON MEASUREMENTS

Discrepancies between simulation and experimental
observations have often been traceable to a poor understanding of
high temperature coronal plasma physics. Not surprisingly,
LASNEX simulations of the higher energy or suprathermal x-ray
spectra observed are often in disagreement with the data.[4]
Deviations are most pronounced in the five to fifteen keV range
with good agreement below and, with a flux limiter parameter
adjusted to fit, reasonable agreement at energies above about
forty keV. LASNEX typically over estimates this mid-range x-ray
flux from high Z targets.[4]
Two instrument packages provide the bulk of the LLNL
suprathermal x-ray spectral information. A sensitive K-edge
filtered four channel photomultiplier system was needed to record
the weak suprathermal emission from 0.35 μm heated
targets.[5,25] Higher intensity 1.06 μm irradiations produced
sufficient flux to measure with a more accurate filter-fluorescer
spectrometer.[5,25] Working together, these two spectrometers
produced the spectra in Figure 18. By fixing intensity on target
of approximately 3×10^{14} W/cm^2, the suprathermal spectra from
Au plasmas can be compared as the laser wavelength is varied.
The general trend is clear, shorter wavelength irradiations give
rise to significantly fewer, cooler suprathermal electrons.

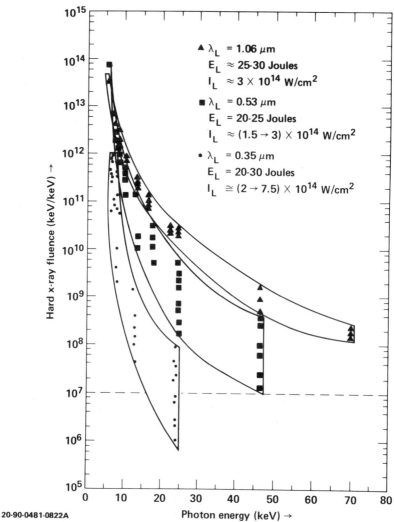

Fig. 18 The fraction of energy in hot electrons decreases
drastically at shorter laser wavelengths.

20-90-0481-0822A

The bremsstrahlung x-rays produced by suprathermal electron
scattering from target ions provide a measure of the suprathermal
electron population. While the data shown in Figure 18 for each
wavelength and Au plasmas scatters over as much as two decades,
the measurements at each x-ray energy do not overlap. That is,
the highest x-ray flux observed in 0.35 μm target irradiations
lies below the lowest x-ray flux detected in 0.53 μm experiments
employing the same average intensity on target. Simulation
calculations typically do not employ the highly structured beam
profiles with which these targets were irradiated. This, more

targets were irradiated. This, more than any other factor,
contributes to the calculational uncertainty, experimental data
scatter and concomitant disagreement between simulations and
experiments.[4,20]

Like the soft x-ray conversion efficiency determinations,
suprathermal x-ray flux measurements are sensitive to incident
laser intensity. Figure 19 bins the data according to intensity

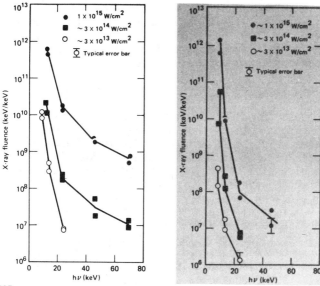

20-90-0282-0725

Fig. 19 Dependence of 2ω and 3ω x-ray fluence on laser-beam
intensity.

and reveals that suprathermal flux increases with increasing
intensity on target. Once again, the scatter in the data is
large probably due to beam quality, but the general trend is
clear. Suprathermal x-ray spectra from plasmas whose heating is
not dominated by resonance absorption, often do not exhibit the
clear two temperature character evident in earlier short pulse
1.06 μm studies.[23] Assignment of unique intensity or
wavelength dependent temperatures is not meaningful for spectra
such as those in Figure 19.

Experimentally, there are two avenues to pursue. On the one
hand, every reasonable effort must be expended to produce and
reliably diagnose more uniform laser beams at the target. At the
same time, the signatures of low density, high temperature
coronal processes which might produce suprathermal electrons must
be searched for in target emissions. Limited by time and optics
availability, the disk experiments reported here did not involve
attempts to significantly improve the laser beam profile.
Optical spectrographs and filtered detectors were available,

however, so scattering measurements were made at frequencies both above and below the target irradiation frequency to look at processes occurring at densities below Nc/4, i.e. Raman light and $3/2\omega_0$ light.[4] If resonant stimulated Raman scattering takes place in the coronal plasma, detectable quantities of light at twice the incident wavelength should be emitted out of the plane of polarization of the incident laser light. Refraction will cause this light to be observed near the target's normal. Non-resonant Raman scattering can occur at electron densities below Nc/4 and involves the decay of the incident photons at ω into electron plasma wave phonons at ω_{pe} and scattered photons at ω_R where energy conservation requires

$$\omega_R = \omega - \omega_{pe} \tag{2}$$

A related process, $2\omega_{pe}$ decay, might also be operative. In this case, incident optical photons give up their energy to two electron plasma waves, both at ω_{pe}, and this can only be important near Nc/4. This latter process, like the Raman process, can give rise to very high phase velocity electron plasma waves whose wavebreaking energies are typical of suprathermal electron temperatures observed experimentally.[4,26] It is difficult to quantify the levels of either of these processes because Raman scattered photons are born very close to their critical densities and hence may be absorbed while $2\omega_{pe}$ decay produces no scattered photons. Evidence for $2\omega_{pe}$ decay can come from electron emission measurements and optical spectroscopy at sum frequencies such as $3/2\omega$.

Time integrated and spectrally integrated Raman emission at 1ω and 2ω are shown in Figure 20.[4] Spectra of $3/2\omega$ emissions are displayed in Figure 21. The strongest Raman emission was observed out of the plane of the incident laser's polarization, as expected.[4] Angular and spectral integrals for Raman emissions from 1.06 μm heated disk targets are plotted against average incident intensity in Figure 22.[4] Like suprathermal x-ray emission, Raman and $3/2\omega$ light levels fall rapidly with reduced intensity on target, reduced plasma size, and shortened wavelength. In some cases, clear evidence of non-resonant Raman sidescattering has been obtained. Figure 23, reproduced from Reference 4, shows wavebreaking energies expected for stimulated Raman scattering in 5 keV electron plasma temperatures. For example, the Raman spectrum displayed in Figure 20b indicates that substantial amounts of Raman energy was sidescattered at electron densities near $0.1N_c$. Raman backscatter may be excluded because at $N/N_c \simeq .05$, $K\lambda_D$ and thus Landau damping would be too large. Figure 24 is the suprathermal x-ray spectrum measured on that same shot. The slope of this spectrum suggests a suprathermal temperature of $\sim$40 keV in good agreement with the Raman sidescatter wavebreaking energy for $0.1N_c$ shown in Figure 23. Data is therefore in hand which ties unique suprathermal electron temperatures to Raman scattered spectra. Where they

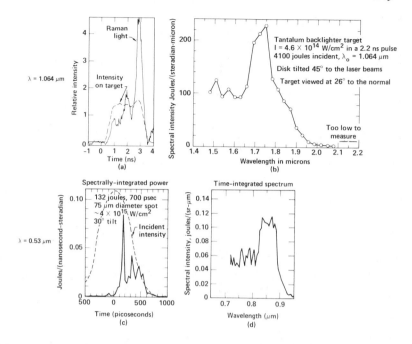

Fig. 20   Raman emission from laser heated high Z disk targets.

(a) Stimulated Raman spectrally-integrated power and
(b) Raman light spectrum for a tantalum disk irradiated
with 4100 J in a 2.2 ns pulse at 4.6 x 10^{14}W/cm^2.

(c) Time-resolved Raman light spectrum for a gold disk
irradiated with 5320 Å light. Spectrally-integrated
power and (d) Time-integrated spectrum.

exist, data such as that in Figure 20 together with the model
used to calculate Figure 23 can be used as a plasma electron
temperature diagnostic.[4] The Raman spectrum can provide an
estimate of low density coronal electron temperature since the
shifts are large compared to the Doppler shifts. As noted with
stimulated Brillouin scattering, lowered coronal temperatures can
be obtained at lowered intensities or shortened incident laser
wavelength. Although Raman light levels from disk target studies
are energetically small, the Raman light and $3/2\omega_0$ levels rise
rapidly as the irradiating pulse duration and hence the volume of
scattering plasma is increased. Figures 20a and c show time
resolved Raman signals which clearly grow with time into the
laser pulse. Raman and $2\omega_{pe}$ instabilities must therefore be

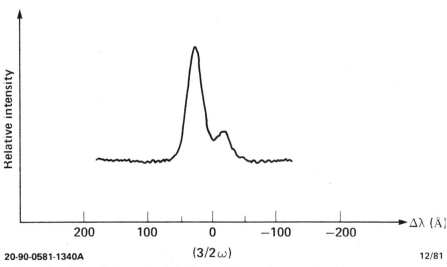

20-90-0581-1340A   (3/2ω)   12/81

Fig. 21   Time integrated 3/2ω spectrum for an incident intensity
of I = 3 x 10^{14}W/cm^2, a Be disk target and laser
wavelength = 1.06 μm. We observe a red peak shifted
≈ 30Å and a blue peak shifted ≈ 20Å.

carefully controlled lest they become important mechanisms for
suprathermal electron production. These studies suggest that
these instabilities can be suppressed simply by employing short
wavelength lasers to drive ICF targets.

ENERGY PARTITION

The proceeding sections provide a basis for determining the
relative fractions of the incident laser energy which is absorbed
and then partitioned into soft x-ray emissions, thermal and
suprathermal electron populations. Table III summarizes these
findings for Au plasmas irradiated at each wavelength at an
average incident intensity of 3x10^{14}W/cm^2. Substantially
cooler coronal plasmas, free of severe electron plasma wave
instabilities which simultaneously radiate efficiently in the
soft x-ray spectral region can be realized by shortening the
irradiating laser wavelength.

Taken together, stimulated scattering spectra and
time-resolved x-ray images provide average fluid velocity
estimates at several electron densities. Figure 25 indicates the
observed relationships for Au disk targets in the range of peak
average powers explored.[4]

The data in hand is sparse and the scalings suggested in
Figure 13 and 25 may not survive future studies. It is
remarkable, however, how well the simple scaling arguments put
forward by Rosen et al describe the plasma behavior.[4,12] The
model assumes that electron thermal conduction transports energy

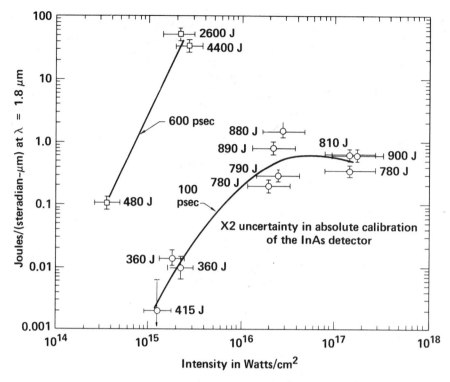

Fig. 22 Stimulated Raman scattering depends on the laser wavelength, the intensity and the size of the underdense plasma.

from the laser absorption region (at critical density, $N_e = N_c \simeq (10^{21}/\lambda^2)$ electrons/cm^3 for λ in μm) to the overdense plasma. Equating absorbed laser flux, I_a, with the heat flux carried off by the plasma, $N_c T_e v_e$, leads to the relation $I_a \lambda^2 \sim T_e^{3/2}$.[12] To use this expression, the model assumes that the plasma is expanding outward (toward the laser) at the local ion sound speed which scales as the square root of pressure divided by density, $v \sim C_s \cong (P/\rho)^{1/2} \sim (\bar{Z}T)^{1/2}$ and that T is proportional to T_e. Table II shows that near $Nc/3$, $\bar{Z}$ varies weakly with Te for Te > 2 keV. Neglecting this $\bar{Z}$ dependence leads to

$$v \sim \left(I_a \lambda^2\right)^{1/3} \tag{3}$$

which may be compared to the velocities inferred from SBS spectra in Figure 25.

Since dense material radiates more efficiently than less dense material at a given temperature, the soft x-ray emission peaks at the highest density reached by the thermal-conduction

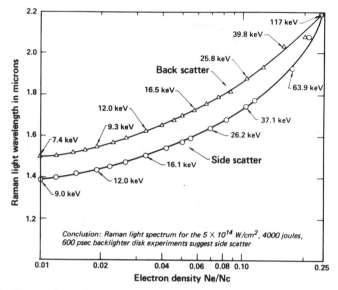

Fig. 23   Comparison of the solutions to the dispersion equations for side and backscatter at 5 keV.

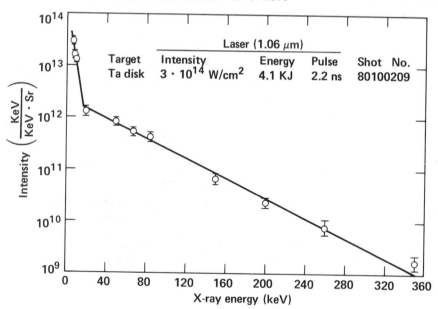

Fig. 24   Suprathermal x-ray spectrum.

TABLE III

Absorbed Energy Partition at $I \sim 3 \times 10^{14} W/cm^2$

Au Disk Targets

	1.06 μm	0.53 μm	0.35 μm
Soft x-ray flux	0.4 ± 0.1	0.5 ± 0.1	0.6 ± 0.1
Coronal hydro	0.6 ± 0.1	0.5 ± 0.1	0.4 ± 0.1
Suprathermal electrons	0.03 to 0.06	0.01 to 0.001	0.001 to 0.0001

front.[4] For example, bremsstrahlung from a single species hydrogen-like plasma may be written

$$P_{Br} \cong 1.6 \times 10^{-34} (T_e^{1/2}\bar{Z}^3N_i^2) \text{ Watts/cm}^3 \tag{4}$$

where N_i is the ion density.[21]

Rosen assumes that the temperature, which is falling rapidly near the thermal front as it propagates into the dense plasma, is proportional the temperature, T_e, in the absorption region. Equations of state for gold indicate that, for T less than 1 keV, $\bar{Z} \approx 1.6 [T(ev)]^{1/2}$ so the sound speed in the peak soft x-ray emission region should scale as $T^{3/4}$ or

$$(I_a \lambda^2)^{1/2}.\tag{5}$$

By assuming that inverse bremsstrahlung is the dominant absorption mechanism, Rosen arrives at a scaling for the temperature in the absorption region in terms of the incident laser intensity, $T \sim I^{0.6}.[12]$ Thus, the peak velocity of the emission region should scale as $v \sim I^{0.45}$ which is within the experimental error range, $v \sim I^{0.44 \binom{+0.22}{-0.17}}.[4]$

The same model crudely estimates that for fixed laser pulse duration the depth of penetration of the thermal conduction front is proportional to temperature, $T_e.[4]$ Time resolved x-ray images such as that in Figure 9 showed the width of the emission region to scale as $I^{0.7\pm0.25}$ in keeping with this model.

Soft x-ray spectra such as those in Figure 10 show the

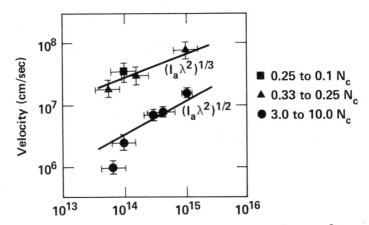

Fig. 25   Fluid flow velocities estimated from stimulated
           scattering and time resolved x-ray images. One
           dimensional current continuity would require
           $N_e v_e \cong$ constant and one dimensional mass continuity
           would require $N_i v \cong$ constant so it was not surprising
           that typical velocities in the subcritical plasma are
           about ten times higher than those in the denser x-ray
           emission region. The solid lines come from the scaling
           arguments of Rosen et al.[4,12]

non-local-thermodynamic equilibrium of these disk plasmas.
Strong Au N-band emission near 800 ev is apparent, particularly
at the higher incident intensities, and the spectra are hardly
Plankian. It is interesting, nevertheless, to apply the same
model to single species bremsstrahlung emission. Emitted
intensity would vary as

$$I_{Br} \sim \bar{Z}^3 N_i{}^2 T_e{}^{3/2}. \tag{6}$$

Given current continuity; i.e. that $N_e v_e = Z N_i T_e{}^{1/2}$ is
nearly constant into the material for each absorption region
temperature, the emitted intensity can be written

$$I_{Br} \sim \bar{Z} T_e{}^{1/2} \sim T \sim I_a{}^{2/3}. \tag{7}$$

Consequently this part of the x-ray conversion efficiency would
scale as

$$\eta_{Br} = I_{Br}/I_a \sim I_a{}^{-1/3} \tag{8}$$

Which suggests that the x-ray conversion efficiency should
decline weakly as I_a increases. Similarly, given mass
continuity, i.e. $N_i v \sim N_i \bar{Z}^{1/2} T^{1/2}$ is nearly constant into

the material, the emitted intensity becomes

$$I_{Br} \sim \bar{Z}^2 T_e^{1/2} \sim T^{3/2} \sim I_a. \tag{9}$$

This scaling suggests that the x-ray conversion efficiency is roughly independent of I_a. Figure 13 shows the total x-ray conversion efficiency scaling as $(I\lambda^2)^{-0.2}$ for $I\lambda^2 > 10^{13}$. Without accurate x-ray line production and transport calculations, however, quantitative modeling of x-ray production is impossible.

Simulation of sub-critical density plasma is an even greater challenge. Fortunately, the data available to us suggests that suprathermal electron production may be made energetically unimportant by short wavelength irradiation. ICF target designers can thus avoid most of their preheat problems by specifying short laser wavelength.[24]

ACKNOWLEDGEMENTS

The experiments reviewed here were carried out by the members of the Experiments Group, an element of H. G. Ahlstrom's Fusion Experiments Program, while the author had the privilege of leading that group. Senior members of the Experiments group during this interval included J. Auerbach, D. Banner, E. Campbell, D. Matthews, D. Phillion, V. Rupert, R. Turner, and F. Ze. Experiments of this scope would have been impossible without the enthusiastic support of the Target Fabrication Program under C. Hendricks, the Diagnostics Group under W. Slivinsky and the Laser Systems Operations Group under O. Barr and J. Hunt. Finally, the design and analysis of these studies was carried out in close collaboration with John Nuckolls' target design program whose simulation calculations are our best guide in the search for practical inertial confinement fusion.

REFERENCES

1. E. Fabre et al, in Proceedings of the Conference on Plasma Physics and Controlled Nuclear Fusion Research, Vienna, Austria, 1980 (International Atomic Energy Agency, Vienna, 1981), Paper No. IAEA-CN-38/1-4.

2. A. G. M. Maaswinkel, K. Eidmann, and R. Sigel, Phys. Rev. Lett. 42, 1625 (1979); D. E. Slater et al, Phys. Rev. Lett. 46, 1199 (1981); W. Seka et al, Bull. Am. Phys. Soc. 25, 895 (1980).

3. M. C. Mead, E. M. Campbell, R. E. Turner, W. L. Kruer, P. H. Y. Lee, B. Pruett, V. C. Rupert, K. G. Tirsell, G. L. Stradling, F. Ze, C. E. Max, and M. D. Rosen, "Laser-Plasma Interactions at 0.53 μm for Disk Targets of Varying Z", Phys. Rev. Lett. 46, 1289 (1981).

4. Section 7 of "Laser Program Annual Report - 1980", Lawrence Livermore National Laboratory, Livermore, California, UCRL-50021-80, pages 7-2 to 7-76 (1981).

5. H. G. Ahlstrom, "Laser Fusion Experiments, Facilities and Diagnostics at Lawrence Livermore National Laboratory", Applied Optics 20, 1902 (1981).

6. J. Nuckolls et al, Nature (London) 239, 139 (1972).

7. J. H. Gardner and S. E. Bodner, "Wavelength Scaling for Reactor-Size Laser Fusion Targets", NRL Memorandum Report 4623, Naval Research Laboratory, (1981).

8. C. E. Max, C. F. McKee, and W. C. Mead, Phys. Fluids 23, 1620 (1980).

9. C. E. Max and K. G. Estabrook, Comments Plasma Phys. Controlled Fusion 5, 239 (1980).

10. D. R. Speck et al, "The Performance of Argus as a Laser Fusion Facility", Lawrence Livermore National Laboratory, Livermore, California, UCRL-52487.

11. D. R. Speck et al, "The Shiva Laser-Fusion Facility", IEEE Journal of Quantum Electronics QE-17, 1599 (1981).

12. M. D. Rosen et al, Phys. Fluids 22, 2020 (1979).

13. K. Estabrook et al, Phys. Rev. Lett. 46, 724 (1981).

14. M. Herbst et al, NRL Memorandum #4436 and BAPS 26, 1024 (1981).

15. Section 5 of "Laser Program Annual Report - 1980", Lawrence Livermore National Laboratory, Livermore, California, UCRL-50021-80, pages 5-23 to 5-31.

16. C. E. Max et al, BAPS 26, 873 (1981).

17. A. Hauer et al, Applied Optics 20, 3477 (1981).

18. V. Rupert, D. L. Matthews and L. N. Koppel, "X-ray Backlighting Sources of 4 to 10 keV for Laser Fusion Targets", presented at the 1st European Conference on Cineradiography with Photons or Particles, UIC, Paris, France, May 1981, UCRL-85592 (1981).

19. J. D. Hares, J. D. Kilkenny, M. H. Key and J. G. Lunney, Phys. Rev. Lett. 42, 1216 (1979).

20. B. Arad and S. Eliezer, Appl. Phys. Lett. 32, 401 (1978).

21. D. L. Book, "NRL Plasma Formulary", NRL Memorandum Report No. 2898.

22. K. R. Manes, V. C. Rupert, J. M Auerbach, P. Lee, and J. E. Swain, Phys. Rev. Lett. 39, 281 (1977).

23. K. R. Manes, H. G. Ahlstrom, R. A. Haas, and J. F. Holzrichter, J. Opt. Soc. Am., 67 717 (1977).

24. R. L. McCrory, BAPS 26, 972 (1981) and references cited therein.

25. Low Energy X-ray Diagnostics - 1981, Edited by D. T. Attwood and B. L. Henke, American Institute of Physics Conference Proceedings No. 75, New York, (1981).

26. Section 6 of "Laser Program Annual Report - 1979", Lawrence Livermore National Laboratory, Livermore, California, UCRL-50021-79 (1980).

GENERATION OF VACUUM ULTRAVIOLET RADIATION
BY PLASMA NONLINEARITIES*

R. L. Carman and C. H. Aldrich
Los Alamos National Laboratory, Los Alamos, NM 87545

ABSTRACT

Numerical simulations of the relevant plasma hydrodynamics indicate that ten (19.3 µm) or more harmonics of an ArF laser should be generated at high efficiency in a high irradiance solid target plasma interaction experiment.

INTRODUCTION

The observation of very high harmonic light in CO_2 laser plasma interaction experiments[1] and the subsequent experimental study of the properties of this radiation[2,3] have led to a revived interest in the theory of harmonic generation in laser produced plasmas. We will begin with a brief review of the key experimental results and the current theoretical situation for CO_2 laser plasmas. We then turn our attention to a theoretical extension of this work to shorter wavelength irradiation systems where we shall predict that optical harmonics should be produced in the vacuum ultraviolet. When we use the ArF laser for a high irradiance source, up to five harmonics are predicted for pulse widths of < 10 ps, and even more harmonics may be possible for pulse widths of < 2 ps. Several uncertainties in the calculations make it difficult to be very precise, but the results are sufficiently interesting to suggest that the ideas be pursued experimentally.

HIGH IRRADIANCE CO_2 LASER-PLASMA INTERACTION RESULTS

In early work, we noted that for CO_2 laser intensities of I_L > 10^{14} W/cm^2, the behavior of the harmonic conversion efficiency with harmonic order changed from an approximate linear fall-off with order[4,5] to one which was almost constant in conversion efficiency to greater than 20 harmonics, with a subsequent rapid fall-off.[1] The number of harmonics appeared to be approximately independent of the solid target material while the angular distribution appeared

*Work performed under the auspices of the U. S. Department of Energy.

to be approximately flat, leading to conversion efficiency estimates of 10^{-4} to 10^{-5} of the incident energy converted to each harmonic order. Repeating the Gemini laser facility experiments at the Helios laser facility, with incident intensities of up to $I_L \sim 3 \times 10^{16}$ W/cm^2, indicated that more than 30 harmonics were efficiently produced with higher conversion efficiency, constant to within 50% of each other, thus suggesting a correlation between the maximum harmonic efficiently produced n and incident laser intensity I_L.

Two dimension particle simulations were carried out[1] under conditions where collisions could be ignored but where Maxwell's equations were solved exactly. For incident laser intensities of $I_L \sim 10^{14}$ W/cm^2 to 10^{15} W/cm^2, we observed a correlation between the number of harmonics predicted and the features of the plasma density profile modification. The Fourier transform of the scattered or reflected light power showed the constant harmonic conversion efficiency over several harmonic orders. Also, the upper shelf density of the profile modification, caused by the radiation pressure balancing the hydrodynamic expansion pressure, was found to nearly coincide with the critical density for the highest harmonic efficiently produced. The lower-shelf density of the profile modification was found to be approximately one-tenth the critical density for the CO_2 light, while the density jump from 10^{19} e/cc to $\sim 7 \times 10^{20}$ e/cc occurred over ~ 1 μm in space when 8 harmonics were efficiently produced. The conclusion drawn was that an approximate phase matching was taking place. When the skin depths for each of the harmonics and the incident CO_2 radiation approximately coincided, the harmonic conversion efficiency was high. However, for harmonics whose corresponding critical density was above the upper-shelf density, this phase matching was no longer satisfied and the conversion efficiency was therefore substantially reduced.

The origin of the plasma nonlinearity was also indicated by the calculations.[1] The plasma profile modification was known to be driven by the resonant absorption process. Here the light polarized in the plane of incidence couples to a surface Langmuir plasma

wave. The high laser intensity causes steepening of this plasma
wave, and its subsequent radiation is the harmonic light, but this
same steepening was also found to be closely associated with the
ejection of hot electrons. Thus, the light wave was impulsively accelerating the electrons, causing some to be ejected, and the collective plasma response was to cause highly nonsinusoidal Langmuir
waves to be excited, from which the harmonic light originates.

This formed a neat package, but was largely still quite specu-
lative. Also, some disagreements with predictions already existed.
For example, the theory required the laser irradiation to be polarized in the plane of incidence while all the experiments were
carried out with the irradiation polarized normal to the plane incidence. Furthermore, the theory suggested that the harmonic light
should be radiated in the spectular reflection direction and not
into all angles in the backward half space. However, the simulations were carried out for a plane wave which does not produce
cratering in the plasma while ripples in the critical surface could
also cause similar effects.

We attempted to check the theory quantitatively in a dedicated
series of experiments at the Gemini laser facility[2] using $I_L <$
10^{15} W/cm^2. Also, one or two more results were obtained at the
Helios laser facility.[2,3] We found that quantitative checks of
the number of harmonics efficiently generated and the upper shelf
density consistent with the radiation-driven density-profile modification could only agree if the cold-plasma temperature T_c were
between 10 and 120 eV. This can be seen from the relationship[2]

$$P_{radiation} = \frac{E^2}{8\pi} (2-A) = P_{hydro} = N_{upper} T_c , \qquad (1)$$

where P is the pressure, A is the absorption fraction and N_{upper} is
essentially the critical density for the last efficiently generated
harmonic n, implying $N_{upper} = n^2 N_{crit}$. Here $N_{crit} \approx 10^{19}$ e/cc is
the critical density for the incident CO_2 radiation. Streak camera
studies of the harmonic light revealed that the visible harmonic
pulse width was typically ~ 200 ps, while absolute timing checks
indicated that the harmonic light was emitted only during the rise

of the CO_2 laser pulse, which was 200 ps. The remaining ~ 800 ps to 1.2 ns falltime of the CO_2 pulse produced no visible harmonics. In contrast, work at Helios, and elsewhere,[6] found that the infra-red harmonic light was generated throughout the whole laser pulse. While greater than 50 harmonics were sometimes observed to be efficiently generated, the laser intensity required was actually less than the maximum available, while at maximum irradiance we had one example of no visible harmonics being produced. High resolution spectra were then taken, an example of which is shown in Fig. 1. This spectrum is of the 16th harmonic of the P(20) transition in the 10.6 µm band, and was obtained at Helios, where the output is principly on the P(20) transition, although P(16), P(18), and P(22) are also present in the output. The frequency bandwidth of all the observed visible harmonics is ~ 100 cm^{-1} and is the same whether only the P(20) rotational-vibrational line is lasing or if four transitions are lasing. The spectrum, of Fig. 1, appears to have both a fine and a more course periodicity. Careful analysis reveals a 1.8 ± 0.05 cm^{-1} uniform periodicity across the entire spectrum, while portions of the spectrum display an ~ 9 cm^{-1} lower frequency modulation. The spectrum is centered to ~ 1 cm^{-1} at the 16th harmonic of P(20). While the 1.8 cm^{-1} periodicity can be explained as a beating of the 4 transitions whose average spacing is 1.79 cm^{-1}, nearly 100 orders are visible in the spectrum suggesting that the associated four-wave mixing phenomena had to be iterated several times.[3] The lower modulation frequency of 9 cm^{-1} is believed to be associated with self-phase modulation. If this interpretation is

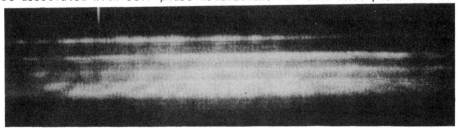

Fig. 1.

High resolution spectrum of 16th harmonic generated in a CO_2 laser plasma experiment. Note the high degree of frequency modulation.

correct, the existence of both of these effects provides indirect evidence for self focusing in the underdense plasma. X-ray pin hole camera data[2] further suggested that self focusing may be occurring, since higher incidence intensity corresponded to more broken-up ~ 1 keV x-ray emission patterns.[2] If self focusing was occurring, it would imply higher intensities could exist locally in the focal region and, therefore, permit T_c to be substantially higher, as inferred from Eq. (1). Since T_c is thought to be nearly 1 keV, an intensity increase of between 10- and 100-fold would thus be inferred. The existence of the high harmonic light only on the rise of the CO_2 pulse then implied that the self focussing only occurred during this time.

In an attempt to rule out other possible explanations to self focussing, we decided to try a second dedicated series of experiments in which the laser risetime was varied while the visible harmonic production was monitored. Here we found, for a CO_2 laser pulse risetime much greater than 200 ps, no visible harmonics were efficiently produced even for peak intensities I_L ~ 2×10^{15} W/cm^2. Similarly, the presence of a prepulse also precluded even the 13th harmonic from being efficiently produced. Since the presence of large quantities of underdense plasma should have improved the probability of self focussing, these results suggested that while self focussing may be occurring, n did not depend on the incombant light intensification. In contrast, for CO_2 laser risetimes of ~ 100 ps (the shortest available) we found that the largest number of harmonics were efficiently produced. In fact, n = 50 for $I_L = 2.3 \times 10^{15}$ W/cm^2 and τ_{rise} = 125 ps, while n = 42 for $I_L = 4.9 \times 10^{15}$ W/cm^2, and τ_{rise} = 160 ps clearly indicates the overall importance of the laser risetime.

TRANSIENT HYDRODYNAMIC THEORY OF HIGH HARMONIC PRODUCTION FOR CO_2 LASER PLASMA EXPERIMENTS

Despite the ability of particle simulations to probe the plasma instabilities and nonlinear coupling associated with this problem, the approach had to be abandoned for several reasons. PIC calculations are usually utilized for collisionless plasmas, while applying the relation $N_{upper} = n^2 N_{crit}$ to problems where n = 50 indicates that $N_{upper} \sim 2\text{-}3 \times 10^{22}$ e/cc. For these densities, collisional effects are clearly very important. Also, with Cray computers, PIC calculations are practical only for time scales of a few hundred optical cycles, or for 10-20 ps, when a range of about 10^3 to 10^4 in plasma density is requested. But what is required is that all densities, from $N < 0.1\ N_{crit} \sim 10^{18}$ e/cc to fully ionized solid densities where $N > 10^{24}$ e/cc, be present for simulation times of 100-200 ps. The presence of large shocks may dictate an even more extreme range of conditions be modeled.

An alternative to the PIC simulation was to use a hydrodynamics code to predict the density profiles and then solve the electro-magnetic problem separately. Since the origin of many of the phenomenological models contained in Lasnex (a hydro code) were PIC simulations, Lasnex seemed to be a logical starting point. In particular, the hot-electron production model, the radiation pressure model and the resonant absorption model in Lasnex came from PIC simulations. However, to accurately describe the relevant range in density requires very fine zoning and long computation times. Also, this use of Lasnex has aggravated a few problems which exist in the code, while causing us to discover a few new problems.

The physics we will be applying is essentially that discussed above, except that N_{upper} is no longer given by radiation-pressure balance. Instead N_{upper} is a dynamic quantity which is given by the density to which the CO_2 laser radiation can penetrate, namely the maximum plasma density which is spatially within one optical skin depth, for the irradiating laser, of its critical density. We first apply this approach to a lower intensity CO_2 irradiation where PIC simulations exist and good data exist. Figure 2 shows the Lasnex 1 results for a 200 μm radius glass microballoon (GMB) overcoated with 15 μm of plastic. However, for Lasnex the hot-electron generation has been turned off, which leads to essentially exponential density profiles;

$$N_e \sim [a \ln(t/t_0) - b]\ e^{z/z_0}, \tag{2}$$

where $a = 9.68 \times 10^{20}$ e/cc, $b = 1.54 \times 10^{21}$ e/cc, $c = 1.6$ μm,

$d = 2.067$, $t_0 = 1$ ps, and $z_0 = c\ t^d$. Now, using our exact
numerical solution of Maxwell's equations in the presence of a steep
density gradient, we can find the skin depth and hence N_{upper}.
For an exponential falloff of plasma density with distance z, we
find a good analytic fit to the skin depth δ, for 10 μm radiation,
namely

$$\delta = 1.24\ Z_0^{0.538}. \tag{3}$$

The assumed laser pulse shape is shown in Fig. 3 (top). Combining
these results, we then arrive at Fig. 3 (bottom), which plots both
I_L and the derived N_{upper} as a function of time. Now, the harmonic light will be efficiently produced over the time scale where
N_{upper} is overdense to that harmonic. Because of the high
power-law dependence of the visible harmonic light on intensity,

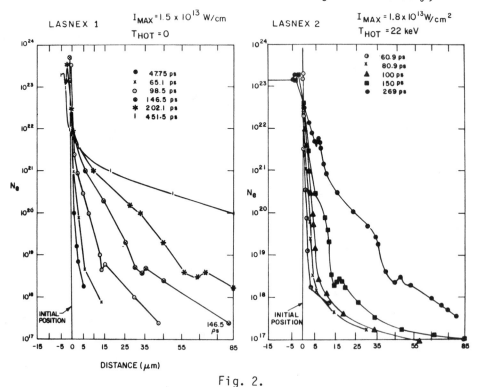

Fig. 2.

LASNEX hydro calculations with (1) and without (2) the presence of
hot electrons. Note the qualitative change in the resulting density
profiles.

we should expect some variation in the generation efficiency with incident laser intensity. No such strong dependence is inferred from the data suggesting that the harmonic generation efficiency is saturated in some sense. We, therefore, do not take this into account. From Fig. 3 (bottom), we would predict a monotonic fall-off of harmonic conversion with n.

Repeating the Lasnex 1 calculation, but with the hot electron generator turned on, yields the Lasnex 2 results of Fig. 2. Note the qualitative changes in the hydrodynamics. In particular, the expansion is slowed markedly because the cold-plasma temperature has been reduced. Also, the radiation-pressure effects are now sufficient, even early in the irradiation causing some steepening of the density profile. Both the Lasnex 1 and Lasnex 2 calculations

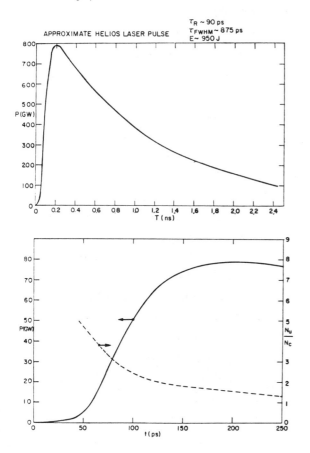

Fig. 3

Plot of the assumed laser pulse shape for all LASNEX calculations (top). Plot of the rising portion of I_L and the inferred variation in N_{upper} for the LASNEX 1 and 2 hydrodynamics, as defined in the text.

were done for $I_{max} = 1.5 \times 10^{13}$ W/cm^2 or $I\lambda^2 = 1.7 \times 10^{15}$ (μm)^2 W/cm^2. In order to understand the effects of increasing the incident CO_2 laser intensity, the Lasnex 4 calculation was performed. Here the laser intensity is now $I_{max} = 2 \times 10^{14}$ W/cm^2. Both the increase in hot-electron temperature and in radiation pressure has cause the density profile to steepen even more than that of Lasnex 2, where

$$T_{hot} \cong 56[I \lambda^2 T_c]^{1/3} \text{ (keV)}. \qquad (4)$$

The units are λ (μm), $I(10^{16}$ W/cm^2), and T_c (keV). This T_{hot} is approximately 4 times the previously published Los Alamos result,[7] and is a better fit to present data for $10^{14} < I_L < 10^{16}$ W/cm^2. To construct the plot of N_{upper} versus time for Lasnex 4, a cubic spline fit to the density profiles is fed into the electromagnetic code to first fine the skin depth and then N_{upper}. The results suggest that the third and fourth harmonic should be produced at about the same efficiency, while the fifth and higher should be produced with a monotonically decreasing efficiency, consistent with both early particle simulations[1] and the experimental data in Refs. 4 and 5.

Lasnex calculations for $I_L \sim 3 \times 10^{16}$ W/cm^2 have also been performed, but severe difficulties developed due to the $T_{hot} \sim 400$ keV near N_{crit}. Basically, the hot-electron transport and the coupling of the hot electrons to both the ion distribution and the cold electron distribution in the underdense plasma are not sufficiently well described to provide for a detailed description of the density profiles. Work is underway to try to remedy these problems.

APPLICATION OF THE TRANSIENT HYDRODYNAMIC THEORY
TO HIGH IRRADIANCE ArF LASER–PLASMA EXPERIMENTS

A significant difference in the techniques for modeling CO_2 laser-plasma experiments and present uv laser-plasma experiments must be taken into account in order to use Lasnex for these uv calculations. The uv absorption experimentally is principly by the inverse-bremsstrahlung process which does not lead to hot-electron production. However, the thermal electron flux must be limited to about 3% of that expected by the Spitzer conductivity. For incident intensities $I_L < 10^{15}$ W/cm^2, the radiation pressure effects for ultraviolet laser will also be neglectable. Another experimental result is that the hot-electron temperature associated with 0.35 μm laser-plasma experiments has a weak dependence on the atomic number Z of the target, namely $Z^{1/4}$, where Z is the state of ionizaton of the material; for example, for helium-like copper, Z = 27. Thus, the higher T_c coupled with the absence of profile modification effects suggest that we should expect the hydrodynamics to proceed similar to that of the Lasnex 1 calculations of Fig. 2.

Guided by the Lasnex 2 and Lasnex 4 results above, we suspect that high-harmonic generation by an ArF laser will become efficient only if hydrodynamic expansion can be greatly diminished. Short pulses with very fast risetimes will satisfy this requirement. For

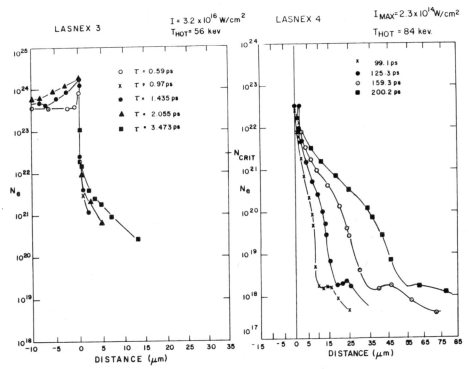

Fig. 4.

Two LASNEX hydro calculations, (3) ArF laser of 0.875 ps rise, 9 ps FWHM, and 2 ps peak intensity, and (4) CO_2 laser of 87.5 rise, 900 ps FWHM and 200 ps peak intensity. Note the high hot-electron temperatures (see text).

$I_L > 10^{16}$ W/cm^2, $I\lambda^2 > 3 \times 10^{14}$ (µm)^2 W/cm^2 and we now expect radiation-pressure effects to be important. Furthermore, when we combine the two above, inverse bremsstrahlung absorption will no longer be as important, since the density gradient scale height will be insufficient. Thus, in analogy to the high irradiance 1.06 µm and 10.6 µm laser plasma results, we will expect resonant absorption to become more important, implying the excitation of surface Longmuir plasma waves, the production of hot electrons, and the potential for producing harmonic light.

The Lasnex 3 calculation of Fig. 4 gives the density profiles associated with an ArF target irradiation at $I_L = 2 \times 10^{15}$ W/cm^2. The target is a 15 µm copper overcoated 200 µm radius GMB. $T_{hot} = 56$ keV is obtained from Eq. (4), where $T_c = 11$ keV and we have also multiplied by $(Z)^{1/4}$ with $Z = 27$. While this T_{hot} is higher than would be inferred from present target experiments using 0.35 and 0.26 µm lasers, they are carried out under conditions which are substantially different than those described above. The assumed irradiation conditions here more closely approximate those associated with CO_2 laser-plasma experiments at Los Alamos. As a further justification for using this T_{hot}, we note that several plasma instabilities could be driven at such high intensities, while many of these instabilities produce hot electrons.

When the results of Lasnex 3 are processed as described above, we predict a substantial conversion to the first seven ArF harmonics. The total absorption expected by the calculation is 39% of which 37% is by resonant absorption. Note that the solid electron density in an uncompressed condition corresponds to $N_e \sim 3 \times 10^{23}$ e/cc, so that the surface material has been shocked up to approximately 4 times solid density. From this result we thus see how to further increase the number of potential harmonics, namely to further shock the surface material. In principal, it is possible to shock the material up to ~ 25 times normal density or $N_e \sim 7.5 \times 10^{24}$ e/cc, which could lead to n = 16. We have not been able to achieve this result using Lasnex. Rather, the highest electron density we have been able to achieve at the surface is $N_e < 3 \times 10^{24}$ e/cc, and this

required an $I_L \sim 5 \times 10^{16}$ W/cm^2 focused on a molybdenum target.
Since we are again approaching plasma conditions where Lasnex is less
likely to predict correctly the hydrodynamics, we regard this result
as even more spectulative. At this point, we wish to stress that the
lack of adequate laser-produced plasma data and models applicable to
this wavelength, intensity, and pulse width regime have forced
assumptions to be made which may not be justified. On the other
hand, no assumptions have been made which are inconsistent with our
present understanding.

In conclusion, we believe that short pulse, high irradiance
laser-plasma experiments utilizing an ArF laser have the potential
for producing several vacuum ultraviolet wavelengths down to possibly
19.3 nm. The use of this harmonic light as a flashlamp, a holographic source, or a probe might be potentially interesting.

The authors wish to thank Eldon Linnebur, David Forslund, Jerry
Blackbill, John Lindl, Keith Boyer and Charles Rhodes for very helpful discussions during the progress of this work.

REFERENCES

1. R. L. Carman, D. W. Forslund, and J. M. Kindel, Phys. Rev.
   Lett. 46, 29 (1981).
2. R. L. Carman, C. K. Rhodes, and R. F. Benjamin, Phys. Rev. A
   24, 2649 (1981).
3. R. L. Carman, "Factors Affecting Target Irradiation Symmetry
   in CO_2 Laser Fusion Experiments," Proc. Topical Conf. on
   Symmetry Aspects of Inertial Fusion Implosions (to be
   published); also Los Alamos National Laboratory Report
   LA-UR-81-1566 (1981).
4. N. H. Burnett, H. A. Baldes, M. C. Richardson, and G. D.
   Enright, Appl. Phys. Lett. 31, 172 (1977) and references
   therein.
5. E. A. McLean, J. A. Stamper, B. H. Ripin, H. R. Griem, J.
   McMahon, and S. E. Bodner, Appl. Phys. Lett. 31, 825 (1977)
   and references therein.
6. P. A. Jaanimagi, G. D. Enright, and M. C. Richardson, IEEE
   Trans. Plasma Sci. P57, 166 (1979).
7. See, for example, D. W. Forslund, J. M. Kindel, and K. Lee,
   Phys. Rev. Lett. 39, 284 (1977) and references therein.

A SOFT X-RAY SOURCE BASED ON THE MAGNETIC CONFINEMENT AND COMPRESSION OF A LASER PRODUCED PLASMA

E.A. Crawford, A.L. Hoffman, R.D. Milroy and D.C. Quimby*
Mathematical Sciences Northwest, Inc.
Bellevue, Washington 98004-1495

ABSTRACT

We have demonstrated high beta magnetic confinement of a laser produced plasma. A scheme for efficient production of soft x-rays by compression of these plasmas will be discussed.

I. INTRODUCTION

Laser heating of dense, magnetically confined plasmas can provide high energy density sources which are unique in their spatial uniformity and ability to be temporally resolved. Using axial magnetic fields for radial confinement, plasmas with pressures in the kilobar range can be produced. These plasmas have applications for basic atomic physics studies and as a medium for soft x-ray laser experiments.

The 10.6μ wavelength of carbon dioxide lasers is ideally suited to heating multiply ionized plasmas to several hundred eV temperatures at densities which can be magnetically confined. Several early experiments have employed CO_2 lasers to heat hydrogen plasmas confined in quasi steady solenoidal fields,[1] and in fast theta pinches.[2] In one early experiment at MSNW,[3] neon was laser heated in a theta pinch and the role of multi-stage ionization and radiation in the plasma energy balance was investigated.

It is both more convenient and more flexible, in terms of available atomic species, to produce a radially confined plasma by irradiating a solid target in a solenoidal magnetic field. Recent experiments at MIT[4,5] and at MSNW have demonstrated the radial confinement of CO_2 laser produced plasmas.

The MSNW experiments were done using solid carbon targets which were irradiated with CO_2 laser pulses of up to 500 J and 50 nsec (FWHM) duration. The targets were immersed in an axial magnetic field of 7.5 T. The resulting plasma is confined radially and expands as a plume toward the heating laser. Plasma conditions in the plume were investigated using ruby laser holographic interferometry to determine electron density profiles and spatially resolved Thomson scattering to determine electron temperture profiles. The resulting data was found to be in good agreement with a high beta pressure balance plasma model. Typical plasma conditions in the plume are T_e = 100-150 eV, and

*Supported in part by the National Science Foundation

$N_e = 1.0 - 1.2 \times 10^{18}$ cm^{-3}. The electron temperature is extremely uniform due, in part, to radial heat conduction, but primarily to the strong temperature dependence of the laser absorption coefficient. Uniformity of the plume is also promoted by matching the target diameter to the laser focal spot size, although there is some loss of heating efficiency due to unfavorable refraction for these small targets.

We have closely coupled our experimental programs with computer modeling of present and contemplated experiments, using various MHD codes with a collisional-radiative atomic physics package. A one-dimensional radial code, DYNASOR,[6] successfully predicted the observed radial profiles in the solid target experiments, while a one-dimensional, axial code, CORK,[7] has given reasonable agreement with the axial structure of the plume.

We have recently begun to investigate theoretically the adiabatic z-pinch compression of these plasma plumes as a possible efficient source of soft x-rays for microlithography applications. A z-pinch can compress, for example, an aluminum plasma to pressures of hundreds of kilobars where it would be an efficient generator of K-shell x-rays in the 1.5 to 2 keV range. We have used DYNASOR to predict the performance of these adiabatic compressions. Preliminary results demonstrate efficient production of K-shell x-rays from an aluminum plasma when the electron temperature exceeds 400 eV. Compressed plasma plumes of this type, with densities in the 10^{19} cm^{-3} range and temperatures of 400 eV - 1 keV may also be attractive pump sources for soft x-ray laser experiments.

II. LASER PLUME GENERATION EXPERIMENTS

Solid targets were irradiated with the 10.6μ radiation from a large e-beam sustained, 200 liter CO_2 laser. A nitrogen free mix and rapid electrical switching were employed to limit the output to a single gain switched 100 nsec long Gaussian shaped pulse of 500 J maximum energy. An unstable oscillator provided good optical quality and the output beam could be focused to a 3 mm diameter (50 percent energy point) with f-15 optics.[8]

The experimental arrangement is sketched on Fig. 1. A 7.5 kG axial magnetic field was provided by a quasi-steady (3 μsec half cycle time) solenoidal magnet. A 3 mm diameter carbon rod was inserted inside the magnet. Vacuum access was provided in four directions for x-ray pinhole photography, ruby laser holographic interferometry, Thomson scattering, filtered x-ray diodes, and grazing incidence spectroscopy. An x-ray pinhole photograph in Fig. 1 shows the approximate size of the plasma plume. The radiating plasma only extends a few millimeters in front of the target without the confining magnetic field.

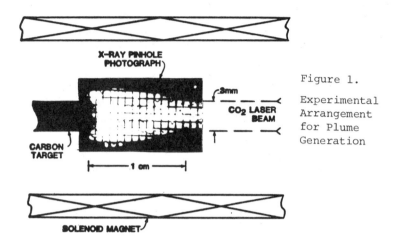

Figure 1.

Experimental
Arrangement
for Plume
Generation

The temperatures measured 6 mm in front of the target surface
for 450 J laser irradiations are shown on Fig. 2. Both the x-ray
diode measurement (based on differential transmission through two
thicknesses of beryllium foil) and the Thomson scattering data
indicate a peak electron temperature of between 130 and 150 eV.
This is somewhat lower than the 230 eV predicted by a CORK
calculation. The lower plasma temperatures are probably due to a
decrease in the actual laser intensity striking the target due to
unfavorable refraction in the plasma plume. Beam focusing rather
than defocusing could be produced by irradiating larger diameter
targets. This would result in hotter control temperatures, but only
at the expense of also incurring a dense, colder outer annulus.

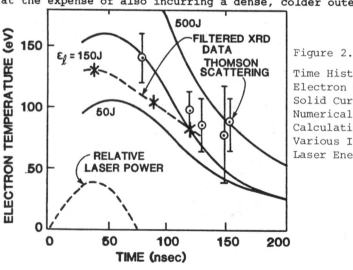

Figure 2.

Time History of
Electron Temperature
Solid Curves are
Numerical
Calculations for
Various Incident
Laser Energies

The intensity actually reaching the carbon target can be estimated to be one third the incident value by comparing the measured temperatures with the the results of the three CORK calculations also shown on Fig. 2.

The radial uniformity of the plasma plume was investigated using temporally and spatially reduced interferometry and Thomson scattering. An experimental profile measured 6 mm from the target face and 120 nsec after the start of laser heating, is shown on Fig. 3. These measurements are compared with DYNASOR calculations based on the measured electron line density and the intensity inferred from the CORK calculations. Within the data scatter, there is excellent agreement with the computed temperature uniformity. The calculations also predict formation of a highly flux excluding $\beta=1$ plasma column. The measured temperatures and densities confirm this prediction.

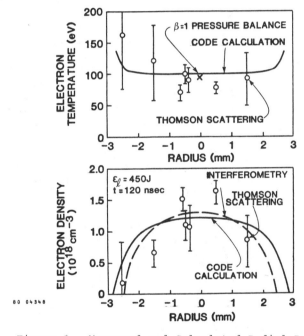

Figure 3.   Measured and Calculated Radial Profiles

III.   ADIABATIC Z PINCH CALCULATIONS

The plasma plumes can be increased in temperature and density through z pinch compression to values where where intense aluminum K-shell radiation will be produced. A simple experimental arrangement, encompassing the laser generated, magnetically confined plume, is shown on Fig. 4. The plume itself can be used as the

240

switch to trigger the axial current discharge.

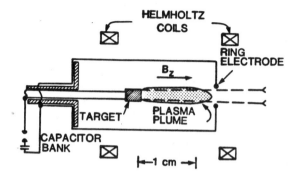

81 04532r

Figure 4. Basic Laser Initiated Z Pinch Scheme

The rate of radiation from aluminum (coronal calculation by Post, et. al.[9]) is shown on Fig. 5 to illustrate the advantage of the combined laser heating, z pinch approach. Laser heating is most efficient at low temperatures, and can rapidly heat the aluminum plasma above the L-shell radiation barrier at 100 eV. K-shell radiation rates will be maximized at 500-600 eV temperatures where the ions are stripped down to the He-like (Z=11) state. A z pinch can provide the high fields necessary to confine these hot, dense plasmas and also supply energy through adiabatic compression and joule heating to reach the high temperatures.

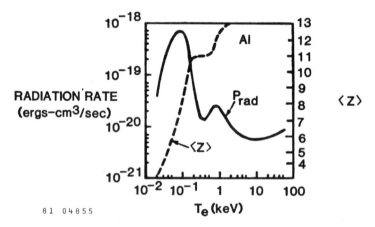

81 04855

Figure 5. Coronal Radiation Rates and Average
Ionization for Aluminum

The z pinch compression can be analyzed by expressing the

azimuthal magnetic field as

$$B_\theta = 0.2 \, \frac{I(\text{kA})}{r(\text{mm})} \text{ tesla} \tag{1}$$

and using an adiabatic relationship

$$p \propto n^{5/3} \propto r^{-10/3} \tag{2}$$

Combining these two equations yields the ratio of final plasma radius to initial plume radius r_o

$$r/r_o = \frac{11.2}{\beta^{3/4}} \left[\frac{B_{zo}(T) r_o(\text{mm})}{I(\text{kA})} \right]^{3/2} \tag{3}$$

where B_{zo} is the initial confinement field and β is the final ratio of plasma pressure to magnetic confinement pressure.

Table I shows an example where a 100 eV, 3 mm diameter plasma column is compressed by a factor of 4 and raised to a temperature of 600 eV. The radiation rate from such a column, as given by Fig. 5, is 50 MW/mm^3, or 250 MW per centimeter of column length. A 2 cm long plasma column could then provide 50 J of K-shell x-rays in a 100 nsec pulse.

Table I Plasma Plume Example Before and After Z Pinch
(I = 200 kA, β = 0.5)

	Before Z Pinch	After Z Pinch
B_{zo}	7T	
r	1.6 mm	0.4 mm
B_θ	—	100 T
n_e	10^{18} cm^{-3}	1.6×10^{19} cm^{-3}
T_e	100 eV	600 eV

The predictions of this simple scaling model were investigated by using the radial DYNASOR code. The pinch current in these simulations is governed by an external circuit equation which also accounts for the variable plasma inductance. Plasma radiation is calculated through the use of an optically thin collisional-radiative model for aluminum ions. Since the motivation for this work is to determine the useful soft x-ray output for photoresist exposures, distribution of radiated power is calculated in 200 eV energy bins, and the useful output is defined as the radiated energy between 1.5 and 2.2 keV. the results of a typical calculation are shown in Fig. 6. Based on a 200 J laser, 0.1 μf capacitor charged to 250 kV, and a 50 nh external inductance, the useful output energy is 50 J for a 2 cm long plasma column.

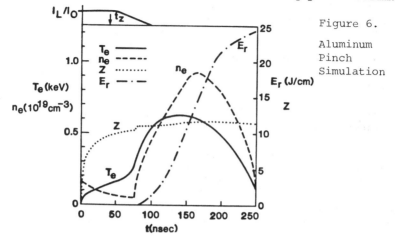

Figure 6.

Aluminum
Pinch
Simulation

REFERENCES

1. T.K. Chu and L.C. Johnson, Phys. Fluids **20**, 1460 (1975).

2. A.L. Hoffman, Appl. Phys. Lett. **23**, 693 (1973).

3. A.L. Hoffman and E.A. Crawford, J. Appl. Phys. **49**, 3219 (1978).

4. W. Halverson, N.G. Loter, W.W. Ina, R.W. Morrison and C.V. Karmendy, Appl. Phys. Lett. **32**, 10 (1978).

5. N.G. Loter, W. Halverson and B. Lax, J. Appl. Phys. **52**, 5014 (1981).

6. D.C. Quimby and L.C. Steinhauer, Phys. Fluids **23**, 1426 (1980).

7. R.D. Milroy and L.C. Steinhauer, Phys. Fluids **24**, 39 (1981).

8. A.L. Hoffman, J. Magnetism and Magnetic Mat. **11**, 350 (1979).

9. D.E. Post, R.V. Jensen, C.B. Tarter, W.H. Grasberger, and W.A. Lokke, "Steady State Radiative Cooling Rates for Low-Density High-Temperature Plasmas," Princeton Plasma Physics Laboratory Report PPL-1352 (July 1977).

A LASER PRODUCED PLASMA AS AN EXTREME ULTRAVIOLET CONTINUUM SOURCE FOR ABSORPTION EXPERIMENTS

E. Jannitti, P. Nicolosi and G. Tondello
Centro Gas Ionizzati, C.N.R., Università di Padova,
35100 PADOVA, Italy

ABSTRACT

An experiment for absorption spectroscopy of highly ionized species in the XUV is described. Two plasmas produced by laser beams act one as a background continuum source and another as the absorbing medium. The performance of the laser produced plasma as a continuum source was optimized and the output absolutely measured. Similarly the absorbing plasma was studied for minimizing spurious broadening mechanisms. An example of absorption spectra of Be II and Be III is shown.

INTRODUCTION

Several experiments have already demonstrated the usefulness of laser produced plasmas for applications other than the inertial confined fusion [1]. In particular plasmas produced by lasers can be used as intense sources of extreme ultraviolet (EUV) or soft X-ray (XUV) radiation. Spectrally both continua or discrete line emission can be emphasized depending on laser-target parameters [2]. Already explored applications for laser produced plasmas as sources of XUV radiation range from photolitography [3], X-ray diffraction studies [4], pulsed EXAFS [5] to plasma diagnostics [6], to more fundamental use in ionic absorption spectroscopy.

As known absorption spectroscopy of ionized species is still quite limited in range especially for the difficulty of producing a suitable column of ionized medium. Recent results in this field are the absorption spectra of singly or doubly ionized species with the technique of laser induced ionization [7].

By using two laser produced plasmas the absorption spectrum of Li^+ has recently been studied down to 70 Å [8]. This technique appears promising and quite simple. The same laser is utilized for producing two plasmas: one emitting a continuum spectrum acting as the source and another for the absorbing medium. By a proper choice of the laser-target interaction mechanisms, and an optimal transfer and dispersing optical system, it is possible to optmize the overall performance, arriving to a rather versatile experimental set-up for absorption spectroscopy of highly ionized species. In the following we will describe the experiment and report measurements on both the emitting laser produced plasma, the absorbing one and some preliminary results on the absorption spectra of Be^+ and Be^{2+} in the range 110-60 Å.

THE EXPERIMENT

Figure 1 shows the experimental system. The output of a Q-switched ruby laser (10 J of energy, 15 ns duration) is split in two beams: the first one (beam 1) carrying 70-90% of the energy is focused on target T_1 with the aspheric lens L_1 after deflection on mirror A,

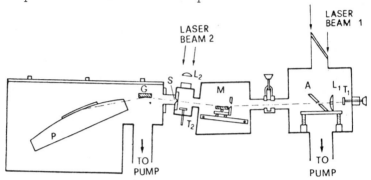

Fig. 1. The experimental system

while the second one (beam 2) is focused on target T_2 with the lens L_2. The two generated plasma plumes will be called respectively the radiating and the absorbing plasmas. The radiating plasma is observed through two small holes in the mirror A and in the lens L_1 by the toroidal mirror M, working typically at an angle of incidence of 87°, which collects and focuses the XUV radiation. The radii of mirror M are such as to produce a first astigmatic image on the entrance slit S of a 2 m grazing incidence spectrograph and a second one at the position of a given wavelength λ diffracted by the grating G on the focal Rowland cylinder P[9]. In such a way the absorbing plasma, produced close in front of the spectrograph entrance slit, is crossed with a narrow blade of XUV radiation reflected by the toroidal mirror. Since the absorbing plasma expands roughly on the normal direction to target T_2 this configuration minimizes the effects of the spatial gradients on the absorption process. The delay between the two beams as well the irradiated section of the absorbing plasma can be properly adjusted. The grating G was or a 600 lines/mm or a 2400 lines/mm platinum coated working at an angle of incidence of 86°.

Photographic detection is used mainly on the spectrograph but some measurements were made also with a scanning exit slit followed by a scintillator-photomultiplier combination. Table I summarize the relevant parameters of the experiment.

THE CONTINUUM EMISSION FROM LASER PRODUCED PLASMAS

An important requirement for an absorption experiment is to have a continuum emitting background source as strong as possible. The continuum must be also clear of both emitting or absorbing features, re-

TABLE I. Optical parameters

Distance	source-mirror	576 mm
	mirror-entrance slit	161 mm
Angles of incidence	on the mirror	87°
	on the grating	86°
Radii	of the mirror in the dispersion plane	5900 mm
	of the mirror in the sagittal plane	26 mm
	of the grating	2000 mm
Entrance slit of the spectrograph		5 μm
Observed portion of absorbing plasma (height x width)		0.250 mm x 2 mm
Grating		600 1/mm, 2400 1/mm
Detection		photografic films, photomultipliers

producible and spectrally confined to the absorbing interval in order to minimize problems related to overlapping orders in the spectrograph. A considerable effort was spent in optimizing the emission of continuum from laser produced plasmas [10].

As known the continuum emission from an ionized plasma can be of free-free or free-bound type. Only for the very simple H-like or He-like ionic configurations the spectrum can be calculated with sufficient accuracy. In these cases the intensity emitted E_{ff} (free-free) and E_{fb} (free-bound) can be written:

$$\frac{d E_{ff}}{d\lambda} = n_e \, n^{z+} \, F_{ff} \, (\lambda, \, T_e) \quad \left[W \, cm^{-3} \, \overset{\circ}{A}^{-1} \, sr^{-1} \right] \qquad (1)$$

$$\frac{d E_{fb}}{d\lambda} = n_e \, n^{z+} \, F_{fb} \, (\lambda, \, T_e) \qquad " \qquad (2)$$

where n_e and n^{z+} are respectively the electron and fully stripped ion densities and the functions $F_{ff}(\lambda, \, T_e)$ and $F_{fb}(\lambda, T_e)$, $\left[W \, cm^3 \, \overset{\circ}{A}^{-1} \, sr^{-1} \right]$ describe the dependence of the emitted intensity on wavelength λ and electron temperature T_e. As known [11] the free-free emission function $F_{ff} \, (\lambda, \, T_e)$ is a smoothly peaked function of wavelength reaching the maximum at $\lambda_m (\overset{\circ}{A}) = 6200/T_e$ (eV), decaying at long wavelengths as λ^{-2} and at very short as exp $\{-h\nu/KT_e\}$. The free-bound emission function $F_{fb} \, (\lambda, \, T_e)$ has the same dependence on λ but exhibits a step-wise behaviour corresponding to each of the discontinuities in the spectrum associated with the ionization potential X_{in} from the i=1,2,3,.... levels.

For ionization stages other than H-like or He-like, and taking into account the fact that several ionization stages are simultaneously present especially for plasmas produced on high Z targets, the dependence of the continuum is very difficult to calculate. In practice the step-like structure is lost and the continuum appears smooth even at rather short wavelengths.

This effect has already been exploited in plasmas produced by focusing laser beams on rare hearth targets [12] where a very smooth continuum was produced.

As evident from eqs. (1) and (2) the intensity depends on the square of the electron density. The strongest emitting regions are the regions of the highest electron density. In the interaction of a laser beam with a plane target such region is near the ablation region where there is a layer of critical density, i.e. the density at which the plasma frequency equals the laser frequency. In plane targets this region is near or below the unperturbed zero position of the target. It is for this reason that by looking inside the ablation region a considerable gain in emitted intensity can be realized with respect to the intensity emitted by the expanding plasma (x6) [13].

In order to test such facts careful measurements on the emissivity of the laser produced plasma especially for low Z materials have been performed. On one hand plasma parameters like the electron temperature T_e, the electron n_e and ion n^{z+} densities, the physical dimensions of the emitting layer and the duration of the emission have

TABLE II. Parameters of the Be and C plasmas inside the crater.
Laser pulses: 6J, 14 ns - Power density $\sim 5 \times 10^{12}$ W/cm^2

PARAMETER	Be	C	DEDUCED FROM
T_e (eV)		100	Slope of the free-bound continua
n_e (cm^{-3})	2.5×10^{21}	1.5×10^{21}	Fitting the Lyman lines $(\gamma, \delta, \varepsilon)$ with theoretical profiles
n^{z+} (cm^{-3})	4.2×10^{20} (z=4)	$6. \times 10^{19}$ (z=6) 10^{20} (z=5)	Ratio of free-bound H-like and He-like continua and the neutrality equation
d (cm)	10^{-2}	10^{-2}	Stigmatic spectra in the XUV
Δy (cm)	1.3×10^{-3}	3×10^{-3}	Ratio of intensity of the continua and the electron and ion densities inside and outside the crater
Δt (ns)	30	30	Time resolved measurement with photomultiplier.

been measured for plasmas of beryllium and carbon: for the former the only species contributing to the continuum emission being the Be^{4+} and for the latter both the C^{6+} and the C^{5+} ions. Table II summarizes the various parameters with a brief indication of the method used in their determination [10]. On the other hand the spectroscopic system has been absolutely calibrated: the absolute efficiency of a 600 lines/mm grating has been measured [14] at 8.3 and 44 Å and the whole spectrograph calibrated in situ with the branching ratio technique. For the latter suitable lines in the visible between n=6,7,8 levels of CVI, CV, Be IV and Be III have been absolutely measured and the calibration transferred to the grazing incidence spectrograph in the range 25-83 Å. Good agreement was found between the two methods. Fig. 2 shows the calculated and measured spectra of a beryllium plasma and Fig. 3 the same for a carbon plasma. For both spectra the resonance lines are visible over the continuum and the step-like structure of the recombination continua is quite pronounced. Good agreement is found between calculated and observed spectra with a peak intensity of over 10^8 W cm^{-2} sr^{-1} Å^{-1} for a carbon plasma.

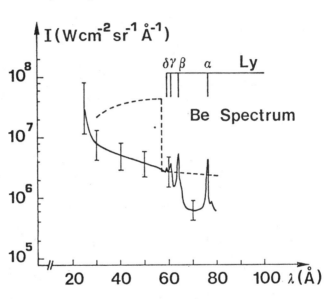

Fig. 2. Calculated (---) and measured spectra (——) of a beryllium plasma.

Spectra emitted by higher Z elements tend to present much less line emission and a smoother continuum. Fig. 4 shows the intensity emitted by aluminium, chromium, copper and uranium plasmas. Particularly for the latters the smoothness of the continuum is noteworthy . Only at very high dispersion one can notice the presence of few very narrow absorption lines.

The intensity of the continuum emitted can be somewhat increased (x2) by firing on a previously built cavity on the target. Practically the cavity can be built with a laser shot at reduced power and with a low aperture lens. Indeed the spectra reported in Fig. 4 were produced with this technique. The main effect of the cavity is to limit the expansion of the plasma. The difference in the plasma characteristics in the two cases can be seen in Fig. 5 where the profiles of the Ly α line of Be IV recorded with the set-up of Fig. 1

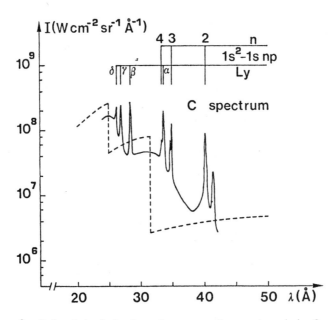

Fig. 3. Calculated (---) and measured spectra (—) of a carbon plasma.

are shown with (dotted li
ne) and without (solid li
ne) the cavity. For the
case without the cavity
the profile is clearly ma
de up of two components:
a wide one emitted by the
plasma in the ablation re
gion and a narrow one emit
ted by the expanding pla-
sma viewed end-on by the
spectrograph. By the way
a numerical modelling of
the two plasmas complete-
ly explains the observed
profile and its asymme-
try[15]. Completely different is the profile emit-
ted when the plasma is
produced in a cavity. The
two components aspect is
almost absent and the li-
ne present a pronounced
self-absorption dip asym-
metrically located. It
is also narrower (about
a factor of two) than

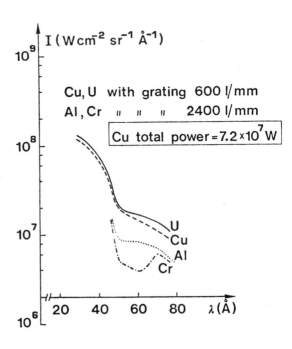

Fig. 4. Intensity emitted by aluminium, chro
mium, copper and uranium plasmas.

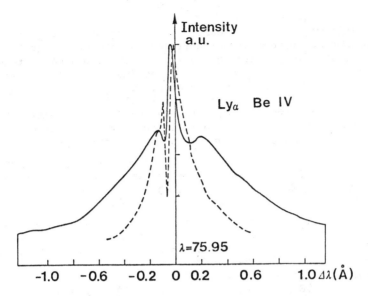

Fig. 5. Profiles of the Ly α line of Be IV with (---)
and without (——) the cavity.

that observed without the cavity. In addition the region emitting
such profile is wider than without the cavity. A profile with relati
vely wide wings and self-absorption is observed even at 0.3 mm out
of the central plane; on the contrary in the case without the cavity
at 0.1 mm only a narrow line was observed. Typical diameters of the
cavity apertures were 200 μm.

It is also very interesting to estimate the total power radiated by the plasma in the XUV and to have indication of the conversion efficiency with respect to the incident laser power of the plasma. For doing this the volume of the emitting plasma has been estimated and the measured intensity integrated in the region   below
100 Å where most of the power is emitted. This results in the most
significant case of a Cu plasma Fig. 4 in a total power emitted of
about 70 MW assuming an isotropic emission. This is of the order of
10% of the laser power reaching the target. Obviously only a very
small fraction of this radiated power is practically usable. However
good continua had been registered with only one shot with a 2400 li-
nes/mm grating and an entrance slit 5 μm wide on kodak 101 emulsions.
This indicates the potential of such technique even with lasers of
quite modest power output.

THE ABSORBING PLASMA

In an absorption experiment ideally the absorbing medium should
be as homogeneous as possible, of controlled and stable density and
free from perturbing effects. These requirements can be fulfilled
only partially when the absorbing medium is ionized and particularly
where strong temporal and spatial gradients are present as in laser
produced plasmas.

The main perturbing mechanism in the absorbing plasma is the streaming motion of the plasma in its expansion in the vacuum. The other usual effects perturbing the ions in these plasmas (Stark, thermal Doppler, opacity) are largely reduced due to the low density and temperature of the actual plasma regions. Consequently the Doppler broadening tends to limit the spectral resolution achievable both in emission and absorption experiments. By choosing the geometry of absorption shown in Fig. 1 it is particularly important to reduce the Doppler component of the streaming in the direction parallel to the absorption target. As known [16] the expanding velocity is a function of the power density on target, so it is important to operate just at the limiting value for producing a given desired species.

Experiments made in emission with plasmas produced on plane targets [17] have indicated that very effective in reducing the Doppler broadening were: 1) the use of focusing lenses of relatively low numerical aperture and 2) the reduction of the plasma expansion produced by the cavity on the target as treated previously. In Fig. 6 the profiles of the line Hα of N VII at λ = 133.8 Å recorded in emission with a 600 lines/mm grating in the second order of the spectrograph are shown for two cases: one (dotted line) using an aspherical lens of f/1 numerical aperture and the second (solid line) using an f/5 doublet and firing inside a previously built cavity. Both profiles refer to the plasma at a distance of 0.4 mm from the surface of the target. The difference in the resolving power is very pronounced: the second profile starts to show the fine structure of the Balmer doublet whose theoretical pattern is marked in the same figure. From similar measurements a resolving power of ≈ 3000 below 100 Å is estimated as the convolved contributions of residual Doppler and instrumental broadening.

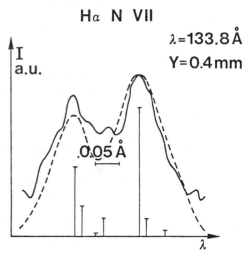

Fig. 6. Profiles of the Hα line of N VII with f/1 aspheric lens (– – –) and with f/5 doublet firing in a cavity (———).

PRELIMINARY RESULTS OF THE ABSORPTION SPECTRUM OF BERYLLIUM

As a test of the system performance we chose to study the absorption spectrum of beryllium, because it has relatively low Z, it is stable and simple to handle and the spectrum of the He-like Be III, quite interesting from an atomic point of view, is in a relatively convenient spectral region. Targets of beryllium metal were irradiated with a line focus approximately of 0.25 x 6 mm area produced by a sphero-cylindrical lens. The power density on the target was $\approx 3.2 \times 10^9$ W/cm^2. The continuum was produced with a Cu target using $\approx$ 6 J of energy. A delay of 15 ns was used between the continuum source and the absorbing plasma observed at a distance of 0.35 mm from the target. A grating with 2400 lines/mm has been used on the spectrograph with an entrance slit 5 μm wide. From 2 to 3 shots were necessary for a well developed photograph. The recorded spectrum is shown in Fig. 7 where the absorption coefficient in relative units

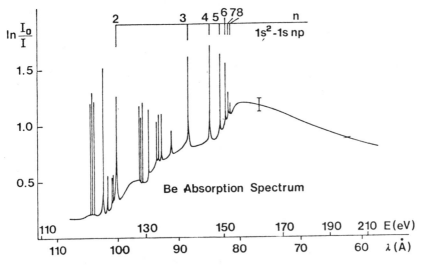

Fig. 7. Absorption spectrum of beryllium

is plotted in the region from 110 to 60 Å. The absorption coefficient has been arbitrarly set to zero at the long wavelength limit; in addition, due to the limited exposure latitude of the films used, the peak absorption on the lines could be underestimated. The spectrum consists of both lines and continua; only the strongest lines are actually shown on Fig. 7. Lines belonging to both doubly excited states of Be II and to the resonance series $1s^2$-1snp of Be III converging to the limit at $\lambda = 80.6$ Å are present. In Table III preliminary measurements on the observed lines together with some identifications are shown. Previously similar spectra had been observed in an absorption experiment utilizing two vacuum sparks[18].

The photoionization continuum $Be^{2+} + h\nu \rightarrow Be^{3+}$ is evident as well as additional continuum absorption at longer wavelengths probably due to some absorption from Be^+.

Work is in progress on identifying the relevant parameters af-

TABLE III. Be II - III absorption lines

Configuration	λ (Å)[a]	Intensity arb.un.	Configuration	λ (Å)[a]	Intensity arb.un.
$1s^2 2p\ ^2P° -$				94.76^d	5.7
$-1s2p^2\ ^2D$	104.62^b	10		93.41^d	3.4
$1s^2 2s\ ^2S -$				93.14^d	4.1
$-1s(2s2p\ ^3P)\ ^2P°$	104.37^b	11		92.60^d	4.1
$1s^2 2p\ ^2P° -$				92.47	1.5
$-1s2p^2\ ^2P$	103.98^b	10.2		91.06^d	2.2
$1s^2 2s\ ^2S -$			$1s^2\ ^1S -$		
$-1s(2s2p\ ^1P)\ ^2P°$	102.47^b	13	$-1s3p\ ^1P°$	88.314^c	7.9
$1s^2 2p\ ^2P° -$			$1s^2\ ^1S -$		
$-1s2p^2\ ^2S$	101.74^b	3.2	$-1s4p\ ^1P°$	84.758^c	8.3
	100.94	2.4	$1s^2\ ^1S -$		
	100.77	2.4	$-1s5p\ ^1P°$	83.202^c	6.8
$1s^2\ ^1S -$			$1s^2\ ^1S -$		
$-1s2p\ ^1P°$	100.255^c	9.3	$-1s6p\ ^1P°$	82.377^c	5.1
	96.25	6.7	$1s^2\ ^1S -$		
	96.02	5.8	$-1s7p\ ^1P°$	81.891^c	2.2
$1s^2 2p\ ^2P° -$					
$-1s2p3p\ ^2D$	95.62^b	7.2			

a: preliminary measurements.
b: identifications from ref. 12
c: from ref. 18.
d: observed also in ref. 19

fecting the absorption and particularly for trying to discriminate between adiacent stages of ionization, by varying the power density on the absorbing target and the delay. With the present set-up however already considerable extension of the known Be II spectrum was produced as well the Be III spectrum. We believe that the technique used, implemented with more sophisticated detectors like Multichannel Optical Analizer, could be very useful for the study of absorption spectra of moderately ionized species.

REFERENCES

1. Y.Couturie, J.M. Forsyth: Anual Meeting of Opt.Soc.Am. Chicago, Ill. (1980)
2. P.G. Burkhalter, F.C. Young, B.H. Ripin, J.M. McMahon, S.E. Bodner, R.R. Whitlock, D.J. Nagel: Phys.Rev. A15, 1191 (1977)

3. D.J.Nagel, R.R. Whitlock, J.R. Greig, R.E. Pechacek, M.C. Pecke-
    rar, SPIE Proceedings, 135, 46 (1978)
4. R.D.Frankel, J.M. Forsyth: Science 204, 622 (1979)
5. P.J. Mallozzi, R.E. Schwerzel, H.M. Epstein, B.E. Campbell, Scien
    ce 206, 353 (1979)
6. M.H. Key, C.L.S. Lewis, J.G. Lunney, A. Moore, T.A. Hall, R.G.
    Evans, Phys.Rev.Lett. 41, 1467 (1978)
7. T.B. Lucatorto and T.J. McIlrath , Appl.Opt. 19, 3948 (1980)
8. P.K. Carroll and E.T. Kennedy, Phys.Rev.Lett. 38, 1068 (1977)
9. G.Tondello, Optica Acta 26, 357 (1979)
10. P. Nicolosi, E. Jannitti and G.Tondello, Appl.Phys. B26, 117
    (1981)
11. J. Cooper: Rep.Prog.Phys. 29, 35 (1966)
12. P.K. Carroll, E.T. Kennedy and G.O. Sullivan: Appl.Opt. 19, 1454
    (1980)
13. G.Tondello, E. Jannitti, P. Nicolosi and D. Santi, Opt.Commun.
    32, 281 (1980)
14. R.J. Speer, J.Spect.Soc.Jpn. 23 Suppl. 1, 53 (1974)
15. E. Jannitti, P. Nicolosi and G. Tondello, J.Phys. to be publised
16. P. Mulser, R. Siegel, S. Witkowski, Phys.Lett. 6C, 187 (1973)
17. E. Jannitti, P. Nicolosi and G. Tondello, XV European Conference
    on Laser Interaction with Matter, Munich (1982)
18. G. Mehlman-Balloffet and J.M. Esteva, C.R. Accad.Sc. Paris 267,
    173 (1973).
19. G. Mehlman and J.M. Esteva, Astrophys.J. 188, 191 (1974)

254

Current Developments in High Resolution X-Ray Measurements[*]

D.T. Attwood, N.M. Ceglio and H.M. Medecki
Laser Fusion Program, Lawrence Livermore National Laboratory

H.I. Smith and A.M. Hawryluk[a]
Sub-Micron Structures Laboratory, Radiation Laboratory
of Electronics, MIT

T.W. Barbee, Jr. and W.K. Warburton[b]
Department of Materials Science, Stanford University

J.H. Underwood
Jet Propulsion Laboratory, California Institute of Technology

B.L. Henke
Department of Physics and Astronomy, University of Hawaii

T.H.P. Chang, M. Hatzakis, D.P. Kern, P.J. Coane,
W.W. Molzen and A.J. Speth
Electron Beam Lithography Group, IBM-Watson Research Center

G.L. Stradling
Physics Division, Los Alamos National Laboratory

D.W. Sweeney
School of Mechanical Engineering, Purdue University

Abstract

Recent developments in high resolution x-ray measurements are reviewed. Among the more interesting developments are sub-1000Å x-ray images, multilayer x-ray interference coatings, free standing transmission gratings, 20 psec soft x-ray streak cameras, normal incidence x-ray mirrors, and tabulated atomic scattering factors.

(a) Also, Physics Department, University of Florida.
(b) Stanford Synchrotron Radiation Laboratory.

*Work performed under the auspices of the U.S. Department of Energy under contract No. W-7405-Eng-48.

ISSN:0094-243X/82/900254-23$3.00 Copyright 1982 American Institute of Phy

INTRODUCTION

A renaissance is under way in the field of high resolution x-ray measurements that will significantly affect the applications we envisage and undertake. After a dormant period of 30 years, the field is now being driven externally by emerging technologies, and internally by diagnostic interests in such fields as transient hot dense plasmas for nuclear fusion, and sub-1000Å x-ray biological imaging. The new technological contributions come from two primary directions. First, the thrust of the electronics industry to build ever smaller electronic circuits has led to an impressive array of microfabrication techniques which permit the construction of two-dimensional structures, with spatial features as small as several hundred angstroms. Second, materials science groups have recently mastered techniques which permit the synthetic construction of stable multilayer structures, of an ever increasing variety of materials, to thicknesses as thin as a few atomic monolayers. With structures of one, two and perhaps three dimensions on the scale of x-ray wavelengths, it has recently become possible to build a new class of "x-ray optical elements" based on fairly large angle diffractive effects. In this paper we review recent developments in this field,[1] and suggest by way of example, some areas where the field is likely to proceed in the near future.

~1950:	Grazing incidence mirrors
	Grazing incidence microscopes
	Grazing incidence gratings
	Molecular multilayer "crystals"
	Natural crystals
~1980:	Zone plate lenses
	Interference mirrors
	Transmission gratings
	Splitters (soon)
	X-ray streak cameras (~20 psec)
	High resolution microscopes (<1000 Å)
	Normal incidence mirrors
	Fabry-Perot interferometer

Fig. 1. X-ray Measurement capabilities, circa 1950, are compared with new capabilities of the late 1970's to early 1980's.

The recent technological contributions of the electronics industry to microfabrication is based on the exploitation of high resolution electron beam pattern generation, x-ray lithography, and related developments in physical chemistry. In combination, these techniques permit the fabrication[2] of high aspect-ratio, two-dimensional structures. Complex structures with features smaller than 1000Å were achieved several years ago. Today work is in progress to extend these efforts towards one or two hundred angstroms. These structures can be made free standing, without a supporting substrate or membrane, and written with a rather flexible two-dimensional pattern generation capability. Of great importance, they can be made sufficiently thick* to provide significant phase and transmission variations, for x-rays of energy as high as several kiloelectronvolts (keV). The latter is necessary in order to obtain reasonable diffraction efficiency (∿ 10%) in both imaging and dispersive applications.[3] As a consequence of these developments, it is now possible to fabricate diffractive x-ray elements such as zone plate lenses and transmission gratings. These x-ray optical elements have been used in preliminary applications to form high spatial resolution (sub-1000Å) biological images,[4] and to study thermal energy transport processes in transient hot dense plasmas.[6] In the future we see the extension to more efficient structures[7] which produce x-ray phase shifts approaching π radians, thus giving diffraction efficiencies approaching 40% in each of the first order (± 1) directions. With smaller spatial periods and higher efficiencies is clear that these elements will see increased use in dispersion experiments, particularly at the longer XUV wavelengths, and will begin to see use as splitters in such applications as sub-kilovolt interferometers. With spatial resolution directly related to the structure's smallest spatial scale, use in higher resolution, element sensitive x-ray microscopy of live biological samples will continue to move forward.[8-10] In fact the extension to off-axis holographic imaging appears quite feasible, perhaps with modest spatial resolution (∿ 1/2 μm) in early applications.

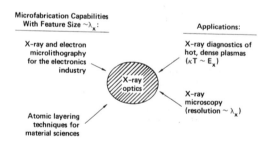

Fig. 2. Primary forces driving a renaissance in high resolution x-ray measurement techniques.

The second major technological breakthrough impacting x-ray optics, is that of atomic monolayering techniques,[11,12] developed

---

* Aspect ratios as large as 4/1 are routine, and perhaps as large as 10/1 are possible in some cases[2]. For certain shadow casting applications, with more modest spatial demands, e.g. five microns, these high aspect ratios permit imaging to be extended to photon energies approaching 100 keV.[5]

by material scientists pursuing the study of interface definition and compound formation. Recently perfected techniques permit the construction of amorphous single layers, of thickness equal to only a few atomic monolayers, laid down in alternating materials of an ever increasing variety. These multilayer microstructures are not only very thin, but are stable, strong and deformable. By utilizing the differing (refractive) scattering properties of low and high atomic number (Z) elements, it is possible to construct x-ray interference coatings which enjoy many of the capabilities of their visible region counterparts, albeit with varying degrees of absorption.[13-16] By constructing coatings whose incremental refractive index varies on a scale of x-ray wavelength dimensions, it is now possible, for instance, to produce coatings that extend present reflective capabilities approximately an order of magnitude higher in energy, to 30 keV or more. In addition, being phase sensitive interference devices, they exhibit a wavelength or photon energy sensitivity, roughly equal to the number of layer pairs (N) playing an effective role. As discussed in the text, mirrors with resolution $E/\Delta E \sim$ 10 to 120, and peak reflectivities of 10% to 80%, are now seeing use in applications covering the spectral region* from 100 eV to 10 keV. These devices are particularly attractive as intermediate resolution monochromators, and as a result have seen early use in applications as energy selectors for synchrotron x-ray studies, and in combination with picosecond (psec) x-ray streak cameras for studying thermal energy transport in short lived hot dense plasmas.[17,18] These devices offer several significant advantages in future applications because of their capability of achieving several angstrom structural sizes. They are additionally valuable because of their very sturdy nature and because the increasing number of material combinations which form sharp, stable interfaces, they can essentially be engineered for use in applications at almost any x-ray wavelength. Applications currently in progress, or being considered, include use as x-ray mirrors or splitters in various monochromator geometries. These applications include higher resolution Fabry-Perot etalons[19], sub-kilovolt interferometry, graded index monochromators[20, 21] energy selective coatings for use with conventional (reflective) x-ray imaging devices[22] and, as reported recently, use as normal incidence x-ray imaging devices.[23]

X-RAY LENSES

X-ray images formed by grazing incidence reflective optics have been in use since the 1950's. Their high reflectance values, approaching 100%, are obtained through total external reflectance of x-rays, at grazing angles of a few degrees, from common materials or coatings whose refractive index is just below unity in

---

* Stable interfaces are currently available with material combinations such as W/C, Ti/C and V/C. Results in which Be or B are used as the low Z material are inconclusive at this point. In general, absorption plays a stronger role at the lower energies. As a consequence the spectral resolution and peak reflectivities are generally less at the lower x-ray energies.

this spectral region.[24] A popular example of that capability, still in wide use today, is the crossed cylindrical mirror pair of Kirkpatrick and Baez[25]. The K-B microscopes, as they are now called, operate conveniently to energies as high as 4 keV, with obtainable resolutions of a few microns, but with relatively small collection solid angle. More sophisticated imaging devices, consisting of concentric conic sections of revolution, initially suggested by Wolter[26], offer larger collection solid angle and, to some degree, better resolution. A review of modern uses of conventional grazing incidence reflection was recently given by Rehn[27], while applications to laboratory and astrophysical imaging applications were given by Price[28] and by Zombeck[29], respectively.

The use of diffractive elements to obtain higher spatial and angular resolution, at x-ray wavelengths, was first addressed seriously in the pioneering paper of Baez[30]. In particular Baez demonstrated the early use of free standing Fresnel zone plates, in which the alternate zones are transmissive and opaque to soft x-rays, albeit using them at longer wavelengths than was his primary interest. The finer the outer zones the higher the diffraction angles and the better the obtainable resolution. Figure 3 shows schematically how such a lens is used as an x-ray focusing element. In the earlier work of Baez, the finest spatial

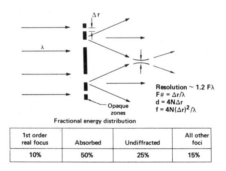

1st order real focus	Absorbed	Undiffracted	All other foci
10%	50%	25%	15%

Fig. 3. The use of a Fresnel zone plate lens for focusing x-rays is demonstrated. For alternately opaque and transmissive zones, 10% of the incident x-rays reach the first order real focus (Ceglio ref. 3; Baez ref. 30).

features were 20 μm wide. As noted in the introduction to this paper, the capability for fabricating structures with ever finer spatial features has advanced several orders of magnitude in recent years. Fig. 4 shows an electron micrograph of a gold Fresnel lens fabricated by Shaver et al.[31]. This lens has an outer zone width of 3200Å, 500 zones, a diameter of 640 μm, and a thickness of 1.3 μm. Figure 5 shows a simple x-ray microscope, employing the above as its objective, to interrogate a resolution test pattern of

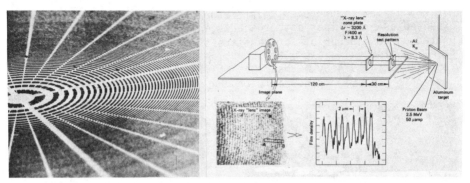

Fig. 4   Fig. 5

Fig. 4. A freestanding gold zone plate lens for x-ray imaging. The outer (500th) zone has a width of 3200 A. The lens is 1.3 μm thick and 640 μm in diameter (Shaver et al., ref. 31).

Fig .5. An x-ray microscope using the 3200 A zone plate of Fig. 4 as its' objective lens. Using aluminum K_a x-rays at 1.5 keV, the one micron lines and spaces of a resolution test pattern are clearly resolved (Ceglio, ref. 32).

1 μm wide lines and spaces. Backlighted by 1.5 keV Al K_α radiation, the 1 μm lines are clearly resolved[32], as one would expect with a 0.32 μm outer zone objective lens.

Figure 6 shows diffraction limited images of biological specimens obtained by Schmahl, Rudolph and Niemann[4], using 280 eV synchrotron radiation and a 1200Å outer zone width objective lens. Spatial features of the diatom indicate that comparable resolution is obtained. More recent work reported by the same group[33], using a lens with a 600Å outer zone, shows that comparable improvements in spatial resolving power are again obtained.

Fig. 6. Biological image obtained by Schmahl et al. (Ref. 4) using 44 Å synchrotron radiation, and a 625 zone Fresnel zone plate lens with a 1200 Å outer zone. Marker in lower right corner indicates one micron scale size.

Additional improvements in Fresnel zone plate imaging are
expected as ever finer patterns are generated, replicated, and
plated up to appropriate thicknesses of gold. Figure 7 shows a
free standing gold zone plate, with an outer zone of 700Å,
fabricated in 1974 by M. Hatzakis et al. of IBM's Watson Research
Center[34]. The Electron Beam Lithography Group at IBM is

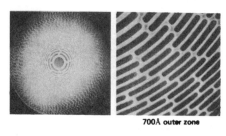

700Å outer zone

Fig. 7

Minimum zone width 350Å

Fig. 8

Fig. 7. A high resolution, free standing, gold Fresnel zone plate,
with a 700 Å outer zone, formed in 1974 by M. Hatzakis et al.
(Ref. 34).

Fig. 8. Progress towards higher resolution zone plate lenses is
demonstrated by this recently generated partial resist pattern,
with outer zone width of 350 Å. Further work remains to generate
the entire pattern, and from that form a free standing gold zone
plate. P.J. Coane, D.P. Kern, A.J. Speth, W.W. Molzen and T.H.P.
Chang (Ref. 35).

currently pursuing e-beam pattern generation which will press the
state of capabilities in that field. Figure 8 shows recent
progress towards generation of a 350Å linewidth zone plate
pattern: a partial resist pattern generated by electron beam
lithography[35]. Complete patterns at that resolution have not yet
been written, nor have they as yet been replicated and plated up to
appropriate thicknesses of gold. In addition to this work, which
has an ultimate goal the fabrication of zone plate patterns with
100Å minimum linewidth, other capabilities are also ripe for
pursuit. For one, finer spatial features may be obtainable
utilizing atomic layering techniques, in rotating cylindrical
geometries, as discussed for instance by Rudolph[36]. In the XUV
and VUV range it may also be possible to produce high resolution
Fresnel phase plate structures (rather than transmission zone
plates), which offer a potential factor of four improvement in
diffraction efficiency[7].

MULTILAYER X-RAY MIRRORS

In this section we review recent advances in the fabrication and utilization of multilayer coatings for use as wavelength selective x-ray mirrors in a variety of x-ray manipulative applications. As discussed in the introduction, it is now possible through feedback controlled sputtering techniques[11], and to a significant extent by evaporative techniques[12], to form multilayer structures, from an ever increasing list of materials, in which the individual layer thicknesses consist of only a few atomic monolayers. For instance, in certain tungsten/carbon multilayer structures, the alternate tungsten layers are only four atomic planes thick, approximately 11Å, over an area of several square inches. These new layering capabilities permit a significant extension of previous x-ray reflective capabilities, in spectral range, incidence angle, energy selectivity, and concomitantly, applications which can be pursued.

The multilayer structures are commonly made of alternately low and high atomic number (Z) materials, such as W/C, V/C, Ti/C, etc. By variations in atomic number, slight changes in the refractive index or local scattering power are introduced, with a spatial modulation corresponding to the periodicity of the layered structure, typically 10Å to several 100Å. In a sense one forms synthetic crystals, amorphous within any plane. Figure 9 shows schematiclly how such a structure is numerically analyzed[15] in terms of its electromagnetic properties at x-ray wavelengths. The

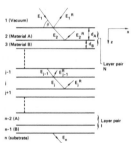

With refractive index n_r given in terms of tabulated atomic scattering factors:

$$n_r = 1 - \delta - i\beta = 1 - \frac{n_a r_e \lambda^2}{2\pi} \, (f_1 + if_2)$$

Fig. 9. Computational model is shown for analysis of multilayer x-ray interference coatings. Refractive index is given in terms of tabulated atomic scattering factors (Underwood, Ref. 15).

incident wave, E_1, at angle θ, passes through the various materials, experiencing small amounts of absorption in each of the thin layers, and producing small reflections at each of the idealized interfaces. Summing these incremental reflections, one finds that with sufficient flatness of the structure, the reflected fields add coherently, giving a sizeable reflection at certain wavelengths, much as their visible region counterparts, except that peak reflectivities are limited somewhat by finite absorption.[12-16] In order to perform the computations one requires knowledge of the complex refractive index or, equivalently, the complex scattering coefficient at relevant wavelengths. Figure 10 shows a sample of atomic scattering factors for gold, from a recent compilation by B.L. Henke of the University of Hawaii.[37] The compilation provides a best fit to available data and theory, for elements from hydrogen to plutonium (Z=94), and for energies spanning the soft x-ray regime.

TABLE I. PHOTOABSORPTION CROSS SECTION, μ
ATOMIC SCATTERING FACTOR, f₁ + if₂

Atomic Weight = 197.0   Z = 79

μ(barns/atom) = μ(cm²/gm) x 327.0   GOLD (Au)
Eμ(E) = 213.5 f, keVcm²/gm

LINE	E(eV)	μ(cm²/gm)	f₁	f₂	λ(A)	LINE	E(eV)	μ(cm²/gm)	f₁	f₂	λ(A)
Na L₂,₃M	30.5	1.28 05		18.23	407.1	Co Lα	776.2	7.95 03	44.05	28.91	16.0
Mg L₂,₃M	49.3	7.25 04		16.74	251.5	Ni Lα	851.5	6.76 03	47.10	26.96	14.56
Al L₂,₃M	72.4	5.24 04		17.77	171.4	Cu Lα	929.7	5.79 03	49.12	25.22	13.33
Si L₂,₃M	91.5	3.34 04		14.31	135.5	Zn Lα	1011.7	4.97 03	50.71	23.53	12.25
Be K	108.5	1.71 04	21.77	8.66	114.	Na Kα	1041.0	4.70 03	51.22	22.90	11.91
Sr M₅	114.0	1.40 04	21.33	7.48	108.7	Ge Lα	1188.0	3.62 03	52.65	20.13	10.44
Y M₅	132.8	8.67 03	18.23	5.39	93.4	Mg Kα	1253.6	3.25 03	52.98	19.09	9.89
S L₁	148.7	7.98 03	15.74	5.55	83.4	Al Kα	1486.7	2.31 03	53.14	16.04	8.34
Zr M₅	151.1	8.02 03	15.38	5.68	82.1	Si Kα	1740.0	1.66 03	51.60	13.55	7.13
Nb M₅	171.7	9.37 03	13.07	7.53	72.2	Zr Lα	2042.4	1.18 03	45.33	11.29	6.07
B Kα	183.3	1.06 04	12.26	8.97	67.6	Nb Lα	2165.9	1.04 03		10.52	5.73
Mo M₅	192.6	1.13 04	11.84	10.22	64.4	Mo Lα	2293.2	3.68 03		39.50	5.41
W N₆N₇	212.2	1.30 04	11.67	12.92	58.4	Cl Kα	2622.4	2.68 03		32.93	4.73
C Kα	277.0	1.52 04	14.20	19.73	44.7	Ag Lα	2984.3	2.27 03		31.78	4.16
Ag M₅	311.7	1.53 04	16.21	22.31	39.8	Ca Kα	3691.7	1.51 03		26.16	3.36
N Kα	392.4	1.54 04	22.36	28.37	31.6	Ba Lα	4466.3	9.53 02		19.93	2.78
Ti L₁	395.3	1.54 04	22.75	28.47	31.4	Ti Kα	4510.8	9.30 02		19.64	2.75
Ti Lα	452.2	1.37 04	27.75	28.99	27.4	V Kα	4952.2	7.37 02		17.09	2.50
V Lα	511.3	1.21 04	31.50	28.92	24.3	Cr Kα	5414.7	5.89 02		14.93	2.29
O Kα	524.9	1.18 04	32.32	28.90	23.6	Mn Kα	5898.8	4.74 02		13.09	2.10
Mn L₁	556.3	1.09 04	33.76	28.47	22.3	Co Kα	6930.3	3.14 02		10.19	1.79
Cr Lα	572.8	1.05 04	34.05	28.21	21.6	Ni Kα	7478.2	2.58 02		9.05	1.66
Mn Lα	637.4	1.02 04	37.23	30.41	19.5	Cu Kα	8047.8	2.14 02		8.07	1.54
F Kα	676.8	9.29 03	39.80	29.44	18.3	Zn Kα	8638.9	1.78 02		7.22	1.44
Fe Lα	705.0	8.79 03	40.87	29.01	17.6	Ge Kα	9886.4	1.26 02		5.84	1.25

ABSORPTION EDGE

M₁	3428	eV	3.616 A
M₂	3150	eV	3.936 A
M₃	2743.9	eV	4.518 A
M₄	2307	eV	5.374 A
M₅	2220	eV	5.584 A

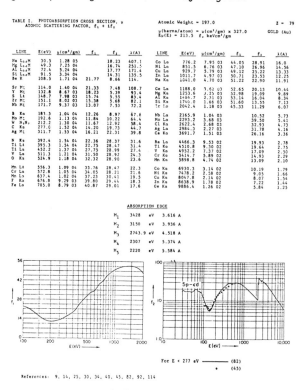

For E < 277 eV ——— (82)
• (45)

References: 9, 14, 25, 30, 34, 40, 45, 82, 92, 114

Fig. 10. Example of tabulated atomic scattering factors for soft x-ray wavelengths and elements from hydrogen to plutonium. Example shown is gold, Z=79 (Henke, Ref. 37).

To give some idea of how strong these undulations are, and as a result, an estimate of the number of layer pair interfaces required to make a sizeable reflection, Underwood of JPL makes the following instructional argument.[38] Consider x-rays incident on a single material interface, from vacuum, as shown in Figure 11. The reflection coefficient in terms of the incremental refractive index variations is as shown, with typical values of δ and β obtained by way of example from Henke's tabulation for carbon and gold at an energy of 109 eV (Be K, λ=114Å). The tabulated results lead to predicted reflection coefficients (fraction of energy reflected) for single surfaces of carbon and gold, of about one part in 10^4, and one part in 10^3, respectively. For higher x-ray energies the reflection coefficients decrease, roughly as wavelength to the fourth power, e.g., to that of a single surface reflection from gold of about 5×10^{-8} at 2 keV ($\lambda=6\text{Å}$). For a succession of interfaces, however, the situation changes dramatically because the fields, which are proportional to the square root of the reflection coefficient ($E_R/E_0=\sqrt{R}$), add coherently. As a consequence a significantly smaller number of interfaces than one might at first suspect is needed to produce a reflection approaching unity. Considering, to first order, that one requires $N \simeq 1/\sqrt{R}$ interfaces (neglecting absorption, incidence angle, polarization, etc.), one sees from the tabulation in Fig. 11 that something like 25 to 75 layer pairs is required for respectable mirror performance. Having some feel for the

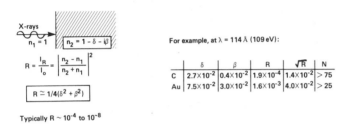

For example, at $\lambda = 114$ Å (109 eV):

	δ	β	R	$\sqrt{R}$	N
C	2.7×10^{-2}	0.4×10^{-2}	1.9×10^{-4}	1.4×10^{-2}	>75
Au	7.5×10^{-2}	3.0×10^{-2}	1.6×10^{-3}	4.0×10^{-2}	>25

Fig. 11. A rough estimate is obtained of the number of layer pairs required to produce a reflection coefficient approaching unity, by first considering a single surface, and then extending the argument to N coherently adding (fields) surfaces. Tabulated values are for carbon and gold at $\lambda=114$ Å (109 eV) (Underwood, Ref. 38).

performance one can expect, it is now interesting to see the results of a full computation, and compare those to experimental measurements made for a specific 30 layer pair tungsten/carbon structure, with periodicity d=32Å (11Å/21Å), illuminated with synchrotron radiation at 5.478 keV and at angles from zero to 2.5^0.[11,15] One observes in Fig. 12 that in addition to the usual grazing incidence reflection at angles below 0.5^0, there is both a predicted and measured Bragg-like interference peak, of greater than 50%, at an angle somewhat larger than 2^0. The major conclusions to be drawn from Fig. 12, are that large reflectances are obtained through the well controlled monolayering techniques, and that good computational models are in hand to aid in the engineering design of desired structures and performance. Figure 13 demonstrates that similar performance is achievable at sub-kilovolt photon energies. In the example shown, a titanium/carbon multilayer structure formed by Barbee of Stanford,

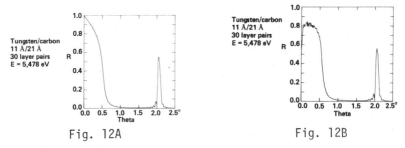

Fig. 12A   Fig. 12B

Fig. 12. The numerically predicted (A) and experimentally verified (B) performance of a 30 layer pair, tungsten/carbon multilayer (11 Å/21 Å), irradiated at 5.478 keV. (Barbee, Underwood and Warburton, Refs. 11 and 15).

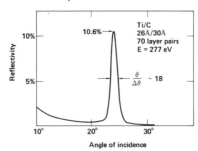

Fig. 13. Sub-kilovolt x-ray mirror performance is demonstrated for a Ti/C multilayer coating. Instrument corrected resolution is $E/\Delta E \simeq 40$. (Barbee, Henke, Stradling, unpublished).

was measured for sub-kilovolt performance by Henke, at the University of Hawaii. Achieving a peak reflectivity of approximately 10% for 277 eV x-rays (C K_α), the energy selectivity of this structure, when corrected for instrument resolution, is approximately $E/\Delta E \simeq 40$. With appropriate filtering, the small angle or low energy reflection component is eliminated, providing a fine intermediate resolution monochromator. Using various materials, it is now possible to construct these intermediate resolution monochromators in applications from 100 eV to 30 keV. In the following section we discuss the work of Stradling[17,18] in which a series of these structures was used in tandem to form a time resolved multichannel analyzer to sample the emission spectrum of a transient hot dense plasma. In other sections we discuss their use as x-ray splitters for phase sensitive measurements. As a concluding point in this section, we show in Fig. 14 results obtained recently by Underwood and Barbee[23], in which for the first time normal incidence x-ray mirrors have been used to form images. In these preliminary demonstration experiments, using multilayer coated mirrors, 100 μm wires and spaces are clearly resolved. Efforts are underway by Henry, Spiller, Kirz and others[39], to extend these results to higher spatial resolution.

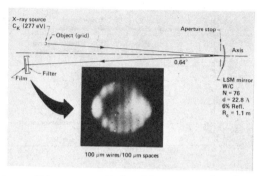

Fig. 14. Imaging with a normal incidence x-ray mirror recently demonstrated by Underwood and Barbee (Ref. 23). Wires and spaces, 100 μm wide, are clearly resolved in the x-ray image. Peak mirror reflectivity is 20% at 274 eV; reflectivity is 6% integrated over the broad carbon line.

X-RAY STREAK CAMERAS

X-ray streak cameras play a major role in the diagnostics of laser created plasmas, and as a result are now reasonably well understood, at least for resolutions down to about 10 picoseconds (psec), and are in use in a variety of applications for photon energies extending from 100 eV to 30 keV.[40-47] In addition to temporal resolutions which can approach a psec, they offer large dynamic range, quantum detection capabilities, broad spectral sensitivity, and a continuous photocathode for combination with spatially or spectrally resolving instruments.[41,48] Dynamic range, which is strongly related to temporal resolution,[49] can exceed three orders of magnitude for resolutions of a few tens of picoseconds. Figures 15 and 16 show typical usage of a soft x-ray streak camera with conventional x-ray mirror/filter pairs, to record sub-keV time histories in observational channels at photon energies of 200 and 700 eV. The data, which was obtained with a modest 1.3 joule, 50 psec Nd laser pulse (λ=1.06 μm), displays the above mentioned performance with regard to temporal resolution, dynamic range (> 10^3), sensitivity, and multi-channel sub-keV recording capabilities.

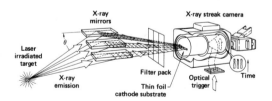

Fig. 15. Schematic of a 20 psec soft x-ray streak camera, shown in combination with a three channel x-ray mirror/filter spectrometer (Ref. 43-46).

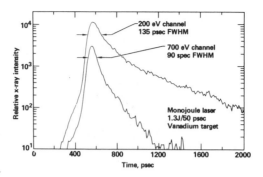

Fig. 16. Typical data obtained with a soft x-ray streak camera, demonstrating large dynamic range (> 10³), good temporal resolutin (20 psec), quantum detection capabilities over a wide spectral range (100 eV to 30 keV), and a continuous photocathode for use in combination with imaging and spectrally dispersing instruments (Refs. 43-46).

An interesting application, combining streak camera technology and multilayer developments discussed in the previous section, is outlined in Figure 17. Here thermal energy transport in laser

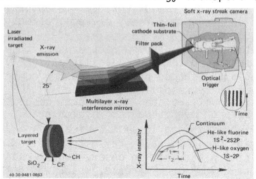

Fig. 17. A five channel multilayer x-ray mirror - soft x-ray streak spectrometer, shown schematically as it would be used to study thermal energy transport in laser irradiated multilayer targets. (Stradling et al., Refs. 17 and 18).

irradiated multilayer targets, is studied with a five channel multilayer-streak spectrometer, in which three of the interference mirrors are engineered to monitor general sub-kilovolt thermal emission, while two channels are keyed to thermally produced heliogenic and hydrogenic atomic configurations of specified target materials. In this example, thermal energy transport is studied, in laser irradiated targets with a five channel multilayer-streak spectrometer. As shown in Fig. 18, the five channels can be chosen to monitor general sub-kilovolt thermal emission, or as is the case in two of the channels, used to monitor emission lines of thermally produced hydrogen-like oxygen (653 eV) and helium-like fluorine (737 eV), which are well identified species in a matched multilayer target designed for such studies. Fig. 19 shows typical streak

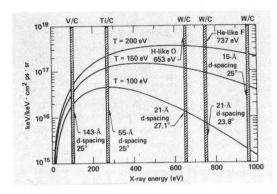

Fig. 18. Spectral location of five multilayer mirror channels as used in Figure 17 (Stradling, unpublished).

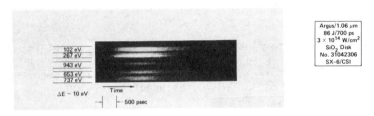

Fig. 19. Streak records of five channel streak spectrometer shown in Figs. 17 and 18. Note that in the sub-thermal spectral range, temporal duration of emission is significantly longer than laser irradiation pulse.

records obtained with this instrument combination, upon irradiation of a simple glass disk target. This work, which will be published elsewhere, is part of a doctoral dissertation soon to be submitted by G.L. Stradling to the University of California at Davis/Livermore.[17,18] Figure 20 shows related studies, by Stradling and his colleagues, of high sensitivity psec x-ray cathodes. In the example shown, CsI is shown to be a factor of ten more sensitive than Au, for photon energies in the 500 to 700 eV range, without significant loss of temporal resolution. The improved sensitivity is related to the dominance of photoelectron conversion to slow secondaries,[50,51] a cathode thickness appropriate to the secondary electron range, and the lower work functions available with certain insulator and semi-conductor materials.[52,53]

MULTILAYER BEAM SPLITTERS

In certain applications it is useful to have available x-ray beam splitters which operate at specified wavelengths. One example would be on a synchrotron beam line in which successive splitters would pick off spectrally distinct portions of the beam, at convenient angles, for independent experimental stations. Another use would be for phase sensitive measurements, such as sub-kilovolt

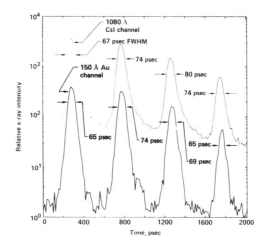

Fig. 20. Intensity vs. time plot showing response of a composite x-ray photocathode (CsI next to Au) to repetitive x-ray pulses from a multi-pulse laser target irradiation experiment. CsI is seen to be far more sensitive in the 500 to 700 eV spectral region (Stradling et al., Ref. 45).

interferometry. Fig. 21 shows a schematic of such a beam splitter, with computational predictions of its performance at 119 eV (λ=104 Å). The system has been numerically optimized by Underwood[54] to give equal split intensities, at 45° incidence angle to a 20 layer pair structure of Mo/B. Values are for σ polarization, as π polarized x-rays experience little reflection at 45°, which is close to Brewster's angle for near unity refractive index.[12] Figure 22 shows predicted performance for the same multilayer with broadband irradiation, and Figure 23 shows how thin (N=20 layer pairs) and thick (N >> 20) multilayers might be combined in a sub-kilovolt x-ray interferometer, which could be designed, or tuned, for use at almost any wavelength of interest and thus extend present interferometric capabilities which are used only in the multi-keV photon energy range[55]. We caution that such multi-layer splitters have not yet been fully developed, and that significant questions remain with regard to achievable surface flatness and thus required longitudinal coherence length.

TRANSMISSION GRATINGS

In our consideration of new developments in x-ray measurement capabilities, a significant advance is that of fabricating free standing x-ray transmission gratings, as recently pursued by Hawryluk[5,6]. Transmission gratings are of particular value in broad band spectral studies of transient, or spatially inhomogeneous plasmas, where their ease of alignment makes combination with temporal or spatial resolving instruments relatively convenient.[6,57] Transmission gratings will also be of

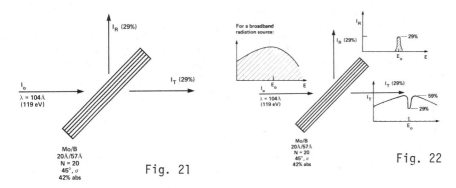

Fig. 21

Fig. 22

Fig. 21. Predicted performance is shown for a multilayer x-ray splitter, optimized for equal reflected and trasmitted intensities at 104 Å (Underwood, unpublished). Such an element might be used in interference experiments.

Fig. 22. Performance of same multilayer splitter as in Fig. 21, but with a broadband radiation source. (Underwood, unpublished).

- Minimal spectral and spatial coherence requirements
- Central wavelength and angles can be engineered

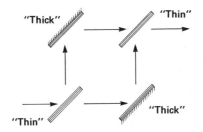

- Thick mirrors set high $\lambda/\Delta\lambda$ in both arms
- Thin last splitter allows larger angular acceptance
- Flatness specifications, coherence length

Fig. 23. A sub-kilovolt x-ray interferometer formed of multilayer splitters and mirrors. The above has minimal spectral and spatial coherence requirements, can be engineered or tuned to almost any sub-keV wavelength, and operate with a relatively large angular acceptance. Performance has not yet been demonstrated; significant questions remain with regard to flatness and required coherence length.

significant value in phase sensitive applications, such as interferometry and off-axis holography, where they can be used to split and recombine x-ray beams with minimal coherence requirements.

Although highly dispersive grazing incidence reflection gratings are available[58] for use in the x-ray range, the small incidence angles render them inconvenient for combination with picosecond streak cameras and microscopes, as is often required for the study of hot dense plasmas. In addition natural[59,60] and synthetic[61] (Langmuir-Blodgett) crystals exist, but cover only limited spectral regions, and suffer from critical and in some situations geometrically inconvenient angular considerations, such that they are not generally combined with spatial or temporally resolving instruments, except in exceptional cases.[48] As an example of the rich data available in sub-kilovolt spectroscopy, figure 24 shows time and space integrated spectral emission from a laser irradiated chromium target, dispersed to film by a lead behenate crystal on a convex substrate.[62] Although the spectral details of the highly ionized species are interesting, it would be valuable in many situations to observe a broader spectrum, say 100 eV to 1000 eV, with several psec time resolution.

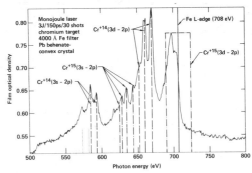

Fig. 24. Line and continuum x-ray emission is shown as typical of laser irradiated disk targets. (Gaines et al., Ref. 62).

As demonstrated by Ceglio[3] and his colleagues, such observations are possible with a streak camera grating combination, albeit with some compromise in spectral detail. It is in the latter sense that available measurement techniques are often complimentary in nature. Figure 25 shows two stages in the multi-step evolution of a high aspect ratio, 3000 Å period freestanding grating, as fabricated and described by Hawryluk[56,57]. After preliminary stages involving pattern generation by interference of 3250 Å He-Cd laser light, and formation of a thin gold pattern mask therefrom for x-ray lithographic replication, a thicker (7000 Å) mask is eventually formed in PMMA, as shown in Figure 25A. This is then used as an electroplating mold to form the relatively thick grating shown in Figure 25B. The use of this grating in combination with a Livermore developed soft x-ray streak camera[40,41,43-46] is then described in Figure 26. In these demonstration experiments, laser driven soft x-ray emisson was for the first time studied with continuous broadband (0.1 to 2 keV) spectral coverage, and high (20 psec) temporal resolution.

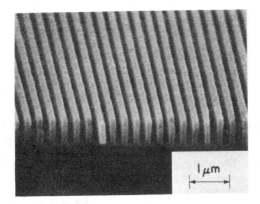

Fig. 25A

Fig. 25B

Fig. 25. Two stages in the evolution of a 3000 Å period free standing gold transmission grating. (A) Grating as replicated in 7000 Å thick PMMA, and (B) after plating pattern up to 6500 Å gold. (Hawryluk, Ref. 56).

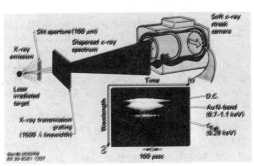

Fig. 26. Time resolved, continuous coverage, soft x-ray spectrometer. (Ceglio, Roth, Hawryluk; Ref. 6).

INTERFERENCE EXPERIMENTS

Figure 27 shows a scheme in which transmission gratings might typically be used in visible light experiments, with sources of minimal spatial and temporal coherence, for use in interferometric and holographic recording schemes.[63] The gratings of the previous section now permit these experiments to be extended to the x-ray regime, where they could be used for similar studies, particularly when the gratings are extended to higher spatial frequencies, and to phase gratings[7] which enjoy a factor of four efficiency over their absorption/transmission counterparts. The extension to higher spatial frequencies, below 1000 Å is possible through either photon[64] or electron beam[65] lithography*, followed by the same steps utilized in high resolution zone plate fabrication. With regard to more efficient grating efficiency, the nominal 10% efficiency, to both plus and minus first order, could as an example be extended to 34% to each of the first orders, at $\lambda \simeq 100$ Å, by fabrication of a 1000 Å period carbon transmission grating, with a thickness of 2000 Å (four to one aspect ratio). Appropriate materials and thicknesses for shorter wavelengths are discussed by Ceglio and Smith[7].

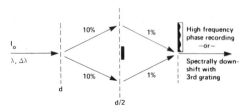

Fig. 27. The use of freestanding gratings as x-ray beam splitters in phase sensitive measurements such as interferometry or holography is suggested. Diffraction efficiencies can be improved to greater than 30% by fabrication of phase gratings. Coherence requirements are related to surface flatness, angular dispersion, properties of the object under study and possible use of a lens.

With regard to using gratings in phase sensitive recording, it is of interest to review some of the literature on x-ray holography[66-72]. A number of very interesting experiments have been conducted over the past ten years using quasi-coherent x-radiation from synchrotrons and characteristic cold line sources. The work of Aoki, Kikuta, and their Japanese colleagues

---

*In fact x-ray and deep UV interference effects have been used to produce 1000Å gratings using spatial-period-division techniques, in which diffraction from a grating of period p, produces a near field pattern of period p/m, where m is n integer.[56,73,74] This technique might be extended in the future by the use of parent phase gratings in the case of x-ray period division, or by the use of coherent UV sources, as they become available with wavelength significantly less than 2000Å[75]

is of particular interest. In Fig. 28 we review one of their experiments,[67] in which an appropriate degree of spatial and temporal coherence was derived from a synchrotron x-ray source by use of a grazing incidence grating and a 7 μm slit. Employing lenseless Fourier transform techniques, they formed an off-axis holographic recording of a nearby object, which consisted of three adjacent 3 μm slits. Resolution was set by the 2.5 μm reference beam slit. Twelve centimeters away they recorded an approximately 30 μm period x-ray interference pattern on film.

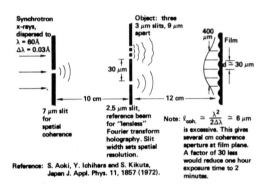

Fig. 28. The off-axis x-ray holography experiment of Aoki, Ichihara and Kikutu (1972) is reviewed (Ref. 67).

Because of the long spatial period recorded, it was then possible to optically reconstruct the images with a visible He-Ne laser. Clear reconstructions of the 3 μm objects are evident in the cited literature[67]. In later work they demonstrated the use of an Al K_α source to record holograms of chemical fibers and red blood cells.[68] It would be interesting to extend these early off-axis experiments utilizing the recently developed x-ray optical components reviewed herein, to broaden our capabilities and applications, using laboratory x-ray sources, and visible or UV reconstruction schemes.

CONCLUSIONS

We have reviewed recent developments in the rapidly evolving field of high resolution x-ray measurements. Capabilities now exist for undertaking applications which could not have been seriously considered even a few years ago, in fields as diverse as physics, biology, chemistry and material science. With further developments in coherent VUV and XUV sources of radiation, likely to be made by participants of this conference, the combined capabilities will see ever more exciting use in applications across the scientific spectrum.

References

1. This review of x-ray optics is excerpted from a recent meeting on x-ray diagnostic techniques, the printed proceedings of which are available as: Low Energy X-Ray Diagnostics-1981 (Monterey), Edited by D.T. Attwood and B.L. Henke (AIP, New York, 1981).

2. H.I. Smith, ibid, p. 223.

3. N.M. Ceglio, ibid, p. 210.

4. G. Schmahl, D. Rudolph and B. Niemann, ibid, p. 225; Ann. N.Y. Acad. Sci. 342, 368 (1980).

5. N.M. Ceglio, G.F. Stone and A.M. Hawryluk, J. Vac. Sci. Technol. 19, 886 (1981).

6. N.M. Ceglio, M. Roth and A.M. Hawryluk, Monterey Proceedings, P. 290.

7. N.M. Ceglio and H.I. Smith Proc. VIII Intl. Conf. X-Ray Optics and Microanalysis (Boston, 1977); UCRL-79966.

8. D. Sayre, J. Kirz, R. Feder, D.M. Kim and E. Spiller, Ultramicroscopy 2, 337 (1977); D. Sayre, Imaging Processes and Coherence in Physics, Edited by M. Schlenker et al. (Springer-Verlag, New York, 1979), p. 229.

9. Ultrasoft X-Ray Microscopy: Its Application to Biological and Physical Sciences, Edited by D.F. Parsons, Ann. N.Y. Acad Sci. V. 342 (1980).

10. E. Spiller, Nucl. Instr. Meth. 177, 187 (1980).

11. T.W. Barbee, Monterey Proceedings, p. 131.

12. E. Spiller, ibid, p. 124; E. Spiller and R. Feder, Sci. Amer. 239, 69 (1978).

13. A.V. Vinogradov and B. Ya. Zeldovich, Appl. Optics 16, 89 (1977).

14. E. Spiller, Appl. Phys. Lett. 20, 365 (1972).

15. J.H. Underwood and T.W. Barbee, Monterey Proceedings, p. 170; Appl. Opt. 20, 3027 (1981).

16. P. Lee, Opt. Commun. 37, 159 (1981).

17. G.L. Stradling, Doctoral Dissertation, Department of Applied Science, Univ. Calif. Davis/Livermore (1982).

18. G.L. Stradling, T.W. Barbee, B.L. Henke, E.M. Campbell and W.C. Mead, Monterey Proceedings, p. 292.

19. T.W. Barbee and J.H. Underwood, to be published.

20. J.V. Gilfrich, D.J. Nagel and T.W. Barbee, Appl. Spectr. 36, 58 (1982).

21. D.J. Nagel, J.V. Gilfrich and T.W. Barbee, Nucl. Inst. Meth., to be published (1982).

22. R.H. Price and T.W. Barbee, unpublished.

23. J.H. Underwood and T.W. Barbee, Nature 294, 429 (1981).

24. A.H. Compton and S.K. Allison, X-Rays in Theory and Experiment (Van Nostrand, New York, 1935).

25. P. Kirkpatrick and A.V. Baez, J. Opt. Soc. Amer. 39, 746 (1948).

26. H. Wolter, Ann. der Physik, 10, 94 (1952).

27. V. Rehn, Monterey Proceedings, p. 162.

28. R.H. Price, ibid, p. 189.

29. M.V. Zombeck, ibid, p. 200.
30. A.V. Baez, J. Opt. Soc. Am. 51, 405 (1961); J. Opt. Soc. Am. 42, 756 (1952).
31. D.C. Shaver, D.C. Flanders, N.M. Ceglio and H.I. Smith, J. Vac. Sci. Tech. 16, 1626 (1980).
32. N.M. Ceglio, Ann. N.Y. Acad. Sci. 342, 65 (1980).
33. G. Schmahl, D. Rudolph and B. Niemann, Proc. Conf. on High Resolution Soft X-Ray Optics, Vol. 316, Edited by E. Spiller (SPIE, Bellingham, WA., 1982); Science 215, 150 (1982).
34. M. Hatzakis, C.H. Ting and N. Viswanathen, VI Int'l Conf. on Elec. and Ion Beam Sci. and Tech., San Francisco, May 1974.
35. P.J. Coane, D.P. Kern, A.J. Speth and T.H.P. Chang, X Int'l Conf. on Elec. and Ion Beam Sci. and Tech., Montreal, May 1982
36. D. Rudolph, and G. Schmahl, Ann. N.Y. Acad. Sci. 342, 94 (1980); D. Rudolph, B. Niemann and G. Schmahl, Proc. Conf. on High Resol. Soft X-Ray Optics, V.316, p. 103, Brookhaven, 1981 edited by E. Spiller (SPIE, Bellingham WA., 1982).
37. B.L. Henke, Monterey Proceedings, p.146 and p.340; Atomic Data and Nuclear Data Tables, 27, 1, (Jan. 1982).
38. J.H. Underwood, T.W. Barbee, and D.C. Keith, Space Optics-Imaging X-Ray Optics Workshop, Ed. by R.C. Chase (SPIE, Bellingham, WA., 1979), V. 184, p. 123.
39. J.P. Henry, E. Spiller, and M. Weisskopf, Proc. Conf. on High Resol. Soft X-Ray Optics, V. 316, p. 166, Brookhaven, 1981, edited by E. Spiller, (SPIE, Bellingham, WA., 1982).
40. C.F. McConaghy and L.W. Coleman, Appl. Phys. Lett. 25, 268 (1974).
41. D.T. Attwood, IEEE J. Quant. Electr. QE-14, 909 (1978).
42. D.J. Bradley, A.G. Roddie, W. Sibbett, M.H. Key, M.J. Lamb, C.L.S. Lewis and P. Sachsenmaier, Opt. Commun. 15, 231 (1975).
43. G.L. Stradling, M.S. Thesis, Physics Department, Brigham Young University, 1981; Lawrence Livermore National Laboratory report, UCRL-52568 (1978, revised 1981).
44. D.T. Attwood, R.L. Kauffman, G.L. Stradling, H. Medecki, R.A. Lerche, L.W. Coleman, E.L. Pierce, S.W. Thomas, D.E. Campbell, J. Noonan, G.R. Tripp, R.J. Schnetz and G.E. Phillips, XIV Int'l Cong. High Speed Phot (Moscow, 1980); UCRL-85043.
45. G.L. Stradling, H. Medecki, R.L. Kauffman, D.T. Attwood and B.L. Henke, Appl. Phys. Lett. 37, 782 (1980).
46. R.L. Kauffman, G.L. Stradling, E.L. Pierce and H. Medecki, Monterey Proceedings, p. 66.
47. P. Jaanimagi, Doctoral Dissertation, Univ. Waterloo (Ontario, Canada, 1981).
48. M.H. Key, C.L.S. Lewis, J.G. Lunney, A. Moore, J.M. Ward and R.K. Thareja, Phys. Rev. Lett. 44, 1669 (1980); P. Cunningham, C.L.S. Lewis et al., Rutherford and Appleton Laboratories Report RL 81-040, p. 4-12.
49. S.W. Thomas and G.E. Phillips, Proceed. of the 13th Int'l Congr. on High Speed Photogr. and Photonics, Tokyo, 1978, pp. 471-475.
50. B.L. Henke, J.A. Smith and D.T. Attwood, Appl. Phys. Lett. 29, 539 (1976).

276

51. B.L. Henke, J.A. Smith and D.T. Attwood, J. Appl. Phys. 48, 1852 (1977).

52. B.L. Henke, J. Liesegang and S.D. Smith, Phys. Rev. B 19, 3004 (1979).

53. B.L. Henke, J.P. Knauer and K. Premaratne, J. Appl. Phys. 52, 1509 (1981).

54. J.H. Underwood, unpublished.

55. D.P. Siddons, Monterey Proceedings, p. 236.

56. A.M. Hawryluk, Doctoral Dissertation, Electrical Engineering Department, MIT (1981).

57. A.M. Hawryluk, N.M. Ceglio, R.H. Price, J. Melngailis and H.I. Smith, Monterey Proceedings, p. 286.

58. E. Kallne, Monterey Proceedings, p. 97.

59. N.J. Peacock, ibid, p. 101.

60. D.M. Barrus, R.L. Blake, H. Felthauser and E.E. Fenimore, ibid, p. 115.

61. B.L. Henke, ibid, p. 85.

62. D.P. Gaines, G.L. Stradling, R.L. Kauffman B.L. Henke and R.T. Perkins, Bull. Amer. Phys. Soc. 25, 1148 (1980).

63. E. Leith, Monterey Proceedings, p. 242.

64. G.C. Bjorklund, S.E. Harris and J.F. Young, Appl. Phys. Lett. 25, 451 (1974).

65. T.H.P. Chang and W.C. Nixon, J.Sc. Instrum. (London) 44, 231 (1967); A.N. Broers, W.W. Molzen, J.J. Cuomo, and N.D. Wittles, Appl. Phys. Lett. 29, 596 (1976).

66. S. Kikuta, S. Aoki, S. Kosaki and K. Kohra, Opt. Commun. 5, 86 (1972).

67. S. Aoki, Y. Ichihara and S. Kikuta, Japan. J. Appl. Phys. 11, 1857 (1972).

68. S. Aoki and S. Kikuta, Japan. J. Appl. Phys. 13, 1358 (1974).

69. B. Reuter and H. Mahr, J. Phys. E. (London) 9, 746 (1976).

70. V.V. Aristov and G.A. Ivanova, J. Appl. Cryst. 12, 19 (1979).

71. V.V. Aristov, G.A. Bashkina and A.I. Erko, Opt. Commun. 34, 332 (1980).

72. A.M. Kondratenko and A.N. Skrinsky, Opt. Spectrosc. (USSR) 42, 189 (1977).

73. D.C. Flanders, A.M. Hawryluk, and H.I. Smith, J. Vac. Sci. Technol. 16, 1949 (1979).

74. A.M. Hawryluk, N.M. Ceglio, R.H. Price, J. Melngailis, and H.I. Smith, J. Vac. Sci. Technol. 19, (1981).

75. A.M. Hawryluk, H.I. Smith, R.M. Osgood, and B.J. Ehrlich, to be published.

POPULATION DENSITY AND VUV GAIN MEASUREMENTS
IN LASER-PRODUCED PLASMAS*

R. C. Elton, J. F. Seely, and R. H. Dixon
Naval Research Laboratory, Washington, DC 20375

ABSTRACT

Recently-measured plasma parameters are used to evaluate vuv-gain produced in a laser-generated carbon plasma by electron-capture pumping of hydrogenic ions. A first attempt towards using such measurements as a gain-gauge for extension towards higher gain at higher pumping flux is described. The possibility of achieving higher gains on lithium-like ions through higher source densities is modeled, based upon presently obtained information. Finally, the analysis and data are applied to selective photon pumping, either for assist or as an independent approach towards higher gain.

INTRODUCTION

Sufficient gain for lasing at very short wavelengths requires an exceptionally high upper state population density N_u, both because of the wavelength scaling of the gain coefficient formula[1,2]

$$G = \frac{\pi^2 r_o \, c\lambda f \, (g_\ell/g_u)}{2} \left(\frac{M}{2\pi kT}\right)^{1/2} N_u I \qquad cm^{-1}, \quad (1a)$$

where I is the inversion factor

$$I = \left[1 - \frac{N_\ell}{N_u} \frac{g_u}{g_\ell} \right], \qquad (1b)$$

and because of the lack of efficient vuv cavities. Equation (1) is written for a Doppler-broadened line of oscillator strength f and wavelength λ originating from transitions between upper and lower states of density and statistical weights N_u, g_u and N_ℓ, g_ℓ, respectively, and for plasma ions of mass M at a temperature T.

Highly-charged ions radiating on optical transitions at short wavelength are accompanied by multitudes of electrons in a quasi-

*Supported by the Department of Energy and the Office of Naval Research.

ISSN:0094-243X/82/900277-10$3.00

278

neutral plasma. At high densities, collisions as well as lower-state radiative trapping reduce the inversion factor I in Eq. (1) to a value near zero, or at least to too small a value to be measured reliably from transitions originating on levels u and ℓ as has been done effectively at lower densities[3]. At such high densities, direct amplification tests[4] are required to indicate laser possibilities. At somewhat lower densities, a finite inversion factor I can be measured so that the absolute gain coefficient can be deduced from Eq. (1), providing the upper state density and temperature can be determined.

This difference in operating regimes is illustrated in the normalized parameter space diagram[1,2] of Fig. 1 for a hydrogen ion 4→3 transition, where two upper-limits on the (reduced) electron density N_e/Z^7 are plotted for inversion factors I ≈ 0 at high density and a measurable I = 0.3 at lower density, respectively. These generalized curves represent only collisional limits in a 3-level (n = 4,3,1) model, with radiative trapping omitted (only in this figure). Hence, at the lower-density finite-I condition, the measured gain coefficient G serves as a valuable gauge for advancing towards higher values in a systematic parameter study. This logical way to proceed is followed below.

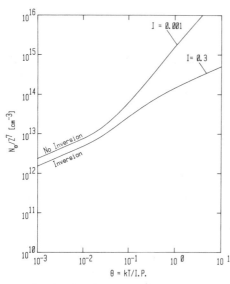

Fig. 1. Normalized gain parameter space for two inversion limits.

An important point bearing on the above discussion is that it is very desirable to maintain ionic conditions as pure as possible, i.e., preferably all ions initially in the particular ionic state from which pumping proceeds, with electrons contributed mainly from the lasant ions (minimal impurities) and with pumping proceeding preferentially into the upper laser state at as high a rate as possible.

In this paper we describe some recent gain coefficient values arrived-at in this manner, as well as the application of some auxiliary measurements reported elsewhere in this volume[5] to a reasonable guiding model for further scaling. While most of the results reported here are for the rather traditional

electron-capture pumping onto stripped (carbon) ions, it is of
interest to inquire if the ionic state purity mentioned above
may be improved by starting with helium-like states; and the
analyses for lithium-like Na and Aℓ are compared to that for
hydrogenic carbon--with more free electrons, however. Also,
it is confirmed with an extended analysis that the selective
upper state pumping mentioned above can be achieved effectively
with matched line radiation from other ions, and that this leads
to greater flexibility in available operational parameter space.

MEASURED GAIN

Most previous data[6] have been obtained on hydrogenic (and
helium-like) ions formed in expanding laser-produced plasmas.
Experiments have shown[3,7] that at expansion distances of about 3 mm
or more, inversion factors $I \approx 0.3$ are possible between n=4 and
n=3 levels, where a lesser degree of lower-state filling by
radiative trapping from the ground state occurs (compared, e.g.,
to 3→2 transitions). At a temperature kT = 10 eV, Eq. (1) then
indicates* that an upper state density $N_u = N_4 \approx 10^{14}$ cm^{-3}
is required for a "threshold" gain coefficient of unity, i.e., a
net gain of exp (1) over a 1 cm path. We have just recently[5]
obtained a time resolved experimental value of $N_4 = 1.2 \times 10^{12}$ cm^{-3}
at 45 ns into the pulse from a measured value of N_7, where the
relative distribution over ion-states can be determined from the
intensities in the soft x-ray Lyman series. This leads to a gain
coefficient of ~2%/cm from Eq. (1). The population density N_7
was obtained from a measurement of the absolute emission from the
7→6 line at 343 nm wavelength in the near-uv spectral region as
described in Ref. (5) herein.

INCREASING THE GAIN -- HYDROGENIC IONS

The next question is how to best increase N_4 and thereby G,
preferably without having the inversion factor I become immeasur-
ably small so that our measurements of G just described can
continue to serve as a gauge of progress. Maintaining generality,
this is accomplished both by increasing the population density
in the initial source reservoir state (C^{6+}) and/or by increasing
the pumping rate into the upper n=4 state. Beyond this point
we must turn to a specific pump example, and free electron capture
is a good starting point in a natural (undoped) quasi-neutral
plasma. Charge transfer from neutrals is considered elsewhere[1,2]
as is photon pumping from the ground state, modeled below. In
steady state, N_4 is related to the initial stripped ion density
N^{6+} by

---

*For thermal Doppler broadening, i.e., streaming is not included for
an elongated medium[5,19].

$$N_4 = N^{6+} \left(\frac{P}{D}\right), \tag{2}$$

where D is the total depopulation rate and P is the mean pumping rate which can be written as[8]

$$P \approx \frac{10^{-31} N_e^2 (n')^6}{Z^6} \left(\frac{Z^2 Ry}{kT_e}\right)^2 \exp \left[\frac{Z^2 Ry}{(n'+1)^2 kT_e}\right] \text{ sec}^{-1}. \tag{3}$$

Here n' represents the collision limit[8,9]. This shows the strong dependence on electron temperature that favors increased pumping with rapid cooling.

Further modeling requires a value for N^{6+} which is sometimes optimistically taken to be $N_e/6$, i.e., a completely-stripped carbon plasma, particularly near the target. At a distance of 3 mm from the target this is clearly not the case because recombination occurs rapidly in the first 100 μm or so, beyond which frozen-ion-state distributions follow in expansion[7]. Again we have been able to recently obtain[5] an experimental value for $N^{6+} = 1.4 \times 10^{15}$ cm^{-3} at a distance from the carbon target of 3 mm, along with $N_e = 3 \times 10^{17}$ cm^{-3} for comparison, i.e., our stripped ion density compared to that for 100% stripping is $N^{6+}/(N_e/6) \approx 0.03$. This is consistent with recombination over the distances involved. This density N^{6+} was obtained from the absolute density N_7 described above, assuming local thermodynamic equilibrium. The localized electron density N_e was obtained from the Stark full width at half maximum $\Delta\lambda$ of the $7 \rightarrow 6$ line at 343 nm wavelength, based on a Lorentzian profile fit to recent detailed calculations of Kepple and Griem[10], which reference also gives the useful approximate relation

$$N_e = \left(\frac{\Delta\lambda}{8.25}\right)^{3/2} \times 10^{18} \text{ cm}^{-3} \tag{4}$$

for a pure carbon plasma and for $\Delta\lambda$ in Å units.

With this value of N^{6+}, Eqs. (1)-(3) can be used in a simple physical model[1,2] to define a gain-parameter region (similar to that begun in Fig. 1) for Z=6 (carbon) and the $4 \rightarrow 3$ hydrogenic transition at 52 nm wavelength. It is also possible now to include radiative trapping in the inversion factor by modifying the lower n=3 level decay rate by an escape factor[11]

$$g(\tau) = 1 - \exp(-\tau)/\tau \quad \text{for } \tau < 2$$

and

$$g(\tau) = 1/\tau(\pi \ln \tau)^{1/2} \quad \text{for } \tau \geq 2, \quad (5)$$

where τ is the (3→1) opacity[8]:

$$\tau = 1.1 \times 10^{-16} \lambda N d f (\mu/kT)^{1/2}, \quad (6)$$

for λ in Å units, μ the atomic mass, kT in eV, and for a depth $d = 0.1$ in cm. Here it was necessary to assume a ground state fractional population density for which we chose[7],[12] $N_1/N_e \approx 0.1$; again, radiative trapping effects are not a major consideration for the 4→3 transition that we chose here. The resulting inversion limit line (labeled by I) and constant-G curves are plotted in Fig. 2 for a fixed inversion factor I=0.3 (see Fig. 1) and for the N^{6+} measured value described above. All of our recent results quoted above were obtained with a Nd:glass laser energy of 7 J in a 10 ns pulse focused to a 500 μm diameter spot for a power density of 4×10^{11} W/cm^2. A fairly typical ion trajectory with distances measured from the target is included dotted in Fig. 2, beginning at the critical density ($N_e = 10^{21}$ cm^{-3}); this indicates the present status in free expansion.

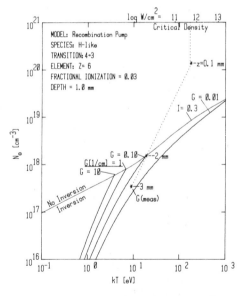

Fig. 2. Parameter space for C^{5+}, 1 mm depth, and measured gain G.

Means of progressing further towards matching plasma conditions with a gain coefficient of unity include: (a) shifting the gain curves in Fig. 2 downwards towards lower N_e by increasing N^{6+}, (b) rapid cooling at a distance of 3 mm by some type of added "barrier" and/or enhanced radiative cooling from higher-Z materials which would increase the pumping rate as seen in Eq. (3), (c) relaxing the requirement that the inversion factor I be as large as 0.3 for measurement which would permit closer-in operation as shown in Fig. 1, but with a net upward shift in the gain curves unless N^{6+} were increased at very close (< 100 μm) distances, and (d) higher-Z materials with closer-in operating conditions (scaling indicated in Fig. 1).

For possibility (a), we have very recently suceeded in increasing the degree of stripping N^{6+}/N_e towards the 1/6 limit at 3 mm distance from the carbon target by increasing the laser power density, i.e., achieving higher temperatures and reduced recombination (Eq. 3). Preliminary results[5] obtained by the time of this conference for up to 6 X 10^{12} W/cm^2 pump flux indicate a dramatic increase of $N^{6+}/(N_e/6)$ by 4.3-times from 0.03 to 0.13 at 1.5 X 10^{12} W/cm^2 at which point an apparent saturation is indicated, the precise origin of which is yet to be determined. The measured n = 7 population density also rose by a factor-of-3, and the electron density increased to 1.4 X 10^{18} cm^{-3}. A corresponding net gain increase will depend on additional cooling either adiabatically with distance [ion states are frozen-in[7]] or artificially, since a rising T_e with pump power will further depopulate the n=4 level by thermalization. This can be roughly visualized in Fig. 2 for N_e = 1.4 X 10^{18}, $T_e \approx 30$ eV and the gain curves shifted upward by 4.3-times for the N^{6+} increase. Hence, extended measurements at different distances and with additional cooling are needed.

LITHIUM-LIKE (& SODIUM-LIKE) IONS

Some measurements of population inversion and gain have been reported[13,14] on lithium-like ions formed by free-electron capture from helium-like ground states. Our hydrogenic analysis above for carbon can be conveniently extended in a Z-2 approximation to lithium-like ion transitions at one higher quantum number, i.e., the 5→4 transition compared to the 4→3 in carbon. Two possibilities emerge:

(a) lithium-like Na^{8+} which is similar to hydrogenic carbon in energy level structure, wavelengths, and inversion limit but with 60% of the ionization potential, so that 30% population in the helium-like ground state could be expected[15] at the same temperature as for carbon in the model. The modeling results for this case are indicated in Fig. 3 and indeed show a similar inversion limit but significantly lower gain, which is not particularly encouraging.

(b) lithium-like Al^{10+} which has a similar ionization potential to hydrogenic C^{6+} but with larger energy level separations, shorter wavelengths, and potential operation at higher densities close to the target, where possibly the fractional concentration of Al^{11+} would be higher. The modeling results are shown in Fig. 4 for the same ionization ratio in the 11+ state as that used above for carbon. It does not appear likely that operation would be possible within a distance of 100 μm where significantly-enhanced stripping would be expected.

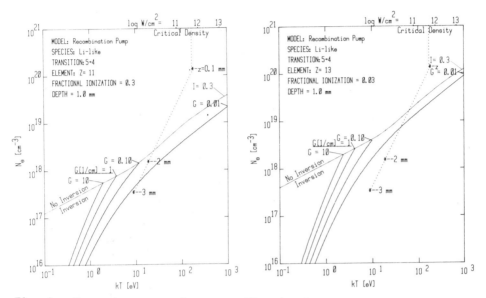

Fig. 3. Parameter space for Na^{8+} (Li-like) on 5→4 transition.

Fig. 4. Parameter space for Al^{10+} (Li-like) on 5→4 transition.

Similar analyses can be done for sodium-like (e.g., Fe^{15+}) ions formed from neon-like ground states and for 6→5 transitions, but the limit for collisional equilibrium begins to dominate the inversion factor at this quantum level.

From this oversimplied model for lithium-like ions (many more levels are involved than for hydrogenic ions) it appears that there is no significant advantage to lithium-like ions for similar laser wavelengths, as long as the currently measured concentrations of stripped ions can be obtained.

PHOTON PUMPING

A very promising additional option for either enhancing the n=4 population density or for supplanting electron-capture pumping altogether is selective photon pumping from the ground state, using a matching line from an intensely-radiating ion in the plasma. We have recently incorporated this in our simple physical parameter-space model illustrated above in Figs. 2-4 for the case of a helium-like neon ion pumped from level 1 to 4 (again, helium-like ground state ions are considered plentiful) by a helium-like sodium resonance line with an almost exact wavelength coincidence at 1.1 nm wavelength[16]. An optically-thick pump line at a temperature T_p is assumed, so that the (1→4) pumping rate

P_{14} is related to the decay rate A_{41} by[17]

$$P_{14} = \frac{1}{2} \left(\frac{g_4}{g_1}\right) A_{41} \exp \left[(h\nu_{14}/kT_p) - 1\right]^{-1} \sec^{-1} \tag{7}$$

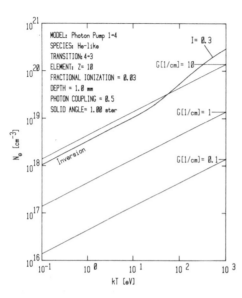

Fig. 5. Parameter space for Ne^{8+} (He-like) $4 \to 3$, photon pumped $1 \to 4$.

for statistical weights g_4 and g_1. Assuming a N^{8+} concentration as measured for C^{6+} above, a solid angle of collection of sodium photons by the neon ions of $\pi/3$, and a spectral coincidence of 0.5, a considerably-enlarged high gain parameter space is indicated in Fig. 5, since the photon source is essentially independent of the plasma neutrality condition, resembling an external flashlamp pumping situation. These results are in good agreement with more detailed rate equation analyses[18] when adjustment is made here for the more-idealized situation in Ref. 18. We are beginning experiments on this approach with similar diagnostics as described above and extending to the higher power densities afforded by the NRL fusion laser. Extension to shorter wavelengths could be obtained with a similar hydrogenic $K \to C\ell$ combination plasma[17].

CONCLUSIONS

In this invited paper we have applied some very recent observations of essential plasma parameters reported elsewhere in this volume[5] to an assessment of presently achieved gain on a $4 \to 3$ transition in a hydrogenic carbon plasma radiating at 52 nm wavelength. We also attempt to show how these measurements can be used as a gauge of progress, as certain variables such as pump power density are adjusted in an attempt to increase the gain towards threshold for lasing. We also extend the analysis to lithium-like ions and conclude that there could be an advantage over hydrogenic ions for enhanced initial-ion concentrations at higher Z's, where complete stripping becomes more difficult than for carbon. Finally, as we advance in our knowledge and gain confidence in this field, it is appealing to consider combination plasmas such as sodium and neon where one (Na^{9+}) serves as a photon source to selectively pump the other (Ne^{8+}) and to apply similar measurement techniques to determine gain in a progressive fashion.

It is clear that we are already measuring gain in laser-produced plasmas that would reach threshold for lasing if cavities were available in the vuv region which were as efficient as those in the visible. Parallel progress towards higher gains with elongated channeled plasma[19], and more efficient cavities could lead to vuv laser operation in the forseeable future with continued research along these lines, followed by continuation isoelectronically towards shorter wavelengths with higher-Z materials.

ACKNOWLEDGMENTS

It is a pleasure to acknowledge the expert technical assistance of J. L. Ford in all phases of the experiment described, including laser operation. Appreciation is expressed to H. R. Griem, both for illuminating technical discussions and for his and P. C. Kepple's permissions to use and quote their recent Stark broadening calculations prior to publication.

REFERENCES

1. R. C. Elton, "Progress and Trends in X-Ray Laser Research", SPIE vol. 279, p. 90 (1981); also J. Optical Engineering, March-April 1982 (in press).
2. R. C. Elton, "Overview and Advances in X-Ray Laser Research", AIP Conf. Proc. No. 75 on Low Energy X-Ray Diagnostics, D. T. Attwood and B. L. Henke, eds., 1981.
3. R. C. Elton and R. H. Dixon, Phys. Rev. Lett., 38, 1072 (1977); also R. H. Dixon, J. F. Seely, and R. C. Elton, Phys. Rev. Lett., 40, 122 (1978).
4. D. Jacoby, G. Pert, S. Ramsden, L. Shorrock, and G. Tallents, Optics Comm., 37, 193 (1981).
5. R. H. Dixon, J. F. Seely, and R. C. Elton, "VUV Calibration and Diagnostic Techniques for Plasma Lasers", postdeadline paper for Topical Meeting on Laser Techniques for Extreme Ultraviolet Spectroscopy, Optical Society of America, Boulder, Colorado, March 8-10, 1982.
6. R. W. Waynant and R. C. Elton, Proc. IEEE, 64, 1059-1092 (1976); also R. C. Elton, "Recent Advances in X-Ray Laser Research", in Advances in X-Ray Analysis, Vol. 21, C. S. Barrett and D. E. Leyden, eds., Plenum (1978); and

R. C. Elton "X-Ray Lasers", in Handbook of Laser Science and Technology, M. J. Weber, ed., CRC Press (1981).

7. F. E. Irons and N. J. Peacock, J. Phys. B: Atom Molec. Phys. 7, 1109 (1974) where low gain coefficients are deduced using low transition probabilities; also J. Phys. B 7, 2084 (1974).

8. R. C. Elton, "Atomic Processes", in Methods of Experimental Physics, Plasma Physics, Vol. 9A, Chapt. 4, H. R. Griem and R. H. Lovberg, eds., Academic 1970.

9. H. R. Griem, Plasma Spectroscopy, McGraw-Hill, p. 160, 1964.

10. P. C. Kepple and H. R. Griem, "Stark-Profile Calculations for the Carbon VI Line n=7 to n=6 at 3434 Å from Dense Plasmas", Phys. Rev. A (submitted).

11. R. W. P. McWhirter, "Spectral Intensities", in Plasma Diagnostic Techniques, Chapt. 3, R. H. Huddlestone and S. L. Leonard, eds., Academic 1965.

12. A. M. Malvezzi, L. Garifo, E. Jannitti, P. Nicolosi, and G. Tondello, J. Phys. B, 12, 1437 (1979).

13. E. Ya. Kononov, K. N. Koshelev, Yu. A. Levykin, Yu. V. Sidel'nikob and S. S. Churilov, Sov. J. Quant. Electron. 6, 308 (1976).

14. P. Jaegle, G. Jamelot and A. Carillon, "Progress in X-Ray Laser Research" in Proc. XVth Int'l Conf. on Phenomena in Ionized Gases, Minsk, USSR (1981).

15. T. F. Stratton, "X-Ray Spectroscopy" in Plasma Diag. Tech., Ch. 8, R. H. Huddlestone and S. L. Leonard, eds., Academic 1965.

16. A. V. Vinogradov, I. I. Sobel'man and E. A. Yukov, Sov. J. Quant. Electron. 5, 59 (1975).

17. R. C. Elton, Ed., "ARPA/NRL X-Ray Laser Program Final Technical Report", pp. 92-114, Naval Research Laboratory Memorandum Report no. 3482, 1977.

18. J. P. Apruzese, J. Davis and K. G. Whitney "Plasma Conditions Required for Attainment of Maximum Gain in Resonantly Photo-Pumped Al XII and Ne IX" (to be published).

19. J. F. Reintjes, R. H. Dixon and R. C. Elton, Optics Letters 3, 40 (1978); also T. N. Lee (to be published).

HOLLOW-CATHODE DISCHARGE FOR XUV LASERS AND RADIATION SOURCES[*]

R. W. Falcone, D. E. Holmgren, and K. D. Pedrotti
Stanford University, Stanford, CA. 94305

ABSTRACT

We have operated a new type of pulsed hollow-cathode discharge that creates high densities of excited-state lithium ions. The discharge operates at high current density (up to 200 amps/cm^2) and high voltage (2000 V) and may be useful for the realization of proposed schemes for XUV lasers and radiation sources.

* * *

Electric discharge excitation of short wavelength (< 1000 Å) lasers requires the development of new types of discharge devices which produce significant densities of highly excited atoms or ions with energies greater than 10 eV. We have designed and studied[1] a novel pulsed hollow-cathode discharge (HCD) which creates densities greater than 10^{12}/cm^3 of highly excited lithium ions, 59 eV above the ion ground state. This device may be useful for the realization of the proposal by Harris[2] for a 200 Å lithium laser, which is discussed in a separate paper in these proceedings,[3] as well as for excitation of other lasers and light sources.

In Fig. 1 we show the design of the HCD. A 2.54 cm diameter stainless steel (alloy 304) tube contains lithium vapor and acts as an anode. Inside and concentric to this outer tube is a 1.27 cm diameter stainless steel tube that functions as the cathode. Two 0.64 cm diameter holes 5 cm apart in the side of the cathode allow lithium vapor and plasma to fill the inside of the tube. The overall length of the device is ~ 50 cm. The central 15 cm length is heated by external Nichrome heaters adjacent to the anode tube.

---

[*]The work described here was supported by the Office of Naval Research.

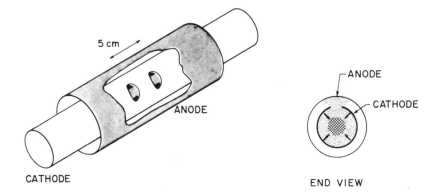

5 cm

ANODE

CATHODE

ANODE

CATHODE

END VIEW

Fig. 1. Diagram of the hollow-cathode discharge tube.
The end view shows the discharge filling the
inside of the cathode; the arrows represent
the paths of the high-energy electrons beamed
off of the cathode.

This device operates as a heat pipe with the lithium vapor confined
to the central region (overlapping the discharge holes) and uses
helium as a buffer gas at both ends of the cell. Stainless steel
mesh on the cathode and the inner anode wall promotes liquid-metal
recirculation in the cell. The cathode is suspended inside the
anode by machinable glass (MACOR) insulators, which are placed at
the ends of the cell. Water cooling rings keep the insulators and
window flanges at room temperature when the central part of the cell
is at its operating temperature of up to 900°C. This design over-
comes a key difficulty in constructing a lithium-vapor discharge
cell: the lack of an insulator capable of maintaining its structural integrity and insulating properties in a hot lithium atmosphere.

The discharge circuitry consists of a dc power supply used to
provide a continuous simmering discharge and a high-voltage high-
current pulser. The pulse circuitry includes a Thyratron switch
(EG & G HY-32), which connects a 75 nF capacitor charged to several
kilovolts across the HCD in a low-inductance geometry.

Voltage and current waveforms measured across the HCD during
pulsing are shown in Fig. 2; the calculated power deposited in the
discharge, shown in the lower half of Fig. 2, is correlated with

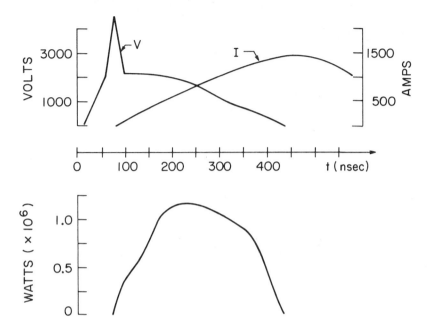

Fig. 2. Electrical characteristics of the hollow-cathode
discharge when a 75 nF capacitor charged to
5000 V is discharged through the HCD.

emission observed from highly excited Li and Li$^+$ levels. We have
observed emission signals originating from inside the cathode to be
10 times larger than emission signals from the region between the
anode and the cathode in pulsed operation of various HCD devices
using both lithium and helium vapor. We conclude that even for
high-current-density $(J = 25 \text{ A/cm}^2)$, short-pulse (100 nsec) operation, the discharge is confined mainly to the interior of the
cathode.

Population densities of both ground state and excited state
species pertinent to the construction of the 200 Å laser in lithium
are given in Table I. These measurements were obtained using the
curve of growth absorption technique with both a continuous, broadband light source (for ground state density) and pulsed, narrowband
dye lasers [for Li $(1s^22p)$, Li$^+$ $(1s2s\ ^1S)$, and Li$^+$ $(1s2s\ ^3S)$ densi-
ties]. The lithium ground-state ion density was estimated from the

Table I. Lithium population densities measured in
hollow-cathode discharge of Fig. 1.

Species	Number Density (atoms/cm^3)	Energy Above Ground State
Li $(1s^2 2s)$	2×10^{16}	--
Li $(1s^2 2p)$	8×10^{15}	1.8
Li$^+$ $(1s^2)$	$< 1 \times 10^{14}$	5.4
Li$^+$ $(1s2s\ ^3s)$	4×10^{12}	59.0[*]
Li$^+$ $(1s2s\ ^1s)$	5×10^{11}	60.9[*]

[*]Energy above ion ground state.

Stark-broadened emission linewidth of the He I 3614 line.[4] (Helium
was present in trace amounts in the discharge zone.) The total
electron density was less than 1×10^{14} cm^{-3}, and we assume that
this is equal to the lithium-ion ground-state density. All these
measurements were made at the peak of the highly excited-state population densities, which occurred ~ 200 nsec after the start of the
current pulse.

We estimate the uncertainty in all measurements to be approximately a factor of 2. However, the absorption measurements actually
imply a product of density times length. We estimate the effective
length of our discharge to be 10 cm by observing cw operation using
the simmering power supply. It is possible that the effective
length of the plasma is in fact much shorter than this, implying
proportionately larger excited-state densities than those given.

A second HCD was designed with a controllable active zone
length. This device is shown in Fig. 3. A 2.54 cm diameter stainless steel tube contained the lithium and acted as an anode as before, but the cathode was now suspended by a support rod in a tee
configuration. Voltage pulses were applied to the cathode through

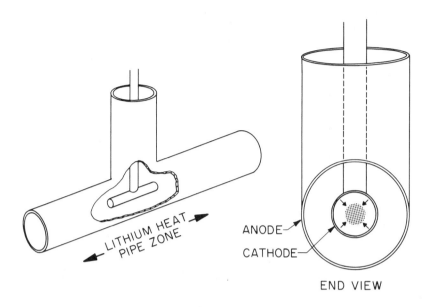

Fig. 3. Diagram of the "tee"-type hollow-cathode discharge.

this conducting support rod. We have operated a 1.5 cm long cathode
with a diameter of 5 mm; population densities obtained in this configuration are given in Table II. As before, these measurements
were obtained using the curve of growth technique except for the
lithium-ion ground-state density which was obtained from the Stark
broadened emission linewidth of the He II 3203 Å line.[4] It was not
possible to measure the Li^+ ($1s2s$ ^3S) population due to an interfering molecular absorption so we estimated triplet ion population
by scaling fluorescence intensities to our previous HCD. Figure 4
shows the temporal behavior of the Li^+ ($1s2s$ ^1S) and Li ($1s^2 2p$)
populations, as well as the current rise after the voltage pulse
was applied at $t = 0$ sec. About 200 nsec after the start of the
pulse the voltage measured across the discharge falls as an arc is
formed.

This "tee"-type HCD gives a factor of 3 or 4 improvement in
both ground state operating pressure and excited state densities

Table II. Lithium population densities measured in
hollow-cathode discharge of Fig. 3.

Species	Number Density (atoms/cm^3)	Energy Above Ground State
Li ($1s^2 2s$)	9×10^{16}	--
Li ($1s^2 2p$)	4×10^{16}	1.8
Li ($1s^2 3d$)	1×10^{13}	3.9
Li$^+$ ($1s^2$)	4×10^{14}	5.4
Li$^+$ ($1s2s\ ^3S$)	$\approx 1 \times 10^{13}$	59.0*
Li$^+$ ($1s2s\ ^1S$)	1.5×10^{12}	60.9*

*Energy above ion ground state.

over our first HCD. It is also more desirable as a laser discharge since it can be scaled to arbitrarily long active zone lengths by replicating the tee structure.

We have preliminary evidence indicating that the plasma created in our pulsed HCD degenerates into a low voltage arc at cathode current densities greater than 200 amps/cm^2. In addition, we observe arcing when the anode-cathode spacing is reduced and the voltage exceeds a critical value. We are exploring these two limitations by varying geometry and introducing insulators into discharge.

Steady-state hollow-cathode analysis[5] suggests that the discharge plasma is a negative-glow region where excitation is dominated by high-energy electrons beamed radially inward from the interior cathode surface. The energy of these beam electrons is nearly equal to the measured anode-cathode voltage, which is dropped principally across the cathode sheath (cathode fall). Although our device operates at high current and high voltage (corresponding to the abnormal-glow region of conventional-glow discharges), as well in a transient or breakdown mode, several experimental observations

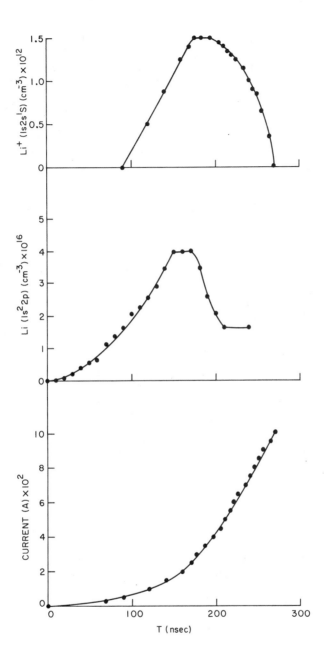

Fig. 4. Temporal behavior of Li$^+$ (1s2s ^{1}S) and Li (1s^{2}2p) populations and current in the "tee"-type hollow-cathode discharge.

indicate that it operates similarly to other hollow-cathode devices. Principally, the discharge is confined mainly to the interior of the cathode. Also, linewidths of optically thin ion-emission lines are narrow (Doppler-broadened at the cell temperature), indicating that the interior of the cathode is reasonably free of fields that could accelerate and heat ions. Finally, sheath formation in pulsed discharges is known to occur in times short compared to our 100 nsec risetime.[6]

Although a complete understanding of this device must await theoretical analysis of the hollow-cathode effect at high voltage and high current and in transient regimes, we have shown that a device operating in this manner is an attractive method of producing high densities of highly excited states.

The new features of this metal-vapor discharge are the operation of an all-metal hollow-cathode discharge tube simultaneously as a heat pipe, to obtain high ground-state densities, and at very high current densities, to achieve appreciable excitation rates of high energy levels. These results may be useful for the realization of proposed schemes for discharge-excited XUV lasers and for short-wavelength light sources based on anti-Stokes Raman scattering from highly excited metastable states.[2,3,7]

ACKNOWLEDGEMENTS

We wish to thank Steve Harris, Jim Young, and Richard Normandin for helpful discussions. R. W. Falcone acknowledges the support of his Marvin Chodorow Fellowship by the Exxon Education Foundation, Varian Associates, Hughes Research Laboratories, and Bell Laboratories.

REFERENCES

1. R. W. Falcone and K. D. Pedrotti, Optics Lett. 7, 74 (1982).
2. S. E. Harris, R. W. Falcone, M. Gross, R. Normandin, K. D. Pedrotti, J. E. Rothenberg, J. C. Wang, J. R. Willison, and

J. F. Young, in Laser Spectroscopy V, A. R. W. McKellar, T. Oka, and B. P. Stoicheff, eds. (Springer-Verlag, New York, 1981).

3. S. E. Harris, J. F. Young, R. W. Falcone, Joshua E. Rothenberg, J. R. Willison, and J. C. Wang, "Anti-Stokes Scattering as an XUV Radiation Source and Flashlamp," in Laser Techniques for Extreme Ultraviolet Spectroscopy, R. R. Freeman and T. J. McIl ath, eds. (AIP, New York, 1982) (to be published).

4. H. R. Griem, Plasma Spectroscopy (McGraw-Hill, New York, 1964).

5. B. E. Warner, K. B. Persson, and G. J. Collins, J. Appl. Phys. 50, 5694 (1979).

6. M. Nahemow and N. Wainfan, J. Appl. Phys. 34, 2988 (1963).

7. J. E. Rothenberg, J. F. Young, and S. E. Harris, Optics Lett. 6, 363 (1981).

SPACE RESOLVED SPECTRA OF 2p-3d MULTIPLETS OF Al XI AND Al X EMITTED FROM LASER IRRADIATED THIN FOILS

M.A. Khan
Department of Physics, University of Petroleum & Minerals,
Dhahran, Saudi Arabia.

G.J. Tallents
Department of Engineering Physics, Research School of Physical
Sciences, The Australian National University, Canberra, Australia.

ABSTRACT

The spectra of 2p-3d multiplets of Al XI ($\lambda \simeq 52$ Å) and Al X ($\lambda \simeq 55$ Å) emitted from plasmas produced by the irradiation of $9\mu m$ thick Al foils with a neodymium laser focused to peak intensities $\simeq 2 \times 10^{15}$ W cm^{-2} have been recorded with spatial resolution$\simeq 30\mu m$. The measured relative intensities show a considerable departure from the expected intensity ratios within the multiplets in the plasma expanding in the rear half-space of the laser irradiated foil. This indicates a partial intermixing of the 3d quantum states of both Al XI and Al X. The role of superthermal electrons in creating the plasma at the rear of the foil is discussed.

INTRODUCTION

Studies of the extreme ultraviolet (XUV) and X-ray emission characteristics of rapidly expanding laser produced plasmas have been of considerable interest in recent years. The classification of spectral lines emitted by highly stripped ions created in laser produced plasmas have been of use to both astrophysicists and laboratory spectroscopists[1-3]. Spectroscopic information has also been of major importance in the determination of plasma parameters in laser-fusion plasmas[4,5]. More importantly, however, an understanding of the intensity behaviour of the XUV and X-ray transitions and the inherent atomic physics could lead to the choice of suitable lines for exciting laser action in this region of the electromagnetic spectrum[6-8].

We report here our studies on the spectra of 2p-3d multiplets of Al XI and Al X emitted from plasmas produced from thin Al foils recorded with spatial resolution $\simeq 30\mu m$. The measured relative intensities show a considerable departure from the expected intensity ratios within the multiplets in the plasma expanding in the rear half-space of the laser irradiated foil. This suggests that a partial intermixing of quantum states may be taking place near the 3d states of Al XI and Al X in the region of plasma being discussed. Furthermore, it is proposed that a measurement of the intensity ratios of the fine structure components within the multiplets 2p-3d

ISSN:0094-243X/82/900296-07$3.00 Copyright 1982 American Institute of Phys of Al XI and Al X could be used as a density and temperature diagnostic in an analogous manner to the current use of hydrogen-like L_α doublet ratios[9-11]. The role of superthermal electrons in creating the plasma at the rear of the foil is also discussed.

EXPERIMENTAL DETAIL

We used a standard Nd: glass laser system delivering upto 10J energy in single pulses of about 180 picosecond (ps) duration[6]. Streak camera measurements showed that the pulse length was reproducible to ±40 ps and that there were no spurious or satellite pulses. In the present experiments, a small pre-pulse of about 20% nominal energy preceded the main laser pulse by 200 ps. This was achieved by inserting a partial mirror in front of the amplifier chain alighment mirror. With this arrangement, an after-pulse with some 14% energy was also produced a further 200 ps after the main pulse. A streak camera recording of the three pulse sequence is shown in Figure 1.

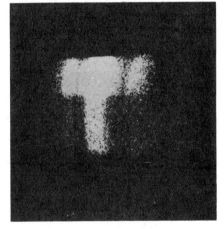

Fig.1.   Image intensifier camera record of the three-pulse sequence at a nominal streak speed of 30 ps/mm. The lower portion of the streak is attenuated to 20% of the upper portion intensity by a filter at the entrance slit.

The XUV emission from the plasma created by focusing the three laser pulses onto a 9μm thick Al foil was recorded photographically using a 2m grazing incidence spectrograph (Hilger & Watts E 580) equipped with a 600 lines/mm grating[6]. The spectrograph was positioned at 90° to the laser axis and a 30μm wide slit placed at a distance of 40 mm from the plasma provided spatial resolution in the direction normal to the foil surface. Kodak-Pathe' SC 5 photographic plates were used in the spectrograph.

The laser beam was focused to focal spots of diameter ≃ 40μm giving a peak laser intensity of ≃ 2 x 10^{15} W cm^{-2} for these experiments. The target foils were mounted on micromanipulators and could be positioned to ±5μm in the focal region of the laser by viewing the target with microscopes mounted on the target chamber[6]. The XUV emission from the plasma expanding in both the forward (opposite to the incident laser beam) and the backward direction was recorded.

Upto five laser shots were superimposed on the same photographic plate for a satisfactory exposure and the target was moved after each shot in order to expose a new surface of the foil.

RESULTS & DISCUSSION

Microdensitometer traces of the spectra of 2p-3d multiplets of Al XI and Al X recorded as a function of distance from the foil target are shown in Figures 2 and 3 respectively. There are three lines in the 2p-3d transitions of Al XI namely $2p(^2P_{1/2}) - 3d(^2D_{3/2})$ at 52.299 Å, $2p(^2P_{3/2}) - 3d(^2D_{5/2})$ at 52.446 Å and $2p(^2P_{3/2}) - 3d(^2D_{3/2})$ at 52.461 Å of which the last one is spectrally forbidden and this gives the transitions the shape of a doublet with lines at 52.299 Å ($^2P_{1/2} - ^2D_{3/2}$) and 52.466Å($^2P_{3/2} - ^2D_{5/2}$) [12]. For an optically thin plasma, the doublet ratio should be $I_{52.299}/I_{52.446} \simeq 0.50$ [12] as is the case if we integrate over the experimental line profiles for the plasma expanding in the forward direction (Figure 2). However, the intensities recorded in the rear half-space are measured to be in the ratio $I_{52.299Å}/I_{52.446Å} \simeq 1.0$ showing a considerable departure from the expected behaviour.

In the case of 2p-3d transitions of Al X, there are six lines (we refer to triplet transitions $2p\ ^3P - 3d\ ^3D$). These lines are also grouped to give the appearance of a doublet [13] with lines $2p\ ^3P_0 - 3d\ ^3D_1$ (55.227 Å), $2p\ ^3P_1 - 3d\ ^3D_2$ (55.272 Å), $2p\ ^3P_1 - 3d\ ^3D_1$ (55.279 Å) having weighted line center at 55.262 Å; and $2p\ ^3P_2 - 3d\ ^3D_3$ (55.376 Å), $2p\ ^3P_2 - 3d\ ^3D_2$ (55.384 Å), $2p\ ^3P_2 - 3d\ ^3D_1$ (55.391 Å) having their weighted line center at 55.377 Å. The doublet intensity ratio for an optically thin plasma should be $I_{55.262}/I_{55.377} \simeq 0.8$ obtained by extrapolating along the Be-like isoelectronic series to lower atomic number ions and using the oscillator strengths given by Wiese et al[14]. This is also evident if we integrate over experimental line profiles for forward expanding plasma (Figure 3). However, once again we observe a significant departure from the expected doublet intensity ratio in the plasma expanding from the rear of the foil.

The above "anomaly" in the intensity behaviour of 2p-3d multiplets may arise due to self-absorption in the relatively cooler outer layers of the plasma expanding in the rear of the foils. However, there is no evidence of self-reversal of line profiles of the type to be expected for optically thick doublets[11].

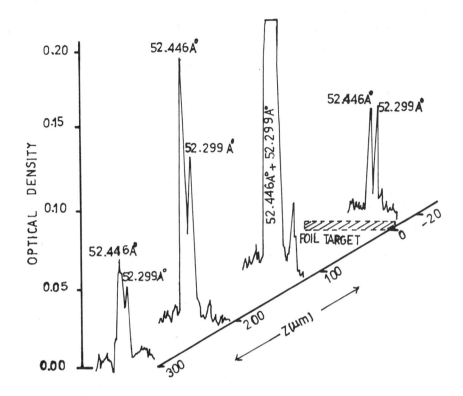

Fig.2: Microdensitometer traces of the 2p-3d transitions of Al XI as a function of distance Z from the target. The position Z = 0 is at the front of the foil target and the laser is incident from positive Z. At Z = 100μm, the doublet lines appear saturated on this optical density scale, but on an expanded density scale, its profile is similar to the shape at Z = 200μ.

The intensity ratios of the fine structure components of the hydrogen-like L_α doublets are currently of interest as a density diagnostic in high temperature plasmas[9-11]. If the two 2p quantum states of the hydrogen-like L_α doublet are in statistical equilibrium, the optically thin intensities should be in the ratio of the corresponding statistical weights of the 2p states i.e. 2:1. However, detailed calculations [9-11,15] have shown that this value is only achieved in the low density (Coronal) or high density (Boltzmann) limit. At intermediate electron densities ($10^{13} < N_e < 10^{21}$ cm^{-3}), there is partial intermixing of quantum states and the intensity

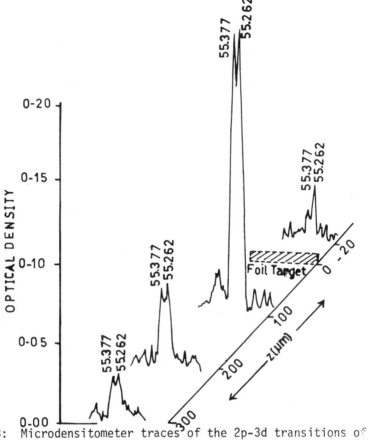

Fig.3: Microdensitometer traces of the 2p-3d transitions of Al X as a function of distance Z from the target.

ratios can be as low as 1.3 - 1.4 : 1. We propose that a similar mechanism of partial intermixing of quantum states near the 3d states is responsible for the intensity anomalies observed by us for Al XI and Al X. As such the measurements of intensity ratios may also serve as density (and temperature) diagnostic of the plasma in a manner analogous to the current use of hydrogen-like L_α doublet ratios[9-11]. The Al XI and Al X 2p-3d transitions are not resonance lines and should be less likely to suffer self absorption which can considerably complicate the interpretation of the L_α doublet ratios[11]. However, the radiative and collisional rate coefficients for Li-like Al XI and Be-like Al X ions are not known to the high accuracy of the rates for hydrogen-like ions and this fact may limit their proposed use.

Intermixing of quantum states near the 3d-states of Al XI is of relevance to schemes for producing optical gain using 3d-4p and

3d-4f lines of this ion[7,16]. Both theoretical and experimental gain deductions from population densities may be in error if the effect of intermixing of nearby quantum states is not included. Intermixing is known to affect the gain values for CVI H_α line deduced from H_α doublet ratios[6].

In order to further investigate the plasma created in our experiment, the laser heating and ionization in the plasma were modeled assuming a self-similar plasma expansion with a complete thermal burn-through of the foil.[17] A peak ionization of 10.4 was predicted by the model which is in satisfactory agreement with time averaged experimental value(from the ratios of resonance line intensities)of average charge of 9.8 near the target surface on the front and 9.3 on the rear. The slight overestimate of the average charge is ascribed to non-symmetric plasma expansion (the dominant expansion being in the forward direction i.e. towards the laser) and 100% absorption of laser energy assumed in the code.

A two dimensional Eulerian hydrodynamic code[17,18] was also employed to simulate the multiple-pulse irradiation of the 9µm thick foil. Contrary to our experimental observation, the code does not predict hot plasma formation ($T_e > 50$ eV) on the rear side of the foil. Experimentally, the role of pre-pulse (and perhaps also after-pulse) appears to be important. In an identical experiment, single pulse irradiation with comparable laser energy failed to produce observable plasma on the rear of the foils. This indicates that superthermal electrons[19,20] which are not modelled in the Eulerian Code[17] may be playing a significant role in creating hot plasma at the rear of the foil. Due to the pre-pulse plasma, the main laser pulse interacts with a longer scale-length plasma. The production of superthermal electrons due to resonance absorption may decrease with a longer scale-length plasma, but it is likely to increase due to Raman scatter and two plasmon decay[21]. The superthermal electrons may pre-heat the target and aid the burn-through[22,23]. The rear plasma may also be formed by the direct bombardment of hot electrons executing large amplitude orbits in the field of the target travelling around the foil edge[22]. The fact that line intensities at the rear surface ($Z = -20$µm) are comparable with intensities at $Z = +300$µm in the front (Fig. 2 and 3) indicates that the volume of the emitting plasma is larger at the rear. This is consistent with the rear plasma being produced by superthermal electrons travelling around the foil edge and striking the rear of the target[22,23].

CONCLUSION

The spectra of 2p-3d multiplets of Al XI and Al X emitted from plasmas produced by focussing $\approx 2 \times 10^{15}$ w cm^{-2} Nd:glass laser radiation on 9μm thick Al foils recorded with spatial resolution of ≃ 30μm have been studied. The relative intensities of components within the multiplets showed a considerable departure from the expected behaviour in the plasma expanding in the rear-half space of the foil which is ascribed to intermixing of quantum states near the 3d states of both Al XI and Al X. The spectra recorded are consistent with the rear plasma being produced by superthermal electrons travelling around the foil edge and striking the back of the foil.

ACKNOWLEDGEMENT

The experimental work reported here was carried out at the Department of Applied Physics, University of Hull, UK. We would like to thank Dr.G.J. Pert for providing the research facilities and his computer codes and for many useful discussions.

REFERENCES

1. E.Ya. Kononov, Physica Scripta 17, 425 (1978).
2. Y.Conturie et al, J. Opt. Soc. Am. 71, 1309 (1981).
3. M.A.Khan, J.Opt.Soc.Am. 72, 268 (1982).
4. J.G.Lunney and J.F. Seely, Phys.Rev.Lett.46, 342 (1981).
5. B.Yaakobi et al, Phys.Rev.A 19, 1247 (1979).
6. D. Jacoby et al, Opt.Comm. 37, 193 (1981).
7. E.Ya. Konovnov et al, Sovt. J.Quant.Electron.6, 308 (1976).
8. V.A. Bhagavatula and B. Yaakobi, Opt.Comm. 24, 331 (1978).
9. A.V. Vinogradov et al, Sovt. J.Pl.Phys. 3, 389 (1977).
10. V.A. Boikov et al, Sovt. J.Pl.Phys. 4, 54 (1978).
11. F.E. Irons, Aust.J.Phys., 33, 283 (1980).
12. W.L. Wiese et al, Atomic Transition Probabilities Vol.II, NSRDS-NBS 22 (1969).
13. C.E. Moore, Atomic Energy Levels, Vol.I NSRDS-NBS 35 (1971).
14. W.L. Wiese et al, Atomic Transition Probabilities Vol.I, NSRDS-NBS 4 (1966).
15. V.A. Boikov et al, J. Phys. B 12, 1889 (1979).
16. E.Ya.Kononov and K.N. Koshelev, Sovt.J.Quant.Electron 4, 1340 (1975).
17. G.J. Pert and G.J. Tallents, J.Phys.B 14, 1525 (1981).
18. G.J. Pert, J.Comput.Phys. (in press).
19. D.C. Slater et al, Phys. Rev.Lett. 46, 1199 (1981).
20. S.P. Saraf and L.M. Goldman, Phys. Rev A 24, 1021 (1981).
21. J.A. Tarvin et al, KMS Fusion Inc. Annual Tech. Report (1980) p.2.16.
22. N.A. Ebrahim et al, Phys. Rev.Lett., 43, 1995 (1979).
23. R.S. Marjoribanks et al, Phys.Rev.Lett. 45, 1798 (1980).

LASER PRODUCED PLASMAS IN Hg:
A SOURCE OF PULSED XUV RADIATION

R. M. Jopson and R. R. Freeman
Bell Laboratories, Murray Hill, NJ 07974

J. Bokor
Bell Laboratories, Holmdel, NJ 07733

ABSTRACT

We have used a laser produced plasma in Hg to generate pulsed XUV radiation which is widely tunable from 900Å to shorter than 300Å. Preliminary estimates of the power exiting the slits of a 1 meter normal incidence XUV monochromator is approximately 5×10^3 photons/shot/Å at 10 pps. at 500Å.

There has recently been a renewed interest in laser-produced plasmas as a source of intense, pulsed VUV, XUV and soft x-rays. Recent work[1,2] has concentrated on the observation of broad band continua from rare earth elements. In most of the previous work, the repetition rate of the system was limited by the need to reposition the solid metal surface because of target deterioration. In this work, we describe the XUV output (from 900Å to less than 300Å) of a laser plasma generated on a liquid Hg metal in which the surface "self-heals" after each shot, presenting a reproducibly clean target to the laser even at 10 pps.

The apparatus consists of a glass tube shaped so as to present a 5° view of the surface of a pool of mercury to a 1 meter normal incidence XUV monochromator. The glass tube and monochromator are filled with approximately 5 torr of pure He; a thin film metallic filter was placed in front of the exit slit to shield the sodium salicylate exit window from scattered laser light. For wavelengths shorter than 800Å, a 0.1μ thick aluminum filter was used, while for wavelengths longer than 800Å a 0.1μ thick indium filter was used. An EMI 9635 photomultiplier recorded the fluorescence from the sodium salicylate window.

A commercial Q-switched Nd:YAG laser doubled to 5320Å was used to create the plasma at 10 pps. Laser energies from 10 mJ/pulse to 200 mJ/pulse, with nominal pulse durations of 5 nsec, were used in these initial experiments, with most of the data taken at ∿60 mJ/pulse. The laser was focused to approximately 75μ on the pool of Hg approximately 1.5cm from the entrance slit of the monochromator for a peak intensity of about 3×10^{11} watts/cm^2.

The resulting plasma was quite intense in the visible, with a well defined hot spot at the focus, and a plume that extended up from the Hg surface. The height of this plume was dependent upon the pressure of He: the lower the pressure the more extended the plume. Visible surface waves were generated in the pool of mercury by the laser light; in addition, high frequency rf was picked up by nearby electronics when the laser plasma was being generated.

ISSN:0094-243X/82/900303-02$3.00 Copyright 1982 American Institute of Physics

For wavelengths longer than 800Å, a spectrum was observed which contained a background continuum, with overlaying strong emission lines which we were able to identify as arising from Hg II and Hg III. The continuum was emitted promptly with the laser (to within the resolution of the PMT, ∿20 nsec), while the line emission was extended over many tens of nanoseconds. The line emission came from the plume, while the continua was generated in the hot focus.

Below 800Å, the spectra revealed mainly continuum with very little structure. This continuum extended down to wavelengths shorter than 500Å, where He absorption in the monochromator cut off the output. When the system was run without the buffer gas, we observed continuum emission as short as 300Å, the expected cutoff of the normal incidence monochromator.

We studied the resolution of the system by examining absorption lines of the He in the monochromator. By careful adjustment of the slits and plasma spot, we achieved a resolution of 1/2Å at 584Å, only a factor of 2 greater than the practical limit for our monochromator with a 600 line/mm grating.

We made preliminary estimates of the intensity at the exit slits of the 1 meter monochromator and found it to be 5×10^3 photons/Å/shot at 500Å, of 2.5×10^6 photons/sec/10% bandwidth at 500Å. We may roughly estimate the conversion of efficiency of the laser to XUV light by assuming a) 3% grating efficiency between 500-800Å; b) an aluminum filter transmission of 1%; c) a calculated solid angle subtended by the slits of 1.7×10^{-4}; and d) a 2π solid angle emission of the light. With these assumptions we find a total emission of 5.6 mW, or 1% of the laser light at 5320Å is converted into XUV between 500 and 800Å.

REFERENCES

1. G. O'Sullivan, P. K. Carroll, T. J. McIlrath and M. L. Ginter, Appl. Optics 20, 3043 (1981).
2. P. K. Carroll, E. T. Kennedy and G. O'Sullivan, Appl. Optics 19, 1454 (1980).

CALIBRATION AND DIAGNOSTIC TECHNIQUES
FOR VUV PLASMA LASERS*

R. H. Dixon, J. F. Seely, and R. C. Elton
Naval Research Laboratory, Washington, DC 20375

ABSTRACT

Absolute intensity calibration and plasma diagnostic techniques
have been developed for vuv and soft x-ray lasers that will use a
highly ionized plasma as the gain medium. The apparatus is
portable and self-contained, and in situ measurements can be made
on a variety of plasmas. The techniques have been successfully
applied to a Nd:glass laser-produced carbon plasma in which
population inversions among the excited levels of the hydrogenic
C^{5+} ion have previously been observed. Using the width and
absolute intensity of the C^{5+} (n=7→6) line at 343 nm wavelength,
the electron and excited-state ion densities are measured as
functions of time and distance from the carbon target. The gain
coefficient for the n=4→3 transition at 52 nm wavelength is
inferred from the bound-state density measurements, and the
dependence of this gain coefficient on the intensity of the Nd:
glass laser is also determined.

I. INTRODUCTION

Population inversions between the n=3 and n=4 levels of the
hydrogenic C^{5+} ion have been observed in a laser-produced plasma
at a distance of approximately 3 mm from the carbon target[1,2].
In these initial observations of population inversion, the
calculated gain for the n=4→3 transition of C^{5+} at 52 nm wavelength was too low for lasing to occur on a single pass through
the inverted medium. Recent analysis has determined the optimal
values of such plasma parameters as electron density and temperature
that result in gain coefficients above the threshold for lasing
on the 52 nm transition[3,4]. In order to achieve a high gain
coefficient experimentally, it is necessary to adjust the plasma
condition by varying, for example, the target geomtry and the
manner in which the target is irradiated, while monitoring the
vital plasma parameters and the gain coefficient itself. Then an
orderly procedure can be developed which leads to optimal conditions
for lasing.

In this postdeadline paper we report the development of
calibration and diagnostic techniques for measuring the gain
coefficient specifically, as well as the plasma parameters required
in the analysis for a carbon plasma produced by irradiating a graphite
target with a Nd:glass laser. The required electron and ion population densities are measured as functions of time, distance from the

*Supported by the Office of Naval Research and the Department of
Energy

ISSN:0094-243X/82/900305-07$3.00   1982 American Institute of Physics

target, and irradiation intensity. From the measured population density of the C^{5+} n=4 level, the dependence of the 52 nm transition gain coefficient on the irradiation intensity is determined. The optical irradiation intensity, which results in the highest value of the gain coefficient, is determined to be 1.5 X 10^{12} W/cm^2.

The measurements that are discussed in Section II and Section III are for a Nd:glass laser intensity of 4.1 X 10^{11} W/cm^2 (8 J, 10 nsec, 500 μm spot diameter). The measurements made at higher irradiation intensities are presented in Section IV and the gain coefficient is evaluated in Section V.

II.   STARK BROADENING OF THE 343 nm LINE FOR N_e MEASUREMENTS

The 343 nm spectral line, resulting from the n=7→6 transition in C^{5+}, is useful for the measurement of electron and ion densities in a carbon plasma[5]. The plasma is optically thin to the 343 nm radiation, and this well-isolated spectral line is readily observed using a normal-incidence spectrograph. The electron density is obtained from Stark broadening and the density of the n=7 level of C^{5+} is inferred from the absolute intensity of the emission. Since the highly-excited n=7 level is in Saha equilibrium with the C^{6+} ionization stage, the C^{6+} density is inferred from the measured n=7 density.

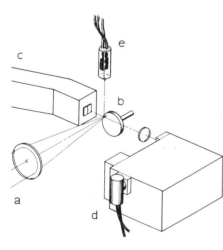

The experimental arrangement is diagramed in Fig. 1. Using a visible light, Czerny-Turner mount spectrometer fitted with a photomultiplier, the time-dependent profile of the 343 nm line is measured at various distances from the target. Within several millimeters of the target surface, the line is typically 1 to 2 nm in width. At distances greater than 5 mm from the target, where the Stark broadening is negligible and the profile is dominated by the instrumental broadening and the Doppler spreading due to the streaming motion of the plasma ions, the profile is nearly Gaussian in shape with a constant full width at half maximum of 0.43 nm.

Fig. 1. Experiment schematic: (a) Nd:glass laser pulse, (b) graphite disk target, (c) grazing incidence spectrograph, (d) near UV-to-visible spectrometer, (e) calibrated deuterium lamp.

The theoretical Stark profiles for the 343 nm line, convolved with a 0.4 nm Gaussian profile, are shown in Fig. 2 for electron densities in

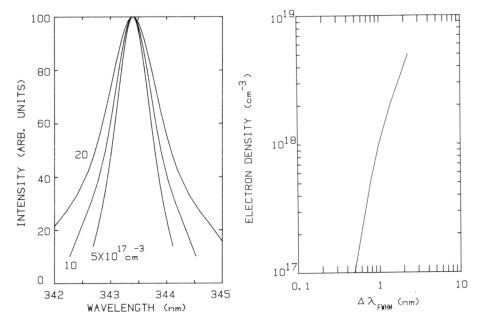

Fig. 2. Convolved Stark profile of the 343 nm line.

Fig. 3. Widths of the convolved Stark profiles.

the range (5-20) X 10^{17} cm^{-3}. The full width at half maximum of the convolved profiles are plotted in Fig. 3 as a function of electron density. These results are scaled from the Stark profile, for an electron density of 10^{18} cm^{-3} and an electron temperature of 20 eV, that has recently been calculated by Kepple and Griem[6]. The accuracy of the calculation is estimated to be 20%.

The 343 nm profile, observed at a distance of 3 mm from the carbon target and at a time 35 ns after irradiation, is compared to the Stark profile for 1 X 10^{18} cm^{-3} in Fig. 4. The electron densities that are measured from the Stark widths are shown in Fig. 5 for the two distances of 1.5 and 3.0 mm from the target. The electron density is highest at the leading edge of the expanding plasma and decreases by an order of magnitude over a period of 20 nsec.

III. ABSOLUTE INTENSITY OF THE 343 nm LINE FOR N^{6+}, N_7, and N_4 DETERMINATIONS

The normal-incidence spectrograph is calibrated using a standard deuterium lamp. This permits the measurement of the absolute intensity of the 343 nm line and the density of the n=7

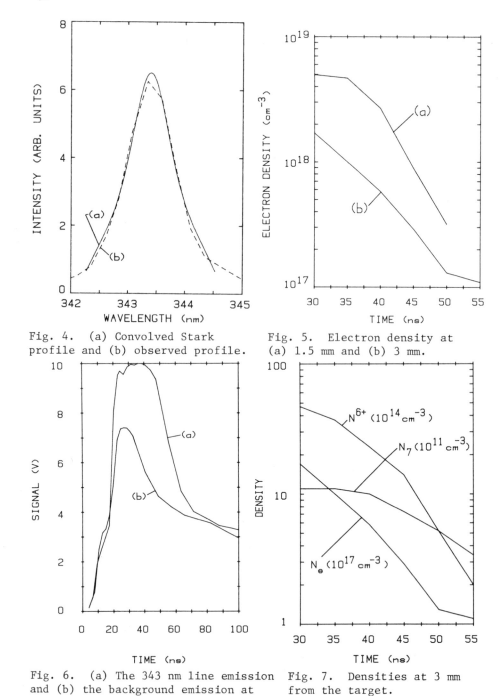

Fig. 4. (a) Convolved Stark profile and (b) observed profile.

Fig. 5. Electron density at (a) 1.5 mm and (b) 3 mm.

Fig. 6. (a) The 343 nm line emission and (b) the background emission at 3 mm.

Fig. 7. Densities at 3 mm from the target.

level of C^{5+}. Typical photomultiplier signals for the 343 nm line and the nearby background are shown in Fig. 6, and the density N_7 of the n=7 level, corrected for the non-linearity of the photomultiplier at the peak of the 343 nm emission, is shown in Fig. 7. The C^{6+} density N^{6+}, obtained from $N^{6+}/N_7 = (g^{6+2}/g_7\,Ne)\,(2\pi mkT_e/h^2)^{3/2}\exp(-\Delta E/kT_e)$ in LTE, is also shown in Fig. 7. As was the case for the electron density, the ion densities are highest at the leading edge of the expanding plasma. However, at later times the C^{6+} density decreases more slowly than the electron density, and the density ratio $6N^{6+}/N_e$ has a peak value of .03 at 45 nsec. Since the excited levels of the C^{5+} ion are populated primarily by electron collisional recombination from the C^{6+} ion, a high density ratio $6N^{6+}/N_e$ (of maximum value unity) is important in the modeling[3,4] of the C^{5+}, n=4→3 line gain coefficient.

IV. HIGHER IRRADIATION INTENSITIES

For a threshold gain coefficient of 1 cm^{-1}, the primary need[4] is an n=4 density N_4 of $\sim 10^{14}$ cm^{-3}, which is achieved in electron-capture pumping by rapidly reducing the temperature in the lasing region for increased recombination into excited states and for preferential population of the n=4 upper laser level. First, however, it is important to maximize the stripped-ion density N^{6+} at a level hopefully $\sim$ 100 X higher than N_4. To achieve this we have recently increased the target irradiance to produce more ions near the target and raise the temperature in the near-target region to increase the ratio $6N^{6+}/N_e$ in the outer (mm's) region.

The measurements described in the previous sections were repeated at the higher irradiation. The energy of the Nd:glass laser was increased to 18 J and the focal spot diameter was reduced to 100 μm. Observations were made at a maximum intensity of 5.7 X 10^{12} W/cm^2 and at four lower intensity levels by using neutral density filters. The photomultiplier signal of the 343 nm line center 3mm from the target is shown in Fig. 8. The 343 nm emission increases with irradiation intensity up to 1.5 X 10^{12} W/cm^2 and decreases at higher intensities.

The electron and ion densities, measured at the time (45 ns) when the 343 nm emission is strongest, is shown in Fig. 9. The maximum values $N_7 = 1.9$ X 10^{12} cm^{-3} and $6 N^{6+}/N_e = 0.13$ occur at an irradiation intensity of 1.5 X 10^{12} W/cm^2. From the relative intensities of the n=7→1 and n=4→1 transitions that are observed using a grazing incidence spectrograph, the density N_4 needed for gain determination[4] is found. This is equivalent to measuring N_4 from the 4→1 soft x-ray intensity obtained by the branching ratio calibration method using 7→6 and 7→1 lines[7], where in this case the same source ion is used in situ.

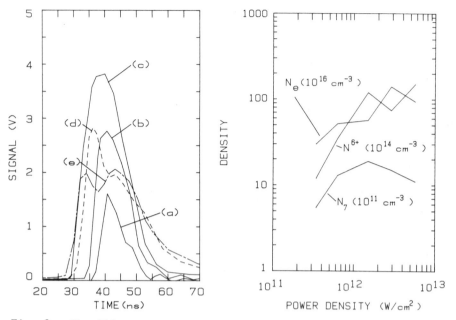

Fig. 8. The 343 nm emission at
irradiation intensities of (a)
.34 (b) .63 (c) 1.5 (d) 2.9 and
(e) 5.7 X 10^{12} W/cm^2

Fig. 9. Densities at 3 mm
and time 35 nsec.

V. GAIN COEFFICIENT

The relative intensities of the n=4→1 and n=3→1 transitions,
corrected for opacity effects, indicate that the fractional population inversion is $[1 - g_4N_3/g_3N_4] = 0.3$. At lower irradiance
of 4.1 X 10^{11} W/cm^2 the ratio N_4/N_7 was measured as 1.63
so that $N_7 = 7.3$ X 10^{11} implied $N_4 = 1.2$ X 10^{12}. For thermal
Doppler broadening at a temperature of 9 eV and 3 mm from the target,
the gain coefficient[4] of the 52 nm transition is ~2%/cm. The
Doppler broadening caused by the streaming motion of the plasma
becomes restricted in a linear plasma produced by a laser beam
focused into a line[8]; therefore, this broadening and the Stark
broadening are small compared to the thermal Doppler broadening
(at least for this irradiance). Absolute gain measurements at the
higher irradiance discussed in Section IV await similar measurements
of N_4, N_7 at those power levels and temperatures.

CONCLUSIONS

In this postdeadline paper, we have shown how basic parameters
vital to vuv-gain measurement and associated analyses are determined
on an absolute basis in laser-produced plasmas. We have also
presented recent preliminary data on optimizing the stripped ion
density N^{6+}, both on an absolute basis for ultimately maximizing N_4
and the gain, and also relative to the electron density in the ratio

$6N^{6+}/N_e$ for competing with electron collisional effects. We find so far that at an irradiance of 1.5×10^{12} W/cm^2 we obtain $N^{6+} = 1.2 \times 10^{16}$ cm^{-3}, which is 100X the density of N_4 needed for a "threshold" gain coefficient $= 1$ cm^{-1}, and a fractional stripping of $6N^{6+}/N_e \approx 0.13$. These values are $\sim$9X and $\sim$4X those obtained[4] at an irradiance of 4.1×10^{11} W/cm^2. Yet to be measured from soft x-ray lines at the higher irradiance is the ratio N_4/N_7, which is not expected to increase with higher irradiance and temperature; however it's value is more important after the pumping is enhanced with rapid cooling[4], now that the stripped density is optimized. How much additional cooling is needed? Starting with these measurements of N^{6+} and assuming LTE with highly-excited states $\left[$e.g., $N_7/N^{6+} \propto T_e^{-3/2} \exp(10/T_e)\right]$, an exponential scaling of the lower (e.g., n=4) bound state distribution, a T_e^{-2} dependent pumping rate, and thermal Doppler broadening (all of which favor increasing N_4 with decreasing temperature), threshold gain could be reached by lowering T_e from approximately 20 eV to 10 eV, a factor-of-2, for the 1.5×10^{12} W/cm^2 irradiance. It is interesting to note that the same factor applies at the lower irradiance $(4.1 \times 10^{11}$ W/cm^2, i.e., 9 eV to 5 eV; however, again, the absolute stripped ion density is marginal at this level.

ACKNOWLEDGEMENTS

It is a pleasure to acknowledge the expert technical assistance of J. L. Ford in all phases of the experiment described, including laser operation. Appreciation is expressed to H. R. Griem, both for illuminating technical discussions and for his and P. C. Kepple's permissions to use and quote their recent Stark broadening calculations prior to publication.

REFERENCES

1. R. C. Elton, T. N. Lee, R. H. Dixon, J. D. Hedden, and J. F. Seely, in Laser Interaction and Related Plasma Phenomena, ed. by H. J. Schwarz, M. Lubin, and B. Yaakabi (Plenum, 1981).
2. F. E. Irons and N. J. Peacock, J. Phys. B. 7, 1109 (1974).
3. R. C. Elton, SPIE vol. 279, p. 90 (1981); see also J. Opt. Engineering, March-April 1982.
4. R. C. Elton, J. F. Seely, and R. H. Dixon, "Population Density and VUV Gain Measurements in Laser-Produced Plasmas", Proceedings Topical Meeting on Laser Techniques for Extreme Ultraviolet Spectroscopy, Optical Society of America, Boulder (this volume), March 8-10, 1982.
5. F. E. Irons, J. Phys. B 6, 1562 (1973).
6. P. C. Kepple and H. R. Griem, Phys. Rev. A (submitted).
7. F. E. Irons and N. J. Peacock, J. Phys. E 6, 857 (1973).
8. G. V. Peregudov, M. E. Plotkin, and E. N. Ragozin, Sov. J. Quantum Electron. 9, 1224 (1979).

SOFT X-RAY POPULATION INVERSION
OBSERVED IN A LINE-FOCUS LASER PLASMA

Y. Conturie, B. Yaakobi, J. Delettrez, and J.M. Forsyth
Laboratory for Laser Energetics, University of Rochester
250 East River Road, Rochester, New York 14623

ABSTRACT

Laser-produced plasmas have proved to be suitable media to achieve soft X-ray amplification and their study benefits directly from the advances in laser development for Inertial Confinement Fusion (ICF) research. In recent years the IR light (1.054 μm) from powerful Nd:Glass lasers has been frequency-tripled into UV light (0.351 μm) with conversion efficiencies around 60%. Such a system, developed because of the superior coupling of short wavelength light to the target, is in operation at the Laboratory for Laser Energetics and has been used to study the feasability of a recombination laser in the 100 Å region. We present results of point-focus and line-focus experiments on step targets, with evidence of population inversion in Helium-like Aluminum and Hydrogen-like Fluorine, respectively. They are supported by a computer model based on the 1-dimensional code LILAC, which includes hydrodynamics, atomic rate equations and radiation transport.

INTRODUCTION

A laser of intensity I is focused on a target of atomic number Z. Depending on the value of I (10^{13} to 10^{15} W/cm^2), a dense, almost fully ionized plasma can be created at the surface of targets of Z-number as high as 13 (Aluminum). In such a plasma the recombination cascade is dominated by collisions, which tend to populate high-lying quantum states of the recombining ion. In particular the three-body recombination rate is proportional to N_e^2 (N_e: electron density) while the radiative recombination rate goes as N_e. Also the probability of a three-body event is strongly dependent on the electron temperature T_e: it increases more rapidly than that of a radiative decay as T_e cools down (1). Therefore the idea is to produce a high density, low temperature plasma from an initially hot plasma (2). One may rely on free expansion for the cooling (3,4) but it is accompanied by a large drop in density, greatly reducing the possibility of significant gain. An alternate solution is to bring the plasma in contact with a heat sink (5): a foil placed a few hundred microns in front of the target allows a rapid cooling of the expanding plasma at an early stage of its recombination, by heat conductivity. The use of the third harmonic (λ=0.351 μm) of a Nd:Glass laser instead of its fundamental (1.054 μm) presents several advantages (6): (i) for intensities of order 10^{14} W/cm^2, the fraction of the light which is absorbed is typically 30% at 1 μm and 70% or more at 0.35 μm, (ii) at short wavelength the laser radiation penetrates to higher densities (10^{22} electrons/cm^3 instead of 10^{21}) where the plasma is more collisional and where

collisional absorption, i.e. inverse bremstrahlung, is more
effective. At 1 μm the dominant absorption mechanism is resonant
absorption, i.e. resonant coupling between light and plasma waves,
and a large fraction of the absorbed energy is transferred into very
energetic electrons instead of being used to ionize the target.
That makes 1 μm light even less effective than the ratio of
absorption would indicate.

EXPERIMENTS

The experimental set-up is shown in Figure 1. The laser is
focused on the target and creates an X-ray emitting plasma. A
crystal spectrograph records the transitions which end on the
ground-state of the Hydrogen-like or Helium-like ion, providing a
measurement of the time-integrated populations of the excited states
of these ions. Spatial resolution is provided (i) along the normal
to the target, by a slit (∼50 μm wide), (ii) in the transverse
direction, by the crystal itself due to its small acceptance angle
at any specific wavelength.
A- Point-focus. Figure 3 shows the spectrum recorded when the UV
laser (20 J, 500 psec) is focused in a ⌀100 μm focal spot at the
surface of an Aluminum slab. The foil, 100 μm thick and located 200
μm from the slab, is of titanium. The choice of material, which
does not affect strongly the result, is largely based on avoiding
spectral overlap. Aluminum X-ray emission is observed mainly in the
focal region at the surface of the target and along the contour of
the foil, with a more diffuse tail as the plasma expands beyond the
foil.

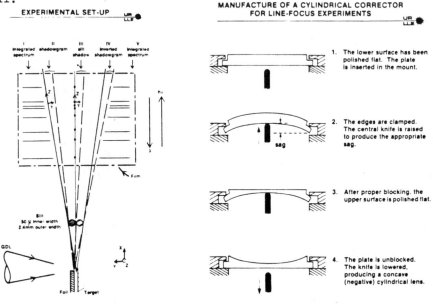

Figure 1.   Figure 2.

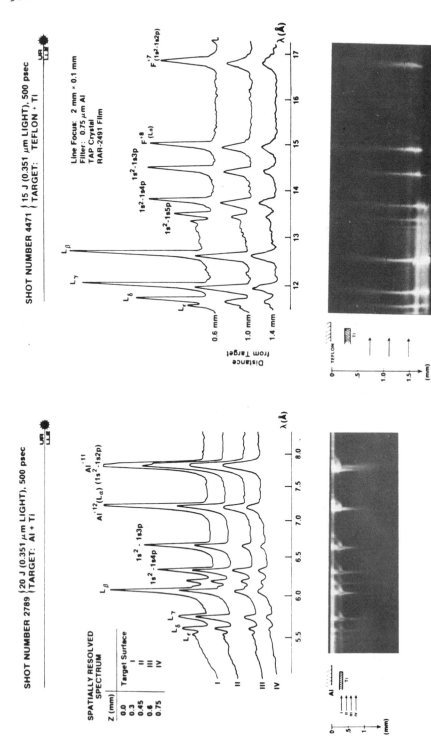

SHOT NUMBER 2789 {20 J (0.351 μm LIGHT), 500 psec
{TARGET: Al + Ti

SPATIALLY RESOLVED
SPECTRUM

Z (mm)	Target Surface
0.0	I
0.3	II
0.45	III
0.6	III
0.75	IV

Figure 3. (Point-Focus)

SHOT NUMBER 4471 {15 J (0.351 μm LIGHT), 500 psec
{TARGET: TEFLON + Ti

Line Focus: 2 mm × 0.1 mm
Filter: 0.75 μm Al
TAP Crystal
RAR-2491 Film

Figure 4. (Line-Focus)

As apparent on the densitometer trace, there is an intensity inversion at distances $\geqslant 600$ μm into the expansion between the $1s^2$-$1s3p$ and $1s^2 4p$ lines of Al^{+11}, showing the possibility of gain at 129.7 Å on 3^3D - 4^3F. Without foil, emission strong enough to be recorded occurs only in the focal region where high density and temperature adequate for line excitation can be found along the gradient from the hot corona to the cold target mass. The expanding plasma, initially at high temperature, is too diffuse by the time it has cooled down to the point where line radiation occurs.

B- Line-focus. An X-ray amplifier requires an elongated medium (i) to increase the length of amplification, (ii) to permit an unambiguous diagnostic of gain from the spatial distribution of the amplified spontaneous emission. A pair of cylindrical, fused silica corrector plates were added to the existing fused silica focussing lens (f/12, f=67") to produce a 100 μm wide line-focus of variable length, up to $L_{(max)}$=4 mm. Because the cylindrical sag required for each lens, $L_{(max)}/16(n-1)f\#$, was small, we used the manufacturing technique described in Figure 2. As the focal area gets larger with a cylindrical lens, the intensity on target decreases and one needs to use materials of lower atomic number to form a fully ionized plasma. Figure 4 shows the Fluorine spectrum recorded when the UV laser is focused on a 100 μm × 2 mm area of Teflon ($(CF_2)_n$). The laser pulse parameters and the Titanium heat sink are the same as before. The plasma expansion is very directional, laminar instead of diverging. As apparent on the film and densitometer traces, the transition Lyman β (1s-3p) is much more intense than Lyman α (1s-2p) almost everywhere, on a time-integrated basis. There are several caveats: (i) the heat sink partially blocks the line of sight of the spectrograph for the radiation emitted near the target, for $\lambda \geqslant 14$ Å, (ii) the entrance window of the spectrograph (0.75 μm Aluminum) transmits about 15% more at 12.64 Å (L_β) than at 14.98 Å (L_α), (iii) the film, RAR-2491, is more sensitive at the shorter wavelength (7). However, even taking those factors into account, there is a strong intensity inversion between the two F^{+8} lines through most of the expansion, showing the possibility of gain at 80.91 Å. There is also some evidence of an inversion between L_γ (1s-4p) and L_β late in the expansion ($\geqslant 1$ mm from the target surface). Gain would be at 231.3 Å on 3d-4f and at 59.95 Å on 2p-4d.

Opacity can contribute to the anomalous line ratios observed, especially in the line-focus experiment, but we find from the simulation that intensity inversion occurs only if there is population inversion between the upper levels of the transitions involved.

COMPUTER SIMULATION

The experiments have been modeled with the 1-dimensional laser fusion code LILAC. In the case of point-focus, the target is represented by a sphere of radius 200 μm, twice as large as the laser focal spot, being subjected to the same irradiance as in the experiment (but a higher total energy). The calculated X-ray intensity is then scaled to the measured laser pulse energy. This

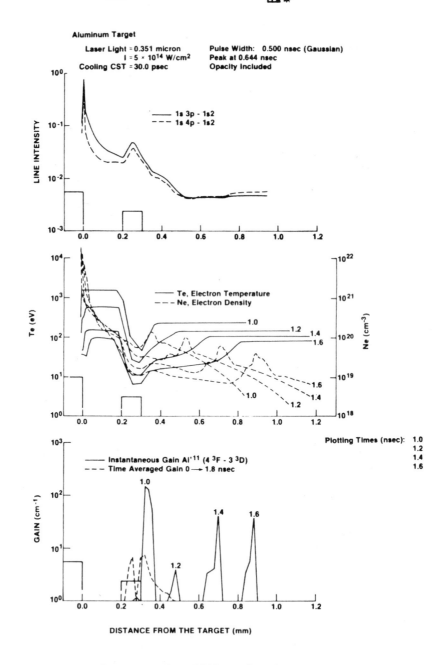

Figure 5. Shot 2789 (Point-focus).

procedure is known to give best results when trying to simulate the inherently two-dimensional behavior of flat target irradiation by using a spherically symmetric numerical simulation. The same method is used in the line-focus case, with a simple intensity scaling. The code computes the hydrodynamic motion of the plasma while providing temperature and density of both ions and electrons at each time step and for each numerical cell. These parameters serve as input for several subroutines:

(i) Rate equations. The calculations by McWhirter and Hearn for Hydrogen-like ions (8) are scaled to Helium-like ions with apparent nuclear charge Z-1.

(ii) Radiation transport. The correction for opacity assumes Doppler-broadened spectral lines (1). The transport simulates the experimental set-up with the spatially resolving crystal spectrograph pointing in a direction perpendicular to that of the laser axis.

(iii) Plasma cooling. The effect of the heat sink is treated by a simple, spatially dependent energy loss term proportional to $T_e(1-\exp(-t/\tau))$. t is the instantaneous time step of the code, τ is a "characteristic cooling time" used as free parameter.

Figures 5 and 6 present the results of simulations for the shots presented on Figures 3 and 4. The main experimental features (increased X-ray emission near the heat-sink, anomalous line ratios) are well reproduced when τ is set around 30 psec and 300 psec in the point-focus and line-focus cases, respectively. There is less cooling in the line-focus geometry because the plasma has a smaller volume of nearby vacuum in which to expand. A periodic spatial variation in the pumping radiation should help solve that problem (9). As shown in Figure 5, the temperature drop near the heat sink produces a pressure gradient. The pressure exerted by the upstream plasma creates a dense region downstream after the foil, appropriate for enhanced recombination. The gain region travels with the velocity of the expanding plasma ($\lesssim 10^8$ cm/sec for shot 2789). While the local gain per cm averaged over the plasma recombination time is small, the instantaneous gain can be fairly high (the gain length product remains small because of the plasma dimensions).

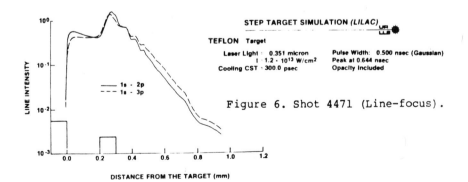

STEP TARGET SIMULATION (LILAC)

TEFLON Target

Laser Light : 0.351 micron   Pulse Width: 0.500 nsec (Gaussian)
I : 1.2 · 10^13 W/cm²   Peak at 0.644 nsec
Cooling CST : 300.0 psec   Opacity Included

1s - 2p
1s - 3p

Figure 6. Shot 4471 (Line-focus).

LINE INTENSITY

DISTANCE FROM THE TARGET (mm)

318

ACKNOWLEDGMENT

This research was partially supported by the following sponsors: Air Force Office of Scientific Research (grant AFOSR-81-0059), Exxon Research & Engineering Company, General Electric Company, Northeast Utilities, New York State Energy Research & Development Authority, Standard Oil Company of Ohio, University of Rochester, and Empire State Electric Energy Research Corporation. Such support does not imply endorsement of the content by any of the above parties.

REFERENCES

1. H.R. Griem, "Plasma Spectroscopy" (McGraw-Hill, N.Y., 1964), p162.
2. L.I. Gudzenko and L.A. Shelepin, Sov. Phys. Dokl. 10-2, 147 (1965).
3. F.E. Irons and N.J. Peacock, J. Phys. B-7, 1109 (1974).
4. D. Jacoby et al., Opt. Comm. 37-3, 193 (1981).
5. V.A. Bhagavatula and B. Yaakobi, Opt. Comm. 24-3, 331 (1978).
6. B. Yaakobi et al., Opt. Comm. 38-3, 196 (1981).
7. R.F. Benjamin et al., Appl. Opt. 16-2, 393 (1976).
8. R.W.P. McWhirter and A.G. Hearn, Proc. Phys. Soc. 82, 641 (1963).
9. W.T. Silfvast et al., Appl. Phys. Lett. 34-3, 213 (1979).

MULTIPHOTON IONIZATION OF ATOMS AND MOLECULES*

R. N. Compton and John C. Miller
Oak Ridge National Laboratory, Oak Ridge, Tennessee 37830

ABSTRACT

We have studied resonantly enhanced multiphoton ionization of
alkali atoms, rare gases, and small molecules using tightly focused
dye laser beams (power densities of 10^9 to 10^{11} W/cm^2). The gas
densities for most species were varied over the range from 10^{-7}
torr to ~100 torr. In the case of alkali atoms, some ionization
signals *appear* as a result of gas density effects (dimers or quasi-
collisions) as previously discovered by Collins and his collaborators.
These have been termed "hybrid-resonances." By contrast, in the case
of the rare gases, certain resonance ionization signals *disappear*
with increasing gas density. At low pressure (10^{-7} to 10^{-5} torr)
we have studied (1) mass spectra, (2) kinetic energy released in
ionic fragmentation, and (3) photoelectron kinetic energy spectra
using time-of-flight mass analysis and a 160° spherical sector
electrostatic energy analyzer. These experiments, combined with two-
color dye laser experiments, can offer an unambiguous and detailed
description of the multiphoton ionization and subsequent fragmentation
events.

INTRODUCTION

The recent invention and development of the pulsed, high-
powered, tunable dye laser is allowing physicists and chemists to
uncover many new nonlinear optical effects which occur when intense
electromagnetic radiation interacts with matter. The extreme
intensity of such light sources allows one to study a multitude of
multiphoton phenomena. Furthermore, since the source is tunable,
very high densities of excited state target species can be produced
leading to collective excitation phenomena. For example, so-called
"energy pooling"[1] and "hybrid resonance"[2] effects have been observed
in alkali excitation studies. It is even possible to perform experiments in which the laser field alters a chemical reaction or makes
the reaction energetically possible. Many recent studies have
shown that intense photon fields which are nonresonant with either
of the separated collision partners can induce energy transfer,[3,4]
collisional ionization,[5,6] and charge transfer.[7] White[8] has recently
observed the absorption of *two* photons during the collision between
barium and thallium ground state atoms, resulting in the simultaneous
excitation of both atoms to excited electronic states. These studies
fall under the general heading of "Collisions in the Field of a
Laser" and are in the early stages of development. There are some

*Research sponsored by the Office of Health and Environmental Research,
U.S. Department of Energy under contract W-7405-eng-26 with the
Union Carbide Corporation.

recent experiments which provide exciting possibilities that one can even study the fleeting transition state of a chemical reaction in absorption[8] or emission.[9]

The use of tunable dye lasers to produce resonantly enhanced multiphoton ionization (REMPI) provides a powerful new method for investigating atomic and molecular energy levels. Recent reviews[10,11] and a paper[12] contain references to work in this rapidly growing field of spectroscopy. The method is particularly useful in discovering and characterizing certain optically forbidden transitions. For example, new electronic states have been observed in iodine,[13] benzene,[14] and ammonia[15] using the REMPI technique. The method is particularly well suited for studying Rydberg transitions in molecules and is experimentally easier than the traditional use of far ultraviolet radiation in conventional spectroscopy. In addition, the use of two or more tunable dye lasers to produce sequential REMPI spectra has significant advantages in signal intensity, spectral simplification, and the ability to probe different Franck-Condon regions of a molecule.[16] Studies of mass-selective two-color (i.e., two different dye laser beams) photoionization of benzene[17] and its clusters[18] have provided new information on high Rydberg states and ionization limits for benzene, and structural information on small benzene clusters. Dual laser experiments have also been used to obtain new information on excited state lifetimes,[18] ionization potentials,[18] spectroscopic constants of free radicals,[19] and dissociation kinetics.[20,21] Such studies have also provided new information on triplet states of carbon disulfide[22] and toluene.[23] This is by no means an exhaustive list of the past and present applications of multiphoton ionization (MPI) techniques but serves to illustrate the vitality and potential of this new field of molecular spectroscopy.

The addition of mass spectroscopy to studies of MPI of polyatomic molecules has shown that considerable fragmentation can accompany ionization. In particular, REMPI of benzene can lead to extensive fragmentation, with C^+ being the dominant ion, while little or no parent ion is observed.[12,24,25] Although at least six photons are energetically required to produce C^+ from C_6H_6, the reported power indices range from 2 to 3.5.[12] The degree of fragmentation is independent of wavelength for benzene; however, for other polyatomics (pyrrole and furan)[24] the degree of fragmentation is highly dependent upon the wavelength (resonant intermediate state). In some regions of the spectrum the parent ion dominates.

Two mechanisms have been invoked to rationalize the production of small fragment ions from benzene and other molecules. In the first, multiple absorption of photons via *neutral* valence, Rydberg, and autoionizing states leads to superexcited states which subsequently autoionize and dissociate to produce the observed fragmentation. The second mechanism assumes that the *parent ion* is initially produced by MPI, and then absorbs further photons to reach dissociative or predissociative ionic states. Evidence that fragmentation occurs via absorption in the parent ion for benzene comes from the two-laser experiments of Beosl et al.[25] and the nonresonant MPI photoelectron spectra of Meek et al.[26] as well as the Oak Ridge

National Laboratory (ORNL) experiments to be discussed below. Conversely, two-laser experiments on azulene[27] indicate that fragmentation results from photoabsorption through autoionizing states. Several theoretical treatments of MPI fragmentation based upon statistical phase-space models have recently appeared.[28,29,30] Recent measurements of photoelectron and ion kinetic energy distributions at ORNL provide direct answers to the above questions and will be discussed in the text.

The interpretation of MPI studies in gases at high pressure (>1 torr) are complicated by the detection techniques, collisional effects, and the possibility of many body or collective phenomena. Payne, Garrett, and Baker[31] recently invoked a collective emission phenomenon to explain some very surprising experimental data in the five-photon ionization of gaseous xenon. In these experiments,[32] it was found that MPI and third harmonic generation (THG) were competing processes and that THG could completely quench the MPI signal at pressures above 0.3 torr. More details will be given below.

The physical mechanisms leading to MPI of alkali atoms has received a great deal of theoretical and experimental attention. A wide variety of ionization phenomena have been observed and the important processes leading to ionization depends upon gas density, laser wavelength, and laser power. When the laser is tuned to a resonance line of the alkali, direct MPI can be observed at low pressure. If the lifetime of the excited states is comparable to the time between collisions, collisional ionization and associative ionization processes are observed. If the laser field is sufficiently intense, laser-induced associative ionization and laser-induced Penning ionization can be important. For somewhat higher alkali density (10^{13}-10^{15} cm^{-3}), electrons which have been heated by superelastic collision with excited atoms can represent the dominant ionization mechanism. Despite the attention devoted to MPI studies of the alkali atoms, the relative importance of the various possible ionization mechanisms and the details of the physics of these processes are only recently being unraveled. Sodium has received the most attention due to the relative ease of handling this alkali and the fact that the sodium D lines occur at a convenient laser frequency. Below we will discuss some of our recent results on MPI of cesium.[33]

EXPERIMENTAL

Our studies of MPI have been performed at pressures above and below ~10^{-3} torr in three different experimental arrangements. MPI of alkali beams at densities of ~10^{-4} are studied in the ultrahigh vacuum beam experiment shown in Fig. 1. Figure 2 shows the apparatus used to obtain MPI photoelectron spectra and mass spectra for effusive beams of atomic and molecular species. A schematic diagram of the high pressure apparatus is shown in Fig. 3. This latter apparatus consists of two parts: one for detecting ionization and the second for detecting third-harmonic light. Pertinent details

322

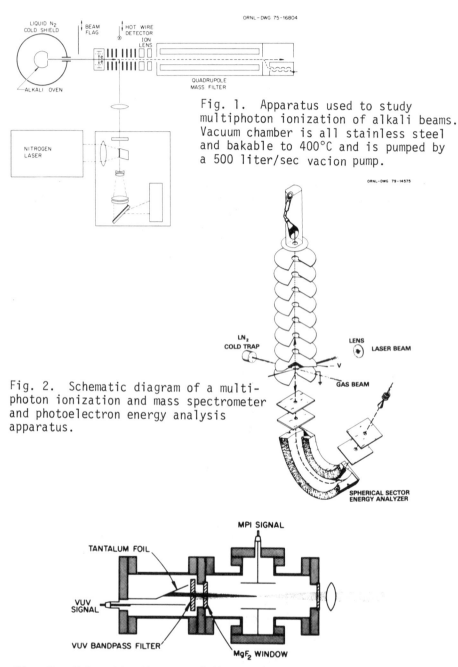

ORNL-DWG 75-16804

LIQUID N₂ COLD SHIELD
BEAM FLAG
HOT WIRE DETECTOR
ION LENS
ALKALI OVEN
QUADRUPOLE MASS FILTER
NITROGEN LASER

Fig. 1. Apparatus used to study multiphoton ionization of alkali beams. Vacuum chamber is all stainless steel and bakable to 400°C and is pumped by a 500 liter/sec vacion pump.

ORNL-DWG 79-14575

LN₂ COLD TRAP
LENS
LASER BEAM
V
GAS BEAM
SPHERICAL SECTOR ENERGY ANALYZER

Fig. 2. Schematic diagram of a multi-photon ionization and mass spectrometer and photoelectron energy analysis apparatus.

MPI SIGNAL
TANTALUM FOIL
VUV SIGNAL
VUV BANDPASS FILTER
MgF₂ WINDOW

Fig. 3. Schematic diagram of the apparatus used to measure the multiphoton ionization spectra and the third-harmonic vacuum ultra-violet radiation. The VUV detector can also be positioned at 90° with respect to the laser beam.

of each apparatus are given in the figure captions. Two different lasers have been used in the studies described below: (1) nitrogen pumped dye laser [Molectron UV24 (1 MW) andDL400] and (2) excimer pumped dye laser (Lambda Physik EMG101 and FL2000). Various focal length lens were used to produce power densities of ~10^9-10^{11} W/cm^2. The bandwidth of either laser was ~0.02 nm.

MULTIPHOTON IONIZATION OF CESIUM

Multiphoton ionization of low density atomic beams of cesium have been studied under the experimental conditions shown in Fig. 1. The cesium MPI spectrum shows a rich array of atomic and molecular features in the near infrared. Figure 4 shows a composite of two dye laser scans showing the MPI signal and the dye laser power. Three photons are required to photoionize cesium over the entire wavelength region shown. The second photon is in near resonance (<100 Å) with the two-photon allowed $6d^2D_{3/2,1/2}$ states of cesium. This near resonance probably accounts for the very strong ionization signals observed in the infrared. Starting at 12,000 cm^{-1} and moving toward larger wavenumbers, the atomic $6p^2P_{3/2}$ transition is observed followed by the $7d^2D_{3/2}$-$6p^2P_{1/2}$ hybrid resonance. The broad continuum like MPI signal observed between these two resonances appears to be molecular in origin but may also be due to nonresonant MPI which is greatly enhanced by both the 6p and 6d states which are near resonant with the first and second photons, respectively. Two photon resonant three photon ionization via the intermediate $7d^2P_{3/2,5/2}$ states is observed next, followed by the second $6p^2P_{1/2}$ fine structure peak. The $^2P_{1/2}$ intensity is considerably stronger than that due to the $^2P_{3/2}$ despite the fact that the statistical ratio should be 2:1, favoring the $P_{3/2}$ state. It is clear that MPI through both of these states is greatly affected by the presence of the 6d states. The last very sharp feature shown on Fig. 4 is due to the dipole forbidden but quadrupole allowed $7p^2P_{3/2}$-$6p^2P_{1/2}$ transition.

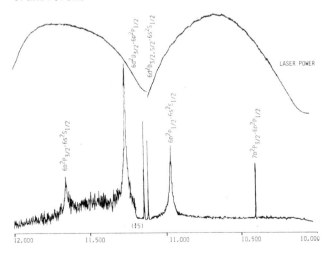

Fig. 4. Near infrared multiphoton ionization spectrum (Cs$^+$ only) of cesium beam at ~100°C. The laser power is also shown.

MPI signals were recorded for many of the one-photon resonant np states (n=7-12) and two-photon resonant s and d states. Figure 5 shows the observed ionization signal as one approaches the two-photon ionization limit at 15703.35 cm^{-1}. The ionization step occurring to the red of the correct ionization limit is due to field (electric and RF from quadrupole) and collisional ionization of high-Rydberg states. The intensity of the ionization signal versus principal quantum number, n, for the data shown closely follows the n^{-8} dependence predicted by simple theoretical considerations (n^{-3} dependence for Rydberg oscillator strengths followed by an n^{-5} dependence on ionization). Red or blue Stark shifts are observed depending upon whether the first nonresonant photon energy lies above or below the intermediate 3p state.

The Cs$_2^+$ dimer ion was observed at the 7p, 8p, and 9p atomic resonances. Since the dissociation limit for Cs$_2^+$ lies between the 8p and 9p levels, all 9p atoms and those 8p atoms with kinetic energy greater than 700 cm^{-1} can participate in the chemi-ionization reaction

$$Cs^* + Cs \rightarrow C_2^+ + e \quad .$$

The exact origin of the dimer ion signal at the 7p resonance is presently unexplained. Chemi-ionization by atoms with high kinetic energy is unlikely due to the small number of atoms in the tail of the Boltzmann distribution. In addition, the dimer ion signal at the 7p level is ~10 times stronger than the signal at the 9p level. The high laser powers (~10^9 W/cm^2) used in these experiments may open additional ionization channels such as ionization by superelastically heated electrons, energy pooling (e.g., Cs* + Cs* → Cs$_2^+$ + e) or laser induced chemi-ionization (i.e., Cs* + Cs $\xrightarrow{h\nu}$ Cs$_2^+$ + e). Two-laser and photoelectron measurements are in progress to distinguish between these possible mechanisms.

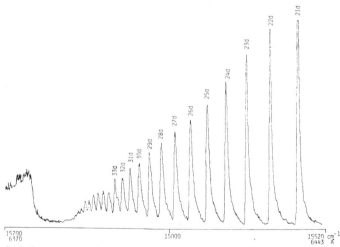

Fig. 5. Multiphoton ionization signal (Cs$^+$ only) in the region near the two-photon ionization limit at 15703.35 cm^{-1}.

MULTIPHOTON IONIZATION PHOTOELECTRON SPECTROSCOPY AND MASS ANALYSIS

Photoelectron spectra and mass analysis following resonantly enhanced multiphoton ionization are carried out with the apparatus shown in Fig. 2. The ionic masses are determined by observing their time-of-flight up the stack of electrodes. Figure 6 shows a typical mass spectra produced by MPI fragmentation of NH_3 molecules with 3866.1 Å laser light. Resonantly enhanced multiphoton ionization photoelectron spectra (REMPI-PES) have been recorded for Xe,[34] I_2,[35] NO,[36] NH_3,[37] H_2S,[38] CH_3I (unpublished), and C_6H_6.[35] Such a spectrum is shown for xenon in Fig. 7 in which the excimer pumped dye laser is tuned to the 6s[3/2],J=1 state. This data is of somewhat higher resolution but is qualitatively the same as that published earlier using an N_2 pumped dye laser.[34] The higher resolution in this data is attributed to three factors: (1) lower operating pressure afforded by the higher power of the new laser, (2) reduced RF noise on the analyzer itself, and (3) a digital data acquisition system which allows detection of a very low ion density in the focal region thus reducing the energy spread due to space charge. The interesting point to note for xenon is that MPI results in leaving Xe^+ in both possible fine-structure states despite the fact that enhancement occurs through the [3/2],J=1 core excitation. MPI-PES contains a great deal of information about the resonant intermediate states as well as the pathways of ion fragmentation. In a recent study of NO (Ref. 37) where fragmentation does not occur, we demonstrated that the distribution of NO^+ vibrational levels was determined primarily by the Franck-Condon factors connecting the resonant intermediate ($A^2\Sigma^+$ or $C^2\pi$) state and the ground $X^1\Sigma^+$ ionic state. Figure 8 gives MPI-PES spectra for the case in which two photons are resonant with either the v=0 or v=1 levels of the $A^2\Sigma^+$ state showing the higher quality data from the excimer pumped dye laser.

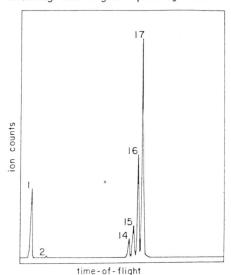

time-of-flight

The peaks at high energy correspond to direct ionization, leaving the ion in the v = 0 and v = 1 vibration levels, respectively. This single peak is expected from Franck-Condon considerations since the $A^2\Sigma^+$ state is a Rydberg state converging upon the $X^1\Sigma^+$ ionic ground

Fig. 6. Mass spectra produced by multiphoton ionization and fragmentation of NH_3 molecules with 386.610 nm laser light as measured by the time-of-flight method. The resonantly excited level is $\tilde{D}$-8. 1:H^+, 2:H^+, 14:N^+, 15:NH^+, 16:NH_2^+, 17:NH_3^+.

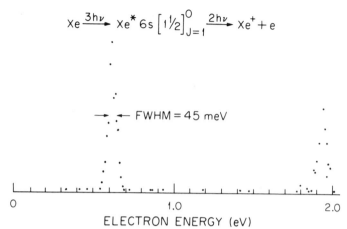

Fig. 7. Three-photon resonant, five-photon ionization photoelectron spectrum of xenon. The two peaks correspond to leaving the Xe⁺ in the $^2P_{3/2}$, $^2P_{1/2}$ final states.

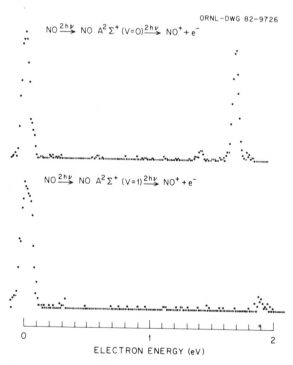

Fig. 8. Resonantly enhanced multiphoton photoelectron spectrum for four-photon ionization resonant with the v = 0 and v = 1 vibrational levels of the $A^2\Sigma^+$ state of NO.

states and hence these two potential curves are almost identical. We also note weaker photoelectron peaks corresponding to leaving the ion in v = 2,3,4, etc. These peaks are probably due to electronic autoionization.

The intense peak at near zero energy is of particular interest. We attribute the peak near zero energy to vibrational autoionization as a result of the admixture of another resonant intermediate state at the third photon energy level. Since *two* additional photons are required to ionize the resonant intermediate state, resonances or near resonances in the third-photon region can allow access to different autoionizing states than direct two-photon ionization from the A state. Thus, the large zero energy peak is attributed to vibrational autoionization from electronic states induced from transitions involving the three-photon excited state. Zero energy electrons are not observed when the laser is tuned to the $C^2\pi$ state *or* v = 3 level of the $A^2\Sigma^+$ state since the third photon is capable of ionization. Further evidence for this interpretation comes from the data in Fig. 9, where we have superimposed the dye laser beam with a beam from the direct output of the XeCl excimer pump (308 nm). The ionization signal resulting from the dye laser alone resonant with the intermediate v = 0 level of the $A^2\Sigma^+$ state is shaded. The large peak near zero energy is due to one photon ionization of the $A^2\Sigma^+$ (v = 0) state by the 308 nm excimer beam. No measurable increase of the zero energy peak is observed when the excimer beam is copropogated with the dye laser beam. This is taken as further evidence for the vibrational autoionization mechanism discussed above (i.e., when the "third photon step" is eliminated by ionizing with one, higher energy, photon, the slow electrons are not observed). The peak at ~2.5 eV is a result of three photon ionization of NO by the 308 nm excimer beam. The signal is resonantly enhanced

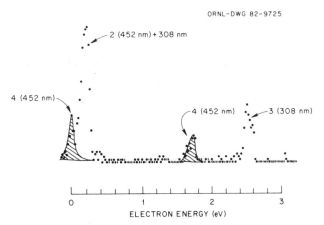

Fig. 9. Multiphoton ionization photoelectron spectrum of nitric oxide using dual laser beams. One laser is tuned to the v = 0 vibrational levels of the $A^2\Sigma^+$ state of nitric oxide (λ = 452 nm). The second laser consists of the XeCl excimer beam at 308 nm. The origin of each photoelectron peak is labeled.

by an accidental two photon resonance with the $H^2\Sigma^+$ (v = 1) state. The electrons occur at $3h\nu-IP-w_0$ ($h\nu$ = 4.02 eV, IP = 9.264 eV, and $w_0 = w_e-w_eXe$ = 0.293 eV; w_0 is the first vibrational spacing of the ion). This is in accord with Franck-Condon factors as discussed above. Finally, further evidence for the proposed mechanism for the zero energy electrons comes from studies of MPI-PES of ^{15}NO where the ratio of slow electrons to fast electrons increases greatly over that shown in Fig. 8 for ^{14}NO. This is taken as further evidence implicating a three photon resonance intermediate.

Iodine, ammonia, and hydrogen sulfide are three simple systems which show intense fragment ions resulting from MPI. In each case, the PES shows directly that the initial ionization step results in the formation of the parent ion, I_2^+, NH_3^+, and H_2S^+. The fragment ions must be produced from the further absorption of photons by the parent ion. Furthermore, in the case of I_2, we were able to measure the kinetic energy of the I^+ ion[35] which corresponded to the reaction $h\nu + I_2^+ \rightarrow I^+ + I$.

For the larger polyatomics, the ion kinetic energies could not provide direct evidence for this model. However, in the case of NH_3 it was possible to show that the extensive fragmentation was due to photon absorption into the first excited state of NH_3^+ which is dissociative. This is shown in Fig. 10 where we have plotted the percent of fragment ions (one particular fragmentation pattern is shown in Fig. 6) as a function of excess internal energy above the ground state of the ion. The data points represent fragment percentages for up to eight vibrational levels of five different electronic intermediate states. The energy onset for fragmentation corresponds exactly with the onset of the first excited state of NH_3^+ as shown in the one-photon photoelectron spectrum (PES). The PES of the two

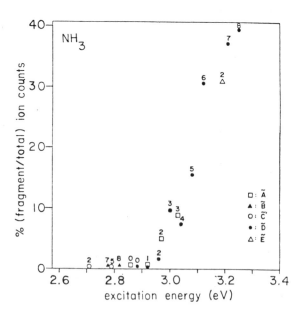

Fig. 10. Extent of fragmentation as a function of excitation energy for a fixed laser power. Squares: $\tilde{A}$, closed triangles: $\tilde{B}$, open circles: $\tilde{C}'$, closed circles: $\tilde{D}$, open triangle: $\tilde{E}$. The numbers indicate the vibrational quanta (ν_2') excited in the particular electronic state.

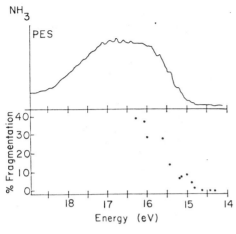

Fig. 11. The photoelectron spectrum of the $(...1e^33a_1^2)^2E \leftarrow (...1e^43a_1^2)$ $^1A_1$ transition $(NH_3^+ \tilde{A} \leftarrow \tilde{X})$ of ammonia (upper panel), and the extent of fragmentation is plotted vs five times the incident photon energy so that the two energy scales coincide. [Adapted from M. J. Weis and G. M. Lawrence, J. Chem. Phys. 53, 214 (1970).]

larger molecules studied at ORNL, CH_3I and benzene, also show that fragmentation follows creation of the parent ion. A major conclusion of our MPI-PES studies is that in every case studied thus far the fragmentation is due to absorption in the ionic manifold. No evidence for absorption in the autoionization manifold has been observed in any molecule but NO where a *minor* ionization pathway involves absorption of one photon from an autoionizing state.[35] The detailed physics of the ionic photodissociation processes is still under investigation.

HIGH PRESSURE RESULTS

Multiphoton ionization studies at high pressures can lead to new and interesting effects, some of which were mentioned in the introduction. These studies of rare gases are detailed in another abstract submitted from ORNL (Miller, Compton, and Cooper, page xx).

REFERENCES

1. V. S. Kushawaha and J. J. Leventhal, Phys. Rev. 22, 2468 (1980); and references cited therein.
2. C. B. Collins, J. A. Anderson, D. Popescu, and I. Popescu, J. Chem. Phys. 74, 1053 (1981); and references cited therein.
3. R. W. Falcone, W. R. Green, J. C. White, J. F. Young, and S. E. Harris, Phys. Rev. A 15, 1333 (1977).
4. P. L. Cahuzac and P. E. Toschek, Phys. Rev. Lett. 40, 1087 (1978).
5. A. V. Hellfeld, J. Caddick, and J. Weiner, Phys. Rev. Lett. 40, 1369 (1978).
6. P. Polak-Dingels, J.-F. Delpech, and J. Weiner, Phys. Rev. Lett. 44, 1663 (1980).
7. W. R. Green, J. Lukasik, J. R. Willison, M. D. Wright, J. F. Young, and S. E. Harris, Phys. Rev. Lett. 42, 970 (1979).
8. P. Hering, P. R. Brooks, R. F. Curl, Jr., R. S. Judson, and R. S. Lowen, Phys. Rev. Lett. 44, 687 (1980).
9. J. C. Polanyi, Faraday Discuss. Chem. Soc. 67, 129 (1979); P. Arrowsmith, F. E. Bartoszek, S.H.P. Bly, T. Carrington, Jr., P. E. Charters, and J. C. Polanyi, J. Chem. Phys. 73, 5895 (1980).

10. D. H. Parker, J. O. Berg, and M. A. El-Sayed, Advances in Chemistry (A. H. Zewail, Ed.), Springer, Berlin (1978).

11. P. M. Johnson and C. E. Otis, Ann. Rev. Phys. Chem. 32, 139 (1981).

12. L. Zandee and R. B. Bernstein, J. Chem. Phys. 71, 1359 (1979).

13. G. Petty, C. Tai, and F. W. Dalby, Phys. Rev. Lett. 34, 1207 (1975); F. W. Dalby, G. Petty-Sil, M. H. Pryce, and C. Tai, Can. J. Phys. 55, 1033 (1977).

14. P. M. Johnson, J. Chem. Phys. 62, 4562 (1975).

15. G. C. Nieman and S. D. Colson, J. Chem. Phys. 68, 5656 (1978).

16. A. D. Williamson and R. N. Compton, Chem. Phys. Lett. 62, 295 (1979).

17. M. A. Duncan, T. G. Dietz, and R. E. Smalley, J. Chem. Phys. 75, 2118 (1981).

18. J. B. Hopkins, D. E. Powers, and R. E. Smalley, J. Phys. Chem. 85, 3739 (1981).

19. T. G. Digiuseppe, J. W. Hudgens, and M. C. Lin, Chem. Phys. Lett. 82, 267 (1981); J. Phys. Chem. 86, 36 (1981).

20. B. H. Rookney and E. R. Grant, Chem. Phys. Lett. 79, 15 (1981); R.J.S. Morrison, B. H. Rookney, and E. R. Grant, J. Chem. Phys. 75, 2643 (1981).

21. G. Radhakrishnan, D. Ng, and R. C. Estler, Chem. Phys. Lett. 84, 260 (1981).

22. R. Rianda, D. J. Moll, and A. Kuppermann, Chem. Phys. Lett. 73, 469 (1980).

23. T. G. Dietz, M. A. Duncan, and R. E. Smalley, J. Chem. Phys. 76, 1227 (1982).

24. C. D. Cooper, A. D. Williamson, J. C. Miller, and R. N. Compton, J. Chem. Phys. 73, 1527 (1980).

25. V. Beosl, H. J. Neusser, and E. W. Schlag, Z. Naturforsch. Teil A 33, 1546 (1978); J. Chem. Phys. 72, 4327 (1980).

26. J. T. Meek, R. K. Jones, and J. P. Reilly, J. Chem. Phys. 73, 3503 (1980).

27. D. M. Lubman, R. Naaman, and R. N. Zare, J. Chem. Phys. 72, 3034 (1980).

28. J. Silberstein and R. D. Levine, Chem. Phys. Lett. 74, 6 (1980).

29. F. Rebentrost, K. L. Kompa, and A. Ben-Shaul, Chem. Phys. Lett. 77, 394 (1981).

30. F. Rebentrost and A. Ben-Shaul, J. Chem. Phys. 74, 3255 (1981).

31. M. G. Payne, W. R. Garrett, and H. C. Baker, Chem. Phys. Lett. 75, 468 (1980).

32. J. C. Miller, R. N. Compton, M. G. Payne, and W. R. Garrett, Phys. Rev. Lett. 45, 114 (1980).

33. W. Christian, R. N. Compton, John C. Miller, J. A. Stockdale, and A. D. Williamson (to be published).

34. R. N. Compton, J. C. Miller, A. E. Carter, and P. Kruit, Chem. Phys. Lett. 71, 87 (1980).

35. J. C. Miller and R. N. Compton, J. Chem. Phys. 75, 2020 (1981).

36. J. C. Miller and R. N. Compton, J. Chem. Phys. 75, 22 (1981).

37. J. H. Glownia, S. J. Riley, S. D. Colson, J. C. Miller and R. N. Compton J. Chem. Phys. (in press).

38. John C. Miller, R. N. Compton, T. E. Carney, and T. Baer, J. Chem. Phys. (in press).

OBSERVATION OF IONIZATION OF LASER EXCITED

SODIUM ATOMS BY SYNCHROTRON RADIATION

J.M. Bizau, F. Wuilleumier and P. Dhez
Laboratoire de Spectroscopie Atomique et Ionique and
Laboratoire pour l' Utilization du Rayonnement Electromagnétique
Université Paris-Sud
F-91405 Orsay, France

D.L. Ederer
Radiation Physics Division
National Bureau of Standards
Washington, DC 20234

J.L. Picqué and J.L. LeGouët
Laboratoire Aimé Cotton
Centre National de la Recherche Scientifique
F-91405 Orsay, France

P. Koch
Yale University
New Haven, Connecticut 06520

ABSTRACT

In a triple, orthogonal crossed beam experiment, we have studied photoionization of excited Na atoms. A cw ring dye laser (few W/cm^2) locked to the D_1 or D_2 absorption lines excited up to 20% of the 10^{13} cm^{-3} ground state atoms to $Na(3p\ ^2P_{3/2})$. Monochromatized synchrotron radiation from the ACO storage ring provided the photoionizing radiation. A cylindrical mirror electron spectrometer was used to measure photoelectron spectra. First measurements of the $2p^63p \rightarrow 2p^53p\varepsilon\ell$ photoionization were obtained. Decay of autoionizing resonances to the $2p^53p$ ionic channel were also observed.

ISSN:0094-243X/82/900331-16$3.00 Copyright 1982 American Institute of Physics

I. Introduction

Photoionization is an important tool for the study of atomic, molecular and solid state systems.[1-3] By the use of the photoionization process, the geometrical and dynamical properties of the electronic motion can be explored.[4] For example, electron correlations can be probed and compared with theoretical models.[5] In fact the continuing progress in the development of theoretical methods has gone hand in hand with the development of experimental techniques to map out photoion-ionization cross sections over a broad energy range. To this end synchrotron radiation has played an important role as a source with a broad spectral distribution. The measurement of the angular distri-bution of the ejected photoelectrons[6] and the spin polarization[7] of the photoelectron have provided additional information about the charac-teristics of the wavefunction needed to describe the photoionization process. These wavefunctions, both their amplitude and phase depend sensitively on the potential used to describe the system. Up to the present time, the vast majority of photoionization measurements have been made when the initial state was the ground state. The information obtained from these measurements, while having great value, is somewhat restrictive because of dipole selective rules. Only a limited class of states can be probed and the initial state is often an ensemble of nearly degenerate levels. This class of measurements have been extended by the observation of photoabsorption from metastable excited states from laser excited states.[8] In most of these cases, the ion-ization was produced by single or multiphoton absorption using lasers. With this mode of excitation, the energy range explored was restricted

to absorption measurements a few volts above the ionization threshold.
The energy range has been expanded recently by pioneering photoabsorption measurements from laser excited states.[9] These measurements
have been made over a broad photon energy range using a pulsed source
of continuum radiation to probe the excited state absorption cross
section of neutral and ionic species.

An important approach we have begun to exploit is to use lasers to
prepare the initial state in a specific way, and to use the photoionization process as a delicate probe of the atom prepared in a specific
initial state. In this paper we wish to report the first successful use
of synchrotron radiation to study photoionization and autoionization
from laser-excited initial states in atomic sodium. The successful
demonstration of this new technique opens many new frontiers for
exploring the geometrical and dynamical properties of atoms, molecules
and solids from states hitherto inaccessable. In addition to the
capability of preparing a system in a specific state, one can obtain
directly the density of excited atoms and study a chain of collision
processes that produce "hot electrons".[10,11] Excitation of an outer
electron can modify the potential of a system and produce a dramatic
change in the atomic structure for atoms with atomic numbers close to
those where a new shell is bound[12] (e.g., the transition elements, the
rare earths, and the actinides). Photoionization cross sections from
excited states can be obtained, and oscillator strengths can be measured
for resonances that decay into channels not accessible from the ground
state by dipole selection rules.

II. Experimental Arrangement

In our experiment (Figure 1) the laser beam irradiated in an effusive, weakly colliminated beam of sodium (density about 10^{13} atoms/cm^3) produced by an oven whose axis was parallel to the axis of a cylindrical mirror analyzer (CMA). The laser beam was perpendicular to the CMA axis and the vacuum ultraviolet (VUV) photon beam was colinear with the CMA axis. The laser beam was produced by an argon-ion pumped ring dye laser, had an intensity of about three watts/cm^2, and was linearly polarized in the horizontal plane. The VUV photon beam was also partially polarized in the horizontal plane at right angles to the laser beam polarization. Synchrotron radiation was the source of VUV radiation. This radiation was produced by the storage ring ACO at the Laboratoire pour l'Utilisation du Rayonnement Electromagnetique (LURE) in Orsay, France, and was monochromatized by a toroidal grating monochromator.[13] The bandpass of the monochromator was typically 0.3 eV for these measurements.

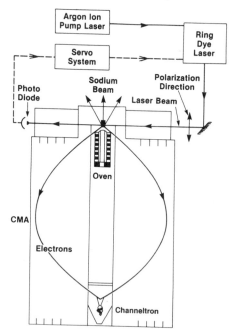

Figure 1. Experimental setup. A cylindrical mirror analyzer (CMA) is used to energy analyze the electrons produced by photoionization of the laser excited sodium vapor. A monochromatized beam of synchrotron radiation (not shown) enters the CMA from the top of the figure along the axis of the CMA.

For an axially symmetric system (i.e., one in which the polarization vectors of the photon beams lie in the same plane) with randomly orientied atoms or molecules in the initial state, the angular deendence of the two photon photoionization cross section is given by:[14,15]

$$\frac{d\sigma}{d\Omega} = A_0(t) + A_2(t) P_2(\cos\theta) + A_4(t) P_4(\cos\theta) \qquad (1).$$

The quantity θ is the angle between the synchrotron radiation polarization direction and the electron emission direction. The functions $P_n(\cos\theta)$ are the Legrendre polynominals of order n and the quantities $A_n(t)$ are coefficients which are products of appropriate photoexcitation cross sections, Clebsch-Gordan coefficients and a time dependent factor that accounts for the mixing and beating of the intermediate and/or final states. This, in general, would be the formula which would govern the angular dependence of the VUV photoionization cross section of laser excited atoms. Under the present experimental conditions, with a sodium density[16] of 10^{13} atoms/cm^3, the alignment of the intermediate state (unmeasured in our preliminary experiments) may have been reduced[17] significantly, at the laser powers used. This reduction in alignment is caused by radiation trapping, which reduces the polarization of the trapped photons relative to that of the incident laser beam. In many circumstances, therefore, a set of operating conditions can be chosen so that the angular dependence of the cross section reduces to a more familiar form:[18]

$$\frac{d\sigma}{d\Omega} = \frac{\sigma}{4\pi}(1 + \beta/2\, P_2(\cos\theta)) \qquad (2)$$

for linearly polarized ionizing VUV radiation. The CMA analyzer[16] used in these experiments transmitted photoelectrons within $\Delta E/E \sim 1\%$ window around a selected kinetic energy E, emitted in cone whose half angle is 54°44' with respect to the photon propogation direction. The total azimuthally integrated count for electrons ejected into this so called "magic angle" is independent of β and is proportional to the partial cross section when the initial state is randomly oriented.

III. Results

A. Continuum Photoionization from Laser-Excited Sodium

A photoelectron spectrum of ground state sodium atoms is shown in the upper frame of Figure 2. The two photoelectron peaks in the figure, obtained with photons of 75 eV, are due to the ionization of one of the 2p electrons of sodium (peak near 37 eV binding energy) and to the ionization of a 2p electron and excitation of the 3s electron to a 3p orbital (peak near 33 eV binding energy). The excitation of this group of ionic levels is due to final state electron correlations. For the $2p^5 3p$ configuration, the resulting six fine structure levels have been tabulated.[19] Five of these levels lie within about 0.5 eV of each other; the sixth is almost one electron-volt lower in binding energy. The relative energy position of these levels is noted on the figure as vertical lines extending below the horizontal line identifying the configuration.

The lower frame of Figure 2 corresponds to a photoelectron spectrum taken with some of the atoms laser-excited to the 3p $^2P_{3/2}$ level. A new peak occurs shifted by 2.11 eV, due to the photoionization of a 2p electron with the outer electron in a 3p orbital. It is interesting to notice that even though they each correspond to the same ionic configuration, the profile of the peak (at 40 eV binding energy) associated with a 2p hole produced in the laser-excited atom is different from the satellite peak at 42 eV binding energy. Studies at higher resolution capable of resolving the fine structure may serve as an important tool to characterize the shake up process.

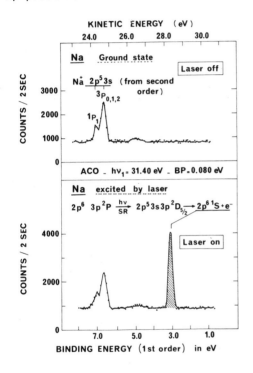

Figure 2. Upper frame: A 2p subshell photoelectron spectrum of atomic sodium obtained with the laser off. The large peak is due to the direct ionization of a 2p electron. The small peak is due to ionization of a subshell 2p electron and excitation of the 3s electron to a 3p orbital.

Lower frame: The 2p subshell photoelectron spectrum obtained with the laser on. In addition to the laser-off spectrum, a new peak appears (hatched area) due to 2p ionization of the laser-excited sodium atoms.

If one kept constant the density of atoms in the laser-excited state and varied the VUV photon energy, the partial photoionization cross section of 2p electrons with the outer electron in a 3p orbital could be obtained relative to that for ground state atoms. It would then be possible to test recent theoretical predictions[20,21] that the partial cross section for ionization of a 2p electron is very nearly the same, independent of the excitation of the valence electrons. If this prediction is true one could use the ratio of areas under the 2p photoelectron peaks for atoms in the ground and laser-excited states, respectively, to determine the relative population of atoms in the two states.

The effective laser optical resolution, the particular transition(s) driven by the laser, and the ground state atom density each influence the achievable steady state of density of 3p atoms. The first depends on the laser linewidth and power and on the Doppler spread for the particular excitation geometry used. The second influences how much optical pumping takes place from one hyperfine component of the ground state to the other. The third influences the amount of radiation trapping.

If the laser were tuned to pump hyperfine components of the 3p $^2P_{1/2}$ state and the source of atoms were a well colliminated beam, a very small population of 3p atoms could be maintained because of the optical pumping of the other hyperfine level in the ground state. In the present experiment, we were able to maintain a fractional population of atoms in the 3p $^2P_{1/2}$ level of a few percent when the laser was tuned to 3s $^2S_{1/2}$ → 3p $^2P_{1/2}$ transition. As explained in the preceeding paragraph, this percentage was determined from the ratio of the innershell 2p photoionization peaks from ground state and laser-excited sodium, respectively, under the assumption that their partial cross section was the same.[20,21] A steady-state 3p $^2P_{1/2}$

population of this magnitude suggests that the Doppler width associated with the effusive beam from the oven was large enough to produce at least a partial overlap of the transitions beginning on each of the hyperfine levels in the ground state, which are separated by 1.7 GHz.

B. Hot Electron Background

In making measurements of this type at photon energies near threshold where the kinetic energy of the electrons is small, one must be aware of the production of low energy electrons by collisional processes involving the laser-excited atoms. The production of these electrons of low kinetic energy could be a troublesome source of background that masks the desired spectrum. In Figure 3, we present an electron spectrum for laser-excited sodium atoms obtained in the absence of synchrotron radiation. In a recent publication[11] the mechanisms responsible for producing these electrons were described. Large numbers of low energy electrons (kinetic energy less than a few tenths of an eV) are produced by associative ionization[10] in collisions of two Na(3p) atoms. These "seed" electrons collide with laser-excited sodium atoms and gain energy by superelastic collisions[22,23] by the reaction process:

$$Na(3p) + e(\underset{\sim}{<} 0.1 \text{ eV}) \rightarrow Na(3s) + e(2.11 \text{ eV}). \tag{3}$$

These "hot" electrons make futher collisions and gain energy in 2.11 eV steps. Electrons produced by associative ionization and making two superelastic collisions are denoted by the peak labled 1 on Figure 3. Similarly, two laser-excited atoms can undergo an "energy pooling" to produce a more highly excited atom and a ground state atom:

$$Na(3p) + Na(3p) \rightarrow Na(3s) + Na(nl) + \Delta E, \qquad (4)$$

where the electron in the nl orbit can be in a 4s, 3d, 4p . . . state,

and ΔE represents kinetic energy obtained from potential energy.

Collisional ionization of these excited atoms produces additional

electrons which can be heated by superelastic collisions. Electrons

from this chain of collisions are identified in the structure

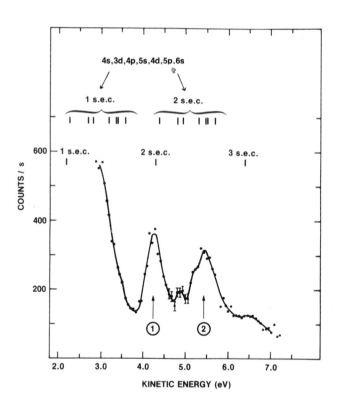

Figure 3. Typical energy spectrum of electrons ejected
from laser-irradiated sodium vapor in the absence of
synchrotron radiation. The upper group of bars indicates
the energy position of electrons created by collisional
ionization from the various excited states and heated
by either one or two superelastic collisions (s.e.c.);
the lower group of bars shows the energy position of
electrons created by associative ionization and having
undergone one, two or three superelastic collisions.

labeled 2 in Figure 3. Photoelectron spectroscopy is a powerful tool for the study of these collisional processes that produce electrons in a laser-excited medium.

C. Even-parity Autoionization Resonances

It was possible to use our apparatus to produce by step-wise excitation (laser plus VUV) even-parity autoionzing resonances. These were detected by electron spectroscopy. We set the VUV monochromator at 31.4 eV, the photon energy corresponding to the excitation[24] $2p^5 3p\ ^2P \rightarrow 2p^5\ (3s3p\ ^1P)\ ^2D_{5/2}$. The upper frame of Figure 4 shows a spectrum we obtained at that energy without the laser. The small peak near 26.5 eV kinetic energy is due to the direct ionization of 3s electrons. The feature between 24-25 eV kinetic energy is due to

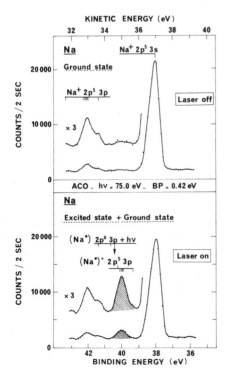

Figure 4. Photoelectron spectra obtained at a photon energy of 31.4 eV with the laser off (upper frame) and with the laser on (lower frame). Photoelectrons from the direct ionization of the 3s electron appear at a kinetic energy of 26.5 eV. Photoelectron due to ionization of the 2p electron by second order radiation hν=62.8 eV) appear with kinetic energy between 24-25 eV. Electrons from the decay of the autoionizing $2p^5 3s3p\ ^2D_{5/2}$ resonances produce the hatched peak in the lower frame.

ionization of 2p electrons by 61.8 eV photons transmitted by the monochromator in second order. The $2p^5 3s$ $^{1,3}P$ states of the sodium ion are nearly resolved. The lower frame of Figure 4 shows the photoelectron spectrum with the laser tuned to the $3s$ $^2S_{1/2}$ → $3p$ $^2P_{3/2}$ transition. The intense feature near 28.5 eV kinetic energy is due to the decay of the even-parity autoionizing resonance excited by step-wise laser plus VUV photon absorption. The area[25] $N_{h\nu}$ of the photoelectron peak for 2p photoionization of from the laser-excited atoms is proportional to the product of the excited state density and the photoionization cross section $\sigma(h\nu)$:

$$N_{h\nu} = KI_{h\nu} E_{h\nu} n_{3p} \sigma(h\nu) \Delta E, \tag{5}$$

where K is the spectrometer constant, $I_{h\nu}$ is the photon flux at energy $h\nu$; $E_{h\nu}$ is the kinetic energy of the photoelectron, n_{3p} is the density of laser-excited atoms and ΔE is the monochromator band pass. Similarly, the integrated area N_R of the photoelectron peak for the autoionzing resonance is given by:

$$N_R = KI_R E_R n_{3p} \int_{\Delta E_R} \frac{df_R}{d\varepsilon} d\varepsilon, \tag{6}$$

where I_R is the photon flux at the resonance excitation energy, E_R is the kinetic energy of the decay electron and ΔE_R is the monochromator bandpass. Since for our experiment, the monochromator bandpass (about 0.3 eV in first order) was much larger than the width of the resonance, and the integral over the oscillator strength density $\frac{df_R}{d\varepsilon}$ in Eq(6) can be replaced by the oscillator strength of the resonance f_R. The ratio of Eq(6) to Eq(5) yields:

$$f_R = \frac{I_{h\nu}}{I_R} \frac{E_{h\nu}}{E_R} [\sigma(h\nu) \Delta E] \frac{N_R}{N_{h\nu}}. \tag{7}$$

All the terms on the right hand side of Eq(7) can be determined experimentally. If we either measure $\sigma(h\nu)$ or assume that it equals the 2p photoionization cross section for ground state sodium we can obtain a numerical value for f_R.

IV. Conclusions

The preliminary experiments reported in this paper, have demonstrated the first union of the techniques of laser, synchrotron radiation and photoelectron spectroscopies to observe inner shell photoionization in an atom whose valence electron was excited with a laser. We have also observed the decay by electron emission of autoionzing resonances that were step-wise excited with laser and VUV photons. Since they are produced by two photon absoption, all final states have the same parity as the ground state; thus, they could not be observed by one photon absorption. A source of low energy electron lines that could contribute background signals in such experiments were shown to be due to collisions of laser-excited atoms. This research establishes the feasibility of combined laser, synchrotron radiation experiments, but this only opens the door to many more detailed experiments that will follow. Technically, the possible combination of laser, VUV and atomic beam geometries must be optimized according to the atomic species studied and the choice of electron spectrometer and detector. We can look forward to detailed measurements of partial photoionization cross sections, angular distribution of photoelectrons, oscillator strength for autoionizing resonances, all as a function of the state of specific valence excitation prepared by the laser, which include alignment or orientation of this intermediate state.

344

V. Acknowledgements

The authors are grateful to the technical staff of the Laboratoire pour l'Utilisation du Rayonnement Electromagnetique for their assistance. This work was supported by the Centre National de la Recherche Scientifique and by NATO Research Grant No. 1735.

REFERENCES

1. M.O. Krause in "Synchrotron Radiation Research," Eds: H. Winick and S. Doniach (Plenum Press, New York, 1980) pp. 101-151.

2. K. Codling, "Topics in Current Physics: Synchrotron Radiation, Technique and Application", (Springer-Verlag, Berlin, 1979) p. 231, and references therein.

3. E.E. Koch and B.F. Sonntag, ibid., p. 270 and references therein.

4. W.R. Johnson, K.T. Cheng, K.-N. Huang and M. LeDourneuf, Phys. Rev. A $\underline{22}$, 989 (1980); H.P. Kelly and R.L. Simons, Phys. Rev. Lett. $\underline{30}$, 529 (1973); and T.N. Chang and R.T. Poe, Phys. Rev. A $\underline{11}$, 191 (1975).

5. A.F. Starace, Appl. Opt. $\underline{19}$, 4051 (1980) and references therein.

6. J.A.R. Samson, "Electron Spectroscopy: Theory, Techniques and Applications", Vol. 4, Eds. C.R. Brundel and A.D. Baker (Academic Press, New York, 1981) p. 361.

7. U. Heinzmann, Appl. Opt. $\underline{19}$, 4087 (1980).

8. S. Ishii and W. Ohlendorf, Phys. Lett $\underline{38A}$, 119 (1972); F.B. Dunning and R.F. Stebbings, Phys. Rev. Lett. $\underline{32}$, 1286 (1974); R.D. Rundel, F.B. Dunning, H.C. Goldwire, R.F. Stebbings, Opt. Soc. Am. $\underline{65}$, 628 (1975); D.J. Bradley, P. Ewart, J.V. Nicholas and J.R.D. Shaw, J. Phys. B: Atom. and Mole. $\underline{6}$, 1594 (1973); H.-A. Bachor and M. Kock, J. Phys. B: Atom. and Mole. $\underline{14}$, 2793 (1981); and A.V. Smith, J.E. M. Goldsmith D.E. Nitz, and S.J. Smith, Phys. Rev. A $\underline{22}$, 577 (1980).

9. T.B. Lucatorto and T.J. McIlrath, Phys. Rev. Lett. $\underline{37}$, 428 (1976); T.J. McIlrath and T.B. Lucatorto, Phys. Rev. Lett. $\underline{38}$, 1390 (1977) and T.B. Lucatorto, T.J. McIlrath, J. Sugar and S.M. Younger, Phys. Rev. Lett $\underline{47}$, 1124 (1981).

10. T.B. Lucatorto and T.J. McIlrath, Appl. Opt. $\underline{19}$, 3948 (1980).

11. J.L. LeGouët, J.L. Picqué, F. Wuilleumier, J.M. Bizau, P. Dhez, P. Koch and D.L. Ederer, Phys. Rev. Lett. $\underline{48}$, 600 (1982).

12. J.P. Connerade, Contemp. Phys. $\underline{19}$, 415 (1978).

13. P.K. Larsen, W.A.M. Van Bers, J.M. Bizau, S. Krummacher, D.L. Ederer, F. Wuilleumier and V. Schmidt, "Proceedings of Conference on Synchrotron Radiation Instrumentation," Nucl. Inst. and Meth. $\underline{195}$, 245 (1982).

14. T. Hellmuth, G. Leuchs, S.J. Smith and H. Walther, "Lasers and Applications", Eds. W.O.N. Guimars, C.-T. Lin and A. Mooradian, (Springer, Berlin-Heidelberg, 1981).

15. J.C. Hansen, J.A. Duncanson, Jr., R.-L. Chien and R.S. Berry, Phys. Rev. A 21, 222 (1980).

16. J.M. Bizau, Thesis, University of Paris Sud (1981), unpublished.

17. A. Fischer and I.V. Hertel, Z. Phys. A - Atoms and Nuclei 304, 103 (1982).

18. J. Cooper and R.N. Zare, "Photoelectron Angular Distributions" Lectures in Theoretical Physics: Atomic Collision Processes, Vol. 11-C, p. 317.

19. C.E. Moore, "Atomic Energy Levels", Vol. 1 (U.S. Govt. Printing Office, Washington, DC, 1971), pp. 91-92.

20. T.N. Chang, private communication.

21. K.D. Chao and S.T. Manson, Phys. Rev. A 24, 2481 (1981).

22. R.M. Measures, J. Appl. Phys. 48, 2673 (1977).

23. D.F. Register, S. Trajmar, S.W. Jensen and R.T. Poe, Phys. Rev. Lett. 41, 749 (1978).

24. J. Sugar, T.B. Lucatorto, T.J. McIlrath and A.W. Weiss, Opt. Lett. 4, 109 (1979).

25. M.O. Krause, Chem. Phys. Lett. 10, 65 (1971).

ELECTRON SPECTRA FOR RESONANT FOUR-PHOTON IONIZATION OF NO

J. Kimman, P. Kruit and M.J. Van der Wiel

FOM-Institute for Atomic and Molecular Physics,
Kruislaan 407, 1098 SJ Amsterdam, The Netherlands

ABSTRACT

Electron energy spectra of (2 + 2)-photon ionization of NO are
presented. The intermediate resonance is $A^2\Sigma^+$ and scans are shown
over the rotational width of v = 0 and v = 1. The vibrational distribution of NO^+ peaks at $\Delta v = 0$ (Franck-Condon for direct ionization), but
also at the highest energetically allowed Δv. We describe evidence
for pre-ionization via a dissociative channel.

INTRODUCTION

The various decay pathways that are open to the NO molecule after
excitation in the near-continuum, are beginning to be unravelled in
detail. A large contribution to this analysis stems from the comparison of photoabsorption and photoionization spectra[1,2], and from the
theoretical basis that was provided recently by the Multichannel
Quantum Defect Theory[3,4]. Moreover, photoelectron spectra have been
reported at various discrete wavelengths[5,6], and a threshold electron
spectrum is available over a continuous wavelength range[7]. In principle, the photoelectron work provides the most detailed information;
in practice, however, limitations are imposed by the use of one or a
few discrete line sources or by the lack of resolution when using
dispersed continuum sources.

In this respect, multiphoton excitation may be considered an
effective alternative for high-resolution VUV excitation[8]. If, in
addition, a resonant intermediate state is excited, we have a technique that is: a) very sensitive; b) sufficiently narrow-band to permit
selection of individual rotational transitions; c) capable of continuous scans over pre-ionizing structure, especially when two dye-
lasers are used.

Here, we are concerned with a 4-photon study of NO. At $\lambda = 430$
and 450 nm, there is a resonance with the $A^2\Sigma^+$, v = 1 and 0 state respectively, and the final states lie at around 2.3 eV above the I.P.
of NO^+. From single-photon electron spectra[5,6] it is already known
that vibrational populations are highly non-Franck-Condon in this
region, which is ascribed to pre-ionization. The work involved measurements at fixed wavelength, where only accidental coincidences
with pre-ionizing states occur. The analysis of Caprace et al.[6] indicates that a possible contribution from pre-ionizing superexcited
valence states accounts for only part of the observed high-Δv transitions. A recent multiphoton study by Miller and Compton[8] provided new
information, but did not lead to a completely satisfactory interpretation, partly because of lack of resolution and sensitivity.

In this paper, we apply the multiphoton electron spectroscopy
technique to perform ~ 3 nm scans over the final energy in the conti-

ISSN:0094-243X/82/900347-05$3.00 Copyright 1982 American Institute of Physics

nuum. Thus, we learn more about the nature of the pre-ionizing state(s) involved.

EXPERIMENTAL

The electron time-of-flight spectrometer used is the subject of a separate paper at this meeting[9], since it is of a novel type suited extremely well for multiphoton ionization studies. It features full collection of electrons over 2π sterad, at an energy resolution of 15 meV. Absolute energy calibration with an accuracy of $\pm$ 5 meV is based on the position of Xe^+ ($^2P_{\frac{1}{2}}$ and $^2P_{3/2}$) 5-photon electron peaks. NO spectra are recorded using 1 : 1 mixtures of NO and Xe at a total pressure of 10^{-3} Pa. The typical data accumulation time for a photo-electron spectrum as shown in fig. 1, is six minutes.

The light source is a Nd-YAG pumped two-stage dye laser with 0.5 cm^{-1} bandwidth. Typical power density in the laser focus is 10^9 W cm^{-2}. The light is linearly polarized along the axis of the spectrometer flight tube. Angular distribution problems are automatically avoided by the 2π-collection.

RESULTS

Figure 1 presents the electron spectrum for the transition:

$$X^2\Pi_{3/2}(v=0,\ J=8\tfrac{1}{2}) \xrightarrow{2h\nu} A^2\Sigma^+(v=0,\ J=6\tfrac{1}{2}) \xrightarrow{2h\nu}$$

$$X^1\Sigma^+(v=0\ \dots\ 6,\ 4\tfrac{1}{2} \le J \le 8\tfrac{1}{2})$$

where $\lambda = 454.2$ nm. The assignment of the 2-photon transition out of the ground state is based on the analysis of Bray et al.[10]. All seven energetically allowed vibrational levels, v = 0 to 6, are populated. The v = 6 peak lies in the energy region below 40 meV, where our conversion from time-of-flight into energy becomes uncertain. We estimate its energy to lie between zero and 20 meV. Relative peak areas are reliable over the full spectrum due to the 2π collection.

In figure 2 a similar spectrum is shown at $\lambda = 430.0$ nm for a transition via $A^2\Sigma^+(v=1)$ at unspecified J. Finally, figure 3 is a recording of the excitation functions of NO^+ in the specific vibrational states v = 1 and v = 8. The spectral range is that corresponding to the rotational width of transitions from the ground state to $A^2\Sigma^+(v=1)$.

DISCUSSION

In both figures 1 and 2 the main aspect of the vibrational distribution is the occurrence of two maxima, one at $\Delta v = 0$ and the second at the highest energetically allowed Δv. Transitions with $\Delta v = 0$ correspond to direct ionization, for which the Franck-Condon overlap is below 10^{-3} for any $\Delta v \ne 0$. The reason is the great similarity of the potential curves of the $A^2\Sigma^+$ and $X^1\Sigma^+$ states. The $\Delta v \ne 0$ transitions therefore require a pre-ionization process.

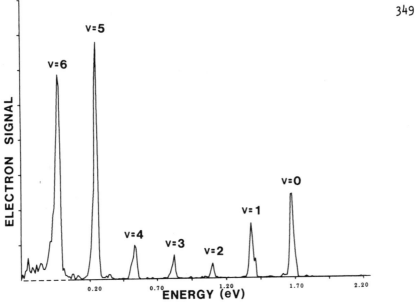

Fig. 1. Electron energy spectrum for 4-photon ionization of NO via the intermediate state $A^2\Sigma^+(v = 0,\ J = 6\frac{1}{2})$ at $\lambda = 454.2$ nm. The calibration of the energy axis is uncertain for energies below 40 meV.

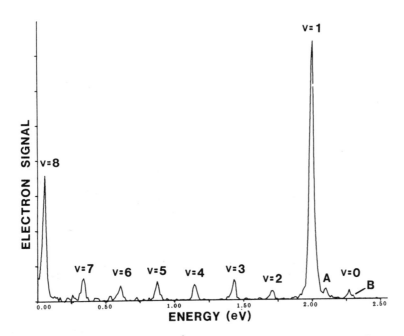

Fig. 2. As figure 1, but for $A^2\Sigma^+(v = 1,\ J$ unspecified) at $\lambda = 430$ nm. Peaks A and B due to fluorescent background of dye laser.

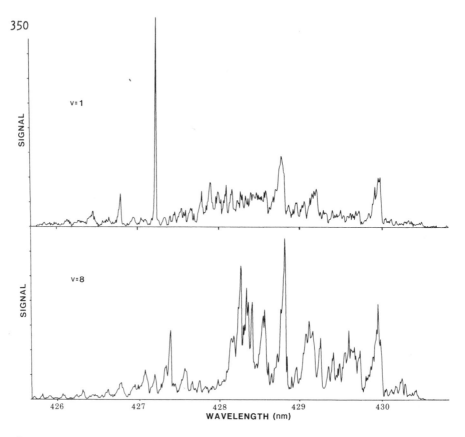

Fig. 3. Excitation functions for the individual vibrational states of the NO^+ ion: $X^1\Sigma^+$, $v = 1$ and 8 respectively. The common structure is due to the rotational branches of the two-photon transition to the $A^2\Sigma^+(v = 1)$ intermediate state.

As to the nature of the pre-ionizing state(s), this should be revealed by the excitation function (fig. 3) for one of the $\Delta v \neq 0$ transitions, e.g. $\Delta v = 8$. Specifically, the $\Delta v = 8$ curve is to be compared to that for $\Delta v = 0$, which represents the population of the $A^2\Sigma^+$ state[10] multiplied by a constant two-photon ionization probability (apart from accidental sharp resonances in the third step). We note that the $\Delta v = 8$ curve contains the same structures as that for $\Delta v = 0$, but a few additional bands or peaks are present.

The latter discrete structure may well be due to superexcited valence states[6], which pre-ionize via vibrational coupling. However, the major contribution apparently stems from an excitation process having a constant transition probability as a function of wavelength, as does the direct ionization. The only possibility is excitation to a vibrational continuum. Such a continuum is available in the form of e.g. the $B^2\Pi$ state, as was pointed out by Giusti-Suzor and by Jungen[4] in the interpretation of the photoionization spectrum around the $v = 4$ threshold. The $B^2\Pi$ is a bound state, of which the dissociation limit lies 0.5 eV below $v = 0$ of the $X^1\Sigma$; it couples electronically with the ionization continuum. The actual Franck-Condon factors for this pre-ionizing transition are a matter for a detailed calculation. However, it has been shown by Mulliken[11], that distributions similar to the ones of figures 1 and 2 may well be ascribed purely to the vibrational overlap of a bound and repulsive curve. Specifically, distributions with a maximum at the lowest electron energy and one or more secondary maxima are predicted.

Finally, the zero-eV peak in figure 1 deserves special attention. At $\lambda = 454.2$ nm, the energy of 4 photons is sufficient only for a $\Delta J = -2$ transition into $v = 6$, $J = 4\frac{1}{2}$ of the ion, while the other rotational transitions, i.e. $-1 < \Delta J < +2$, are energetically not allowed. This may explain the observation that only at this wavelength the highest allowed Δv is not connected with the highest of the $\Delta v \neq 0$ peaks. It is to be noted also, that at no wavelength the direct ionization ($\Delta v = 0$) accounts for more than 50% of the total ionization signal.

This work was subsidized by FOM and ZWO.

REFERENCES

1. E. Miescher, Y.T. Lee and P. Gürtler, J.of Chem.Phys. 68 (1978) 2753.
2. Y. Ono, S.H. Linn, H.F. Prest, C.Y. Ng and E. Miescher, J.of Chem. Phys. 73 (1980) 4855.
3. A. Giusti-Suzor, Invited Paper, XII ICPEAC, Gatlinburg, 1981.
4. Ch. Jungen, Invited Paper, XII ICPEAC, Gatlinburg, 1981.
5. J. Gardner and J.A.R. Samson, J.of Electr.Spectr. 2 (1973) 153.
6. G. Caprace, J. Delwiche, P. Natalis and J. Collin, Chem.Phys. 13 (1976) 43.
7. P. Murray and T. Baer, J.of Mass Spectr.and Ion Physics 30 (1979) 165.
8. J. Miller and R.N. Compton, J.Chem.Phys. 75 (1981) 22.
9. P. Kruit, J. Kimman and M.J. van der Wiel, paper at this conference.
10. R. Bray, R. Hochstrasser and J. Wessel, Chem.Phys.Letters 27 (1974) 167.
11. R. Mulliken, J.of Chem.Phys. 55 (1971) 309.

GENERATION OF NARROW-BAND TUNABLE RADIATION
IN THE 1200 Å SPECTRAL REGION

F. S. Tomkins
Argonne National Laboratory, Argonne, Il. 60439

Rita Mahon
University of Maryland, College Park, Md. 20742

ABSTRACT

Investigation of frequency up-conversion in Hg vapor has been extended further into the VUV, producing continuously tunable narrow-bandwidth radiation from 1220 Å to 1174 Å.

* * *

This paper presents an extension of previously reported work[1,2] on the efficient generation of narrow-band, coherent radiation in the VUV, using a two-photon resonant, four-wave mixing process in mercury vapor.

We have extended the observations to the region from 1221 Å to beyond the first ionization limit at 1187 Å, and show the generation of significant photon fluxes throughout this wavelength range.

The relevant energy levels in atomic mercury are shown in Fig.1. We pump the excited $6s\,7s\ ^1S_0$ level with two ω_1 photons (31964 cm^{-1}), and add a third photon (ω_2) to reach the vicinity of successive $6s\,np\ ^1P$ levels. The process $2\omega_1 + \omega_2$ goes very efficiently near the ^1P levels, and produces tunable radiation in bands approximately 5 Å wide, centered on the ^1P resonances. The earlier work reported generation around the 9p, 10p, and 11p levels, where the generated intensity goes to zero between these bands.

We find, however, in extending the observations to 12p and above, that the generated bands overlap sufficiently that the radiation becomes continuously tunable. As indicated, we also observe the generation of the difference frequency $(2\omega_1 - \omega_2)$ in the range 2000 Å to 2350 Å. The radiation is continuously tunable over this range, but with much reduced conversion efficiency.

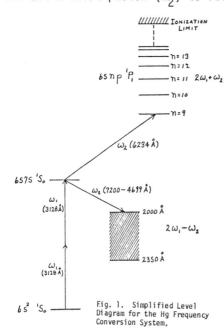

Fig. 1. Simplified Level Diagram for the Hg Frequency Conversion System.

ISSN:0094-243X/82/900352-06$3.00 Copyright 1982 American Institute of Physi

EXPERIMENTAL

The experimental arrangement is shown schematically in Fig. 2.

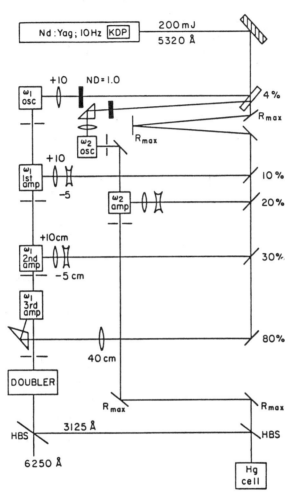

Fig. 2. Block Diagram of Laser
System.

The ω_1 and ω_2 dye laser oscillators are pumped by the second or third harmonic of a Nd:YAG laser, and the outputs amplified as shown. The amplified ω_1 output is doubled, combined with ω_2, and focussed into the mercury cell by a 25 cm quartz-fluorite achromat. For the experiments described here, the mercury cell was operated at a temperature of 140 °C--corresponding to a vapor pressure of approximately 2 Torr.

The ω_1 and ω_2 input energies varied from 0.2 mJ to 10 mJ, depending on the dye and whether ω_2 was pumped by the second or third harmonic of the Nd:YAG laser.

The detection train consisted of a 0.3 meter vacuum monochromator, solar-blind photomultiplier, boxcar integrator, and recorder. Absolute photon fluxes were measured using an NO ionization chamber between the mercury cell and the monochromator.

RESULTS

Table I shows the ω_2 wavelength, the generated wavelength, and the total generated photon fluxes at or near successive np ^{1}P levels. For the 9, 10, and 11p ^{1}P regions, the input energies were approximately 2 mJ and 1 mJ for ω_1 and ω_2 respectively. For the higher ^{1}P levels, coumarin dyes pumped by the third harmonic of the Nd:YAG were required for ω_2. The residual second harmonic was then used to pump ω_1, resulting in considerably lower energy--approximately 0.8 mJ for ω_1 and 0.2 mJ for ω_2.

Table I. Intensities of Generated Radiation.

n	$\lambda(\omega_2)$ (Å)	λ(Generated) (Å)	Photons sec.$^{-1}$
9	6234	1251	
10	5803	1232	$5\times10^{14}-10^{13}$
11	5550	1221	
	5454	1216	3×10^{10}
12	5393	1213	
13	5291	1207	
14	5219	1204	$\sim10^{12}$
15	5165	1201	
16	5129	1199	
.			
.			
.			
Ioniz. Lim.	4395	1188	$\sim10^{11}$
.			
.			
.	4699	1174	

Figure 3 shows the intensity, in arbitrary units, of the radiation generated in the region from 23p ¹P to near the ionization limit, and the enhancement of the generation at the np ¹P levels. The intensity shown where the higher ¹P maxima merge remains essentially constant through and beyond the ionization limit.

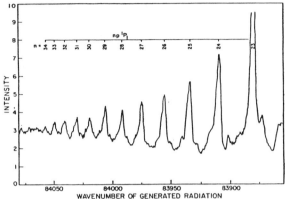

Fig. 3. VUV generation around higher ¹P states.

Figure 4 shows the generation near the 12p ¹P level and the generation at Lyman α, which lies on the short-wavelength wing of the 11p ¹P. The minimum at the indicated position of the 12p ¹P level is due to absorption by the mercury vapor beyond the generating region in the cell.

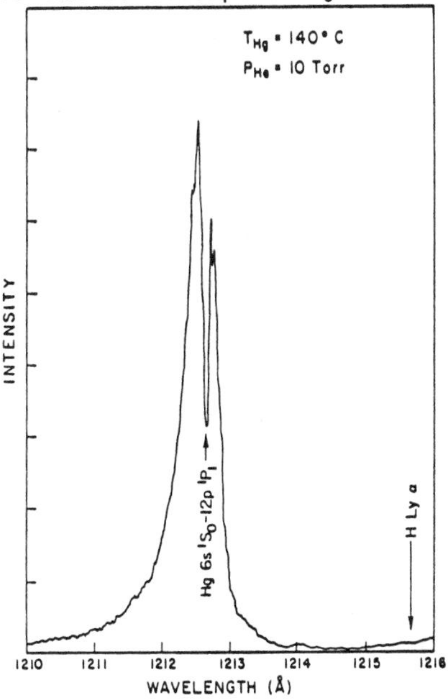

To confirm the generation of tunable radiation in the Lyman α region, we produced atomic hydrogen in a microwave discharge and let it flow through a cell placed between the mercury cell and the monochromator. Figure 5 shows the absorption spectrum produced in a mixture of hydrogen and deuterium as the ω_2 dye laser was scanned through the appropriate region.

Figure 6 is the absorption spectrum of atomic nitrogen, obtained in the same manner, showing the resonance triplet near 1200 Å. In each of the cases the widths of the absorption lines are the room-temperature Doppler widths, 1.03, 0.72, and 0.28 cm⁻¹ for H, D, and N respectively--showing only that the bandwidth of the generated radiation is less than 0.3 cm⁻¹. We expect that it is actually close to 0.02 cm⁻¹, which is the bandwidth of each of the dye

Fig. 4. Generation in the vicinity of 12p ¹P.

lasers.

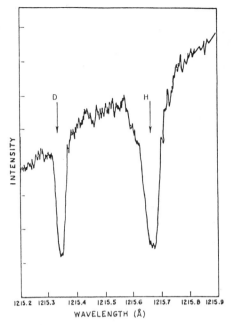

Fig. 5. Absorption spectra of atomic
H and D.

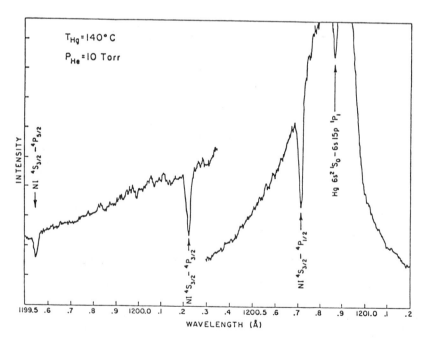

Fig. 6. Absorption spectrum of
atomic nitrogen

Comparing the position of the 15p ^{1}P mercury absorption with that of N I at 1200.714 (a well-determined VUV wavelength standard) we obtain 83273.54 cm^{-1} for the energy of the 15p level. This differs from the Moore table value by 7 cm^{-1}, but is very close to the level classified as 6s12f ^{1}F. It appears that a number of the higher-lying levels in the mercury system have been mis-identified and should be rechecked.

SUMMARY

We show the generation of tunable, coherent, VUV radiation, with a bandwidth of less than 0.3 cm^{-1}, from 1221 Å to 1174 Å, with intensities in every case which are more than adequate for high-resolution spectroscopy, and in more limited regions with intensities high enough for photochemical studies.

REFERENCES

1. F. S. Tomkins and R. Mahon, Optics Lett. 6, 179 (1981).

2. R. Mahon and F. S. Tomkins, to be published in IEEE: J. Quantum Electron., May 1982.

MULTISTEP EXCITATION OF ALKALINE EARTH ATOMS
TO AUTOIONIZING STATES

T. F. Gallagher
Molecular Physics Laboratory
SRI International
Menlo Park, California 94025

ABSTRACT

Using multistep visible laser excitation of autoionizing states, it is possible to excite atoms from the ground state to autoionizing states lying as much as 10,000 cm^{-1} above the ground state. In addition to the fact that the approach allows one to easily reach states lying at high energies, it also considerably simplifies the interpretation of the observed spectra, because typically, one excitation is orders of magnitude stronger than the others. As a result, the Beulter-Fano interference phenomenon associated with the vuv excitation of autoionizing states is absent. We describe the multistep method and its application in a recent study of Ba 6pnd states, which lie $\sim 60,000$ cm^{-1} above the ground state. In addition to illustrating the method, this study shows how the inherent simplicity of the approach allows one to observe the interaction between the $6p_{1/2}nd$ and $6p_{3/2}nd$ series.

INTRODUCTION

One of the most interesting aspects of the spectra of multielectron atoms is the fact that they have autoionizing states, that is, states in which there is enough energy invested in two or more electrons to allow one electron to be removed from the atom. Naturally, the simplest case in which autoionization occurs is He, which has only two electrons, and it is perhaps the most interesting for that reason. In many ways the alkaline earth atoms, with two valence electrons outside a closed shell core, are similar, and the study of such atoms approximates the study of He. Here we shall be concerned mainly with Ba. Let us therefore consider the energy levels of Ba, shown in Fig. 1 which serves not only to give an overall view of the problem but also to give a useful introduction to the multistep excitation scheme.

ISSN:0094-243X/82/900358-10$3.00 Copyright 1982 American Institute of Physic

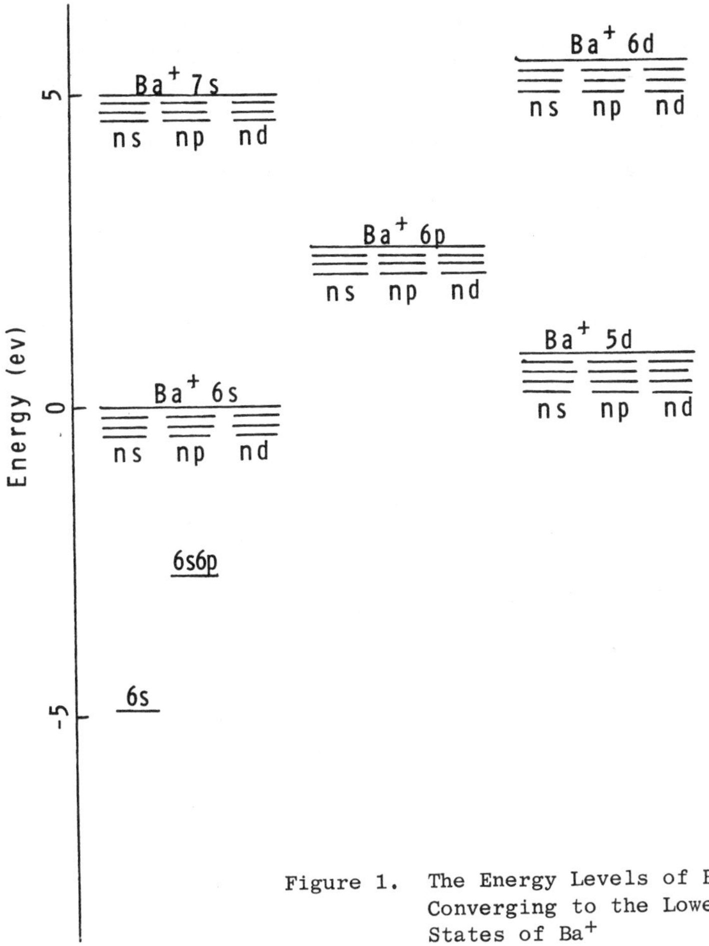

Figure 1. The Energy Levels of Ba
Converging to the Lowest
States of Ba$^+$

One can imagine constructing this energy level diagram in the following way: If we begin with Ba^{++} and add one electron, we have the states of Ba$^+$ [1] shown by the bold lines of Fig. 1. Not surprisingly, the level diagram resembles that of Cs with which Ba$^+$ is isoelectronic. Now let us add the second electron. The lighter lines below each Ba$^+$ state indicate the Rydberg series of the second, or "outer," electron converging to that state (or limit) of Ba$^+$. There are, of course, continua above the Ba$^+$ states. This is an independent electron picture which is a good zeroth approximation for low lying states of Ba$^+$ if the outer electron is in a Rydberg state. Introducing the interaction between the two electrons couples states of Ba converging to higher limits of Ba$^+$ to degenerate continua, and this leads to autoionization.

We note that a critical difference between He and Ba is brought out by Fig. 1. The exicted states of Ba$^+$ are energetically well removed from each other, whereas this is not the case for He$^+$, in which all exicted states of the same n but different ℓ are degenerate. In such a case, there is necessarily a high degree of correlation between the electrons. Crudely, the presence of the outer electron can be thought of as leading to Stark mixing of the He$^+$ states. Experimentally, it is observed that the autoionizing states of He are not represented by simple Rydberg series converging to states of He$^+$, as shown in Fig. 1.[2,3] Thus, in this repsect the autoionizing states of Ba are, in fact, simpler than those of He.

MULTISTEP EXCITATION

The multistep excitation scheme is perhaps unusual in that it is both conceptually and practically simple. The method was first used by Cooke et al.[4] to study the 5pnℓ autoionizing states of strontium, which are analogous to the Ba 6pnℓ states shown in Fig. 1. Since we are considering the excitation of the Ba 6pnd states, we shall use them to illustrate the method.[5] In Fig. 2 we show the relevant energy levels for the excitation of the Ba 6pnd levels.

As shown by Fig. 2, there are three lasers involved in the excitation. The first and second lasers, at fixed wavelengths, excite the Ba atoms from the 6s6s level to the 6s6p level and then to a 6snd level, a bound Rydberg state. The third laser drives the transition 6snd → 6pnd, from the bound 6snd Rydberg state to the autoionizing Rydberg state, which autoionizes yielding an ion and an electron. As the third laser is swept through the 6pnd state, we observe an ionization signal proportional to its spectral density—and essentially nothing else. Why this method works so simply becomes clear if we consider what we are doing to the atom with each laser. This is shown in Fig. 3.

As shown by both Figs. 2 and 3, the first two lasers excite one of the electrons to a Rydberg state. From Fig. 3, though, it is clear that the "outer" Rydberg electron spends most of its time far away from the ionic core near its classical turning point at large orbital radius. The third laser drives the inner electron transition 6s → 6p while leaving the outer electron as a spectator, which makes small adjustments in its orbit as required by any difference in quantum defect of the 6snd and 6pnd states. We note that this is the resonance line of Ba$^+$, $f \sim 1.1$, and it is spread over the autoionization width of the 6snd state, ~ 10 cm^{-1}. This leads to peak optical absorption cross sections for the excitation of the 6snd → 6snd transitions of $\sim 10^{-14}$ cm.^{2} This is to be compared with direct photoionization of the Rydberg state 6snd

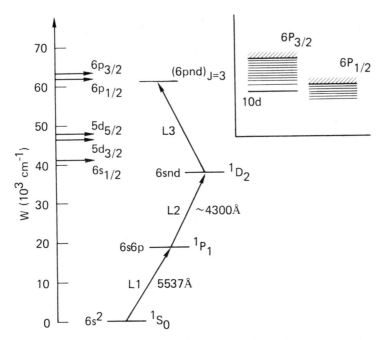

Figure 2. Ba Energy Levels Relevant to the Excitation
of the 6pnd States

(6snd → 6Sεf, for example) by a visible photon. Because of the
gross mismatch between the slow spatial oscillations of the nd
radial wavefunctions and the rapid spatial oscillations of 2 eV
continuum waves, the direct photoionization cross section is
negligible, ~ 10^{-20} cm^2.[6] Since photoionization is effectively
absent, at any photon energy there is only one nonzero transition
amplitude, 6snd → 6pnd. Thus, we observe directly, free of the
usual interference effect,[7] the spectral density of the 6pnd
states. More precisely, we observe the product of the spectral
density and the overlap integral squared between bound and
autoionizing Rydberg states.[8] However, the overlap integral is
quite small for bound and autoionizing states of different n. Thus,
it is a good approximation to assume it is one for states of the
same n and zero otherwise.

The experiments are done in a laser-atomic beam apparatus shown
in Fig. 4.

Three pulsed dye lasers are typically all pumped by the same
pump laser. The dye lasers are modest by any standard, 100 μj
pulse energies, 1 cm^{-1} linewidths, and 5 ns pulse durations. The
three dye laser beams are brought together at a small angle where

362

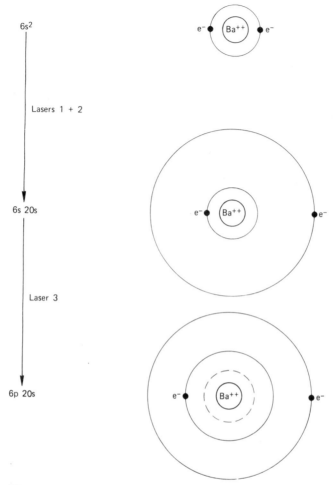

Figure 3.   Physical Picture of the Ba Atom During
Excitation

they cross the atomic beam, and their optical path lengths are
arranged so that the second and third laser pulses come about 5 ns
and 10 ns, respectively, after the first pulse. The atomic beam
apparatus is one 50-cm diameter vacuum chamber at a pressure of
10^{-6} torr. The atomic beam effuses from a resistively heated oven
10 cm from the interaction region and has a density from 10^8 cm^{-3} to
10^{10} cm $^{-3}$ in the interaction region, which consists of a plate and
a grid 1 cm apart. About 1 μs after the laser pulses, a positive
voltage of ∼ 100 V is applied to the lower plate to drive the ions
produced through the grid in the upper plate to the particle
multiplier. The multiplier signal is detected with a gated
integrator and recorded as the "y" signal with an x-y recorder.

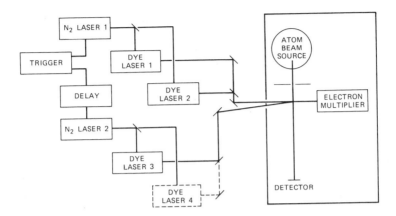

Figure 4. Laser-Atomic Beam Apparatus

In practice then, an experiment consists of setting the first
two lasers to excite a chosen 6snd Rydberg state, tuning the third
laser through the 6pnd state, and recording the ion signal as a
function of wavelength (the x signal for the recorder). An example
is shown in Fig. 5, which is a recording of the 6s16d → $6p_{3/2}$16d
transition. The procedure must, of course, be repeated for each n
state studied.

The preceding discussion outlines the central idea of our
multistep approach. However, it is worth noting some of the less
obvious refinements which have been incorporated into experiments.
The fact that these experiments are (and must be) done in a beam
allows us to use an electron energy analyzer to determine the final
states of the ion and ejected electron subsequent to autoionization.[9] The collision-free environment of the atomic beam also
enables us to study in a controlled fashion the effects of external
fields[10] (and perhaps collisions).

Quite apart from the advantages of the atomic beam are those
which arise from multistep excitation via Rydberg states. First,
controlling the polarization of the three lasers allows us to
separate energetically overlapping autoionizing states using polarization selection rules.[11] Finally, the fact that we use bound
Rydberg states as intermediates enables us to delay the third laser
for ~ 5 μs, during which time we can turn on and off electric
fields to allow the population of otherwise inaccessible high
states.[4]

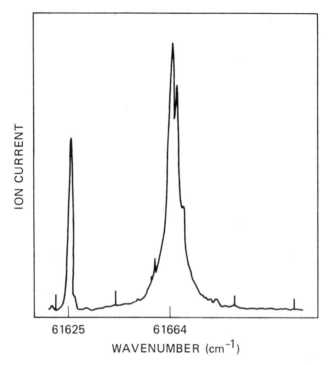

Figure 5. The Excitation Spectrum of the
Ba $6p_{3/2}16d$ State

THE Ba 6pnd STATES

In this study we have restricted ourselves to the $Ba(6p_jnd)_3$ states by circularly polarizing all three lasers in the same sense. Most of the states exhibited excitation spectra similar to the ones shown in Fig. 5. The quantum defects of the Ba $6p_{3/2}nd$ states are very constant, ~ 2.75, and those of the $6p_{1/2}nd$ series are ~ 2.80 except near n = 20.[5] Similarly, the widths W of the series scale as $n*^{-3}$, as indicated by the near constant value of $Wn*^3$, as shown by Fig. 6. From Fig. 6 it is apparentthat for n ~ 20 of the $6p_{1/2}nd$ series, there is a perturbation. The origin of the perturbation can be seen immediately in Fig. 7. The $6p_{3/2}10d$ state is roughly degenerate with the $6p_{1/2}20d$ state, and the configuration, or channel interaction, leads to the evident perturbation of the $6p_{1/2}nd$ series.

Nowhere is the interaction of the series more evident than in the excitation from the 6s10d state to the $6p_{3/2}10d$ state, which is shown in Fig. 7.

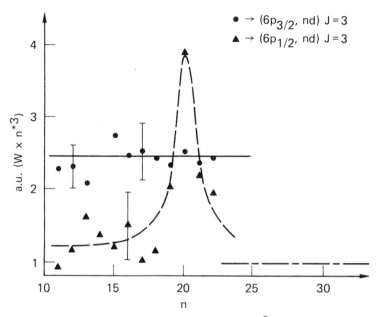

Figure 6. The Normalized Widths, $Wn*^3$, of the 6pnd States

As shown by Fig. 8, the spectrum is a wide envelope of width ~ 100 cm^{-1}, with apparent interferences ~ 20 cm^{-1} apart. The 100 cm^{-1} envelope corresponds to the width of the $6p_{3/2}10d$ state, and the apparent interference structures arise from the $6p_{1/2}nd$ states with $n \sim 20$.

While the structures corresponding to the $6p_{1/2}nd$ states appear to be Beutler-Fano interference structures, in reality they are not. This can be understood by simply writing down the three possible excitations:

$$(1) \qquad 6s10d \rightarrow 6p_{3/2}10d \qquad\qquad (1)$$

$$(2) \qquad 6s10d \rightarrow 6p_{1/2}nd, \; n \sim 20 \qquad (2)$$

$$(3) \qquad 6s10d \rightarrow 6s\varepsilon f. \qquad\qquad (3)$$

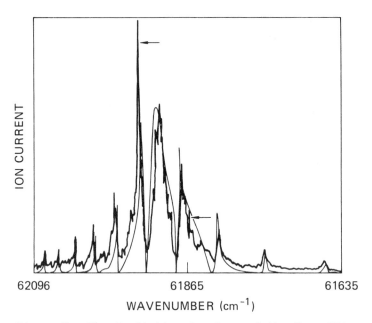

Figure 7. The Excitation Spectrum of the $6p_{3/2}10d$ State

From our previous discussion we know that excitation to the continuum (3) is negligible. Similarly, the overlap integral $< 10d \mid n^*d >$ for n^* (a continuous variable) ~ 20 is $\ll 1$, so that (2) is also negligible. Thus, it is apparent that only (1) gives an appreciable transition amplitude. Therefore, the observed spectrum shown in Fig. 7 cannot arise from an interference between transition amplitudes. Consequently, it must represent the spectral density of the Ba $6p_{3/2}10d$ state. In fact, the light line of Fig. 7 is the spectrum calculated using quantum defect theory[12,13], assuming that the excitation spectrum arises from the spectral density of the $6p_{3/2}10d$ state. We note that a similar spectrum has been observed by Cooke et al.[14], who have made an even more successful fit using a similar approach with two continua. In principle, the same parameters which lead to the calculated spectrum of Fig. 7 should also give good values for the widths (Fig. 6) and the quantum defects. We find this to be the case, indicating consistency in the treatment.

CONCLUSIONS

From the discussion of the approach and its application to the 6p$_j$nd series, it is apparent that multistep excitation is a powerful approach the simplicity of which enables one to unravel potentially very complicated problems, such as the one illustrated here.

ACKNOWLEDGMENTS

The work described here is the fruit of the labor of many, F. Gounand, W. E. Cooke, K. A. Safinya, W. Sandner, R. Kachru, and N. H. Tran, and the ideas presented are those which have survived the ordeal of lively discussion. This work is supported by the National Science Foundation under Grant PHY 80-07041.

REFERENCES

1. C. E. Moore, Atomic Energy Levels, NBS Circular No. 467 (U.S. GPO, Washington, D.C., 1949).
2. R. P. Madden and K. Codling, Phys. Rev. Lett. 10, 516 (1963).
3. J. W. Cooper, U. Fano, and F. Prats, Phys. Rev. Lett., 10, 518 (1963).
4. W. E. Cooke, T. F. Gallagher, S. A. Edelstein, and R. M. Hill, Phys. Rev. Lett. 40, 178 (1978).
5. F. Gounand, T. F. Gallagher, W. Sandner, K. A. Safinya, and R. Kachru (to be published).
6. D. C. Lorents, D. J. Eckstrom, and D. L. Huestis, SRI Report MP 73-2 (unpublished).
7. U. Fano, Phys. Rev. 124, 1866 (1961).
8. N. H. Tran, R. Kachru, and T. F. Gallagher (to be published).
9. W. Sandner, R. Kachru, K. A. Safinya, W. E. Cooke, and T. F. Gallagher (to be published).
10. K. A. Safinya, J. F. Delpech, and T. F. Gallagher, Phys. Rev. A 22, 1062 (1980).
11. W. E. Cooke and T. F. Gallagher, Phys. Rev. Lett. 41, 1648 (1978).
12. U. Fano, Phys. Rev. A 2, 353 (1970).
13. C. M. Lee and K. T. Lu, Phys. Rev. A 8, 1241 (1973).
14. W. E. Cooke, C. L. Cromer, and S. A. Bhatti (to be published).

VACUUM ULTRAVIOLET SPECTROSCOPY OF MOLECULES USING THIRD-HARMONIC GENERATION IN RARE GASES*

John C. Miller and R. N. Compton
Oak Ridge National Laboratory, Oak Ridge, Tennessee 37830

C. D. Cooper
University of Georgia, Athens, Georgia 30601

ABSTRACT

A variety of spectroscopic studies in the vacuum ultraviolet region (100-200 nm) has been performed using tunable, narrow bandwidth (~0.05 Å) light produced by third-harmonic generation in rare gases. These experiments include one-photon absorption and photoionization spectroscopy and photoionization threshold determinations. Two-photon ionization studies of nitric oxide, using the vacuum ultraviolet light to probe Rydberg states and the fundamental light to ionize are also presented. These studies demonstrate the simplicity and utility of a laser-based, tunable high-resolution source for vacuum ultraviolet spectroscopy. As a prelude to these studies we will briefly discuss recent results on third harmonic generation and multiphoton ionization in xenon. Resonant and nonresonant multiphoton ionization involving one vacuum ultraviolet photon and one or two blue photons is shown to be an important loss mechanism for third harmonic generation.

INTRODUCTION

Tunable vacuum ultraviolet (VUV) radiation is an essential element in studies of absorption and photoionization of most atoms and molecules. This region begins at 200 nm (50,000 cm^{-1}) and extends to about 30 nm where soft x rays begin. The region from 200-100 nm (6.20-12.40 eV) is called the VUV region, while the range below 100 nm is usually called the extreme ultraviolet (XUV) region.

The 200-100 nm spectral region is especially important as the ionization potential (IP) of most atoms lies above 6.20 eV, and the IP of almost all organic and inorganic molecules lie between 6.2 and 12.4 eV. Although the spectroscopy of the lowest-lying states of most atoms and molecules is studied in the visible and near UV regions, most Rydberg states lie in the VUV. Due to the difficulty of obtaining and using VUV light, photoionization and Rydberg absorption studies have not been as frequent or as detailed as corresponding studies in the visible and near UV.

The traditional source of tunable photons in the VUV region has been the rare gas or hydrogen discharge lamp coupled with a vacuum monochromator. More recently, synchrotron sources have provided

*Research sponsored by the Office of Health and Environmental Research, U.S. Department of Energy under contract No. W-7405-eng-26 with the Union Carbide Corporation.

relatively more intense, continuum VUV light which also requires a
monochromator for high resolution. Currently, there is intense
interest in developing laser-based sources of VUV light. The aim is
to extend to this region the unique laser properties such as high
intensity, very narrow bandwidth, spatial and temporal coherence,
and short pulse times. These factors along with the extensive
tunability of dye lasers have already revolutionized the fields of
atomic and molecular structure, spectroscopy, and photophysics.

Extension of tunable coherent light sources into the VUV,
however, poses considerable problems as frequency conversion crystals
become opaque in this region. For third harmonic generation (THG)
into the VUV, the rare gases have most often been used as the nonlinear medium. These studies of THG in krypton and xenon have
characterized the properties of the nonlinear process, mostly with
the aim of producing a bright source at the Lyman α and β lines for
hydrogen detection schemes.[1] Tunable THG has been demonstrated in
xenon in the ranges of 140.3-146.9 nm and 125.0-129.6 nm and in
krypton of 120.3-123.6 nm and 110.0-116.5 nm with conversion efficiencies of 10^{-5}-10^{-6} and fluxes in the range of 10^{7}-10^{10} photons/
pulse. The output power, linewidth, and tuning range are, of course,
dependent on the properties of the laser supplying the fundamental
frequencies and whether or not phase matching is attempted.

In addition to THG, however, tight focusing of high powered
lasers into gases leads to other nonlinear processes. In particular,
multiphoton ionization (MPI) occurs when the pump laser wavelength
is such that three or four photons are resonant with atomic levels.
This MPI, which is interesting in its own right, is also an important
loss mechanism for THG in rare gases. Furthermore, the presence of
VUV light from THG has a marked effect on the MPI spectra. Our
experiments, monitoring both THG and MPI as a function of wavelength
and pressure, are discussed following a description of the experimental apparatus. Then a number of applications of the VUV for
molecular spectroscopy will be outlined.

EXPERIMENTAL

The apparatus, shown in Fig. 1, is a further extension of a
cell initially built for MPI studies[2] and previously modified for
THG experiments.[3] It consists of three separately evacuable
chambers, each of which functions as a proportional counter.
Electrons produced in such a counter are accelerated toward an
electrode maintained at positive bias voltages (10-500 V) and under
go ionizing collisions with other gas molecules causing an electron
avalanche. The amplification produced is proportional to the ratio
of electric field to pressure (E/P). The gain is finally limited by
dielectric breakdown. The first chamber (I) consists of a stainless
steel Varian six-way cross connected to an ion pump and capable of
baseline vacuums of 10^{-8} torr. The beam of a N_2 laser (Molectron
UV-24) pumped dye laser (Molectron DL400) is focused in chamber I by
a 3.8 cm focal length lens to a spot $\leqslant$20 μm, giving a power density

370

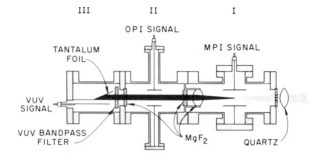

Fig. 1. Schematic diagram of the apparatus used for studies of one-photon absorption and ionization and of two-photon ionization using third-harmonic generation in rare gases. Chamber I is the frequency tripling cell, chamber II is the ionization or absorption cell, and chamber III is a vacuum ultraviolet light detector. Each chamber functions as a proportional counter for detecting electrons. See text for details. Chamber I also functions as a cell for conventional multiphoton ionization studies.

of the order of 5×10^9 W/cm^2. Krypton and xenon gas (Matheson, 99.995 and 99.9%, respectively) were used as the nonlinear medium and any electrons produced by MPI in the tripling gas were detected with the biased flat-plate electrode. The VUV light, produced by THG in the focal spot, is collimated with a MgF$_2$ lens (focal length = 4 cm) and passes through a MgF$_2$ window into the second chamber (II). Again, any electrons resulting from one-photon ionization (OPI) or MPI are amplified in the sample gas and detected at the electrode. The VUV beam exits chamber II through an MgF$_2$ window and passes through a dielectric VUV bandpass filter (Acton Research) to remove the blue pump light. This filter could also be placed between chambers I and II when necessary. Chamber III is a VUV photon detector consisting of a 3.8 cm Varian nipple (8.5 cm long) containing a tantalum foil and a single flat-plate electrode. Photons whose energy exceeds the work function of tantalum (~4.1 eV) eject electrons which are detected in the proportional counter. Signals from charge-sensitive preamplifiers connected to any two of the three chambers were averaged in a dual channel boxcar integrator (Princeton Applied Research, No. 162/165) and displayed on a dual-channel X-Y recorder and oscilloscope as a function of wavelength.

The absolute intensity of the third harmonic light is on the order of 10^7-10^8 photons/pulse depending on the wavelength. The bandwidth of the THG will be at most one-third the bandwidth of the input laser (nominally ~0.1 Å). An additional factor of $\sqrt{3}$ reduction in width results from the I^3 dependence of the THG if a Gaussian pump beam is assumed. The expected bandwidth is thus 0.02 Å at 1470 Å (~0.09 cm^{-1}). Use of an etalon would reduce the laser band-width to 0.01 Å and that of the THG to 0.002 Å (0.09 cm^{-1}) at 1470 Å. This is on the order of the Doppler broadening of light molecules.

THIRD HARMONIC GENERATION AND MULTIPHOTON IONIZATION

For the experiments described below, chamber II was removed from the apparatus (Fig. 1) and chamber III attached directly to chamber I. Aron and Johnson[4] reported the first resonantly enhanced multiphoton ionization (REMPI) spectrum of xenon at pressures ranging from 1 to 100 torr. They noted the complete absence of any ionization signal when tuning through the 6s[3/2]°,J=1 three-photon-allowed intermediate state. At low pressure (10^{-6} torr), we found that MPI through the 6s state was very intense and dominated the spectrum.[5] In later publications,[3,6] we further showed that for a laser beam focused by a 3.8 cm lens to a power density of ~5 × 10^9 W/cm^2, the ionization signal produced via the 6s state rapidly disappeared at pressures above ~0.3 torr. Third-harmonic light generated in the direction of the pump beam was detected as the ionization disappeared. The excitation lineshape of both the MPI and THG signals were similar and were found to shift and broaden asymmetrically to shorter wavelength as the pressure was increased. Although the intensities of the ionization and third-harmonic light are entirely relative, the dramatic decrease of ionization with increase in THG suggested that the two processes were interfering or in "competition."[3,6] The THG light is expected in any negatively dispersing medium and the excitation lineshape is expected to shift to shorter wavelengths as the pressure increases due to phase matching considerations. For wavelengths far off resonance, the normal theory for THG involving the nonlinear susceptibility (X_3) applies. The broadening and shifting of the MPI lineshape along with the quenching of the MPI signal are not easily rationalized in terms of our present understanding of such phenomena. In a recent theoretical study of these results, Payne et al.[7] found it necessary to include terms involving the coherent, collective excitation of xenon atoms. These terms result in greatly enhanced THG near an odd-photon resonance with concomitant, very rapid depletion of the resonant population.

Multiphoton ionization and third-harmonic generation near the 6s'[1/2],J=1 resonance in xenon[6] show similar behavior to that of the 6s[3/2],J=1 state discussed above. Differences in the two sets of data are shown for pressures from 0.5 to 25 torr in Fig. 2 and at higher pressures (50 to 310 torr) in Fig. 3. The ionization signal is seen to broaden and shift at low pressures similar to that near the 6s[3/2],J=1 region. The ionization signal does not disappear however. Also, molecular effects (Xe_2) are seen in the high pressure data (Fig. 3). It is important to note that only four photons are required to ionize xenon in this wavelength region and this may account for the fact that the nonresonant ionization signal does not disappear.

The ionization signal which is observed to broaden and shift in both the wavelength regions near the 6s and 6s' resonant intermediate states can be ascribed to MPI using one third harmonic photon plus two- or one-laser photons, respectively. The fact that both the MPI and THG signal have identical lineshapes at low pressures strongly

ORNL–DWG 81-18464

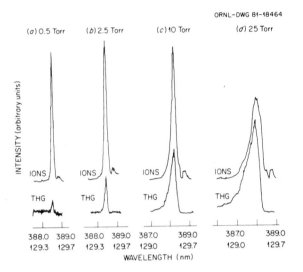

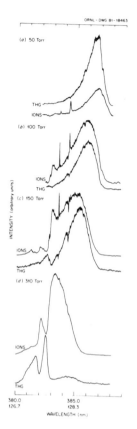

ORNL-DWG 81-18463

Fig. 2. (Top Left) The multiphoton spectra and third harmonic generation lineshapes near the 6s' resonance in xenon recorded at various pressures from 0.5 to 25 torr. The upper wavelength scale corresponds to the laser and the lower scale to the third harmonic generation (one-third of the laser wavelength).

Fig. 3 (Top Right) The multiphoton ionization spectra and third harmonic generation lineshapes near the 6s' resonance in xenon recorded at various pressures from 50 to 310 torr. The upper wavelength scale corresponds to the laser and the lower scale to the third harmonic generation (one-third of the laser wavelength).

supports this hypothesis. However, it does not explain the absence of signal at the resonance line position. This would require some other quenching mechanism.

It is also evident from Fig. 3 (also Ref. 3, Fig. 3 and Ref. 4, Fig. 10) that MPI is an important loss mechanism for THG and ultimately limits its scaling with higher powered lasers. Sharp peaks in the MPI spectrum, which also appear as sharp dips in the VUV, occur at four-photon atomic resonances. At pressures where the VUV occurs at these wavelengths, absorption of one VUV plus one blue photon can also populate these states. One additional blue photon then leads to ionization. Similar absorptions into molecular states are important at higher pressures. Finally, near the 6s', nonresonant two-photon ionization (one VUV plus one blue) limits the THG.

At the present time, new experiments using the higher power of an excimer laser pumped dye laser are underway to test the above models of MPI at high power. Such experiments will include unfocused laser MPI, polarization experiments, focal length dependences, and two-color laser experiments.

Finally, very recent new experiments by Glownia and Sander[8] provide additional support for the competition of THG and MPI. These authors directed counter propogating beams of circularly polarized light into a cell similar to that of Fig. 1. Under these conditions MPI, resonant with the 6s and 6s', is possible using photons from both beams but THG cannot occur due to phase matching considerations. In the absence of THG, the 6s and 6s' MPI spectra are strong, sharp, and unshifted at all pressures.

VACUUM ULTRAVIOLET SPECTROSCOPY

In contrast to the many studies involving *generation* of THG, there have been only a few involving the *use* of tunable VUV as a spectroscopic tool. Wallenstein et. al.[9] have traced the Lyman-α absorption of H and D atoms and a fluorescence excitation band of CO using THG in krypton and argon generated with a Nd:Yag pumped dye laser system. Egger et al.[10] and Rothschild et al.[11,12] in an elegant series of experiments have performed very high resolution studies of H_2 and D_2 Rydberg states in the XUV near 83 nm. These experiments used xenon to triple the frequency of a high resolution KrF laser system. Zacharias et al.[13,14] have studied resonant two-photon ionization of H and D atoms and of CO in a mass spectrometer where the VUV photon was produced by THG in krypton using a Nd:Yag pumped dye laser. The ionizing photon was one of the harmonics of the pump laser.

The third harmonic light generated in the above described experiments is sufficiently intense to be useful in spectroscopic studies of gaseous species in the near ultraviolet region (100-200 nm). Table 1 shows the useful wavelength regions for THG in rare gases. The gaps can probably be filled by using molecular gases as the tripling media. The apparatus, as shown in Fig. 1, was used for these studies. Further details may be found in a forthcoming paper.[15]

Table 1. VUV Tuning Range for Nitrogen Laser Pumped Dye Laser Using Xenon and Krypton Gas as the Tripling Gas

Rare Gas	Laser Tuning Range, nm	VUV Tuning Range		
		nm	cm^{-1}	eV
Xe 6s	420.8-440.6	140.3-146.9	68063-71275	8.44-8.84
Xe 6s'	375.0-388.8	125.0-129.6	77160-88000	9.57-9.92
Kr 5s	360.9-370.5	120.3-123.5	80972-83126	10.04-10.31
Kr 5s'[a]	330.0-349.5	110.0-116.5	85837-90909	10.64-11.27

[a]Accessible only with excimer pumped dye laser or doubled red dye laser.

374

By monitoring the THG intensity in chamber III with a sample gas in chamber II, a VUV absorption spectrum can be obtained. Figure 4 shows an absorption spectrum of the $A^1\pi$ (v = 3) ← $X^1\Sigma^+$ (v = 0) band of CO which is a red degraded band whose head appears at 144.73 nm. The upper wavelength scale is the laser wavelength in air while the lower number is the VUV wavelength corrected for vacuum. This band is partially overlapped by the $e^3\Sigma^-$ (v = 5) ← $X^1\Sigma^+$ (v = 0) band at lower energies. Unfortunately, the spectrum does not show the full resolution of the system due to the mostly unresolved, closely spaced P,Q,R, triplet in each band. The Doppler width for CO at room temperature is 0.0034 Å. Thus, an etalon in the present laser would produce sub-Doppler resolution for CO as was observed by Hilbig and Wallenstein.[1] Absorption spectra were also taken for I_2 and NO in this region which, while complicated, were similar to published spectra. The NO absorption results are shown in Fig. 5 and will be discussed below.

Direct photoionization spectra of benzene, CH_3I, NO, and I_2 were studied in the Xe 6s' and Kr 5s wavelength regions. These spectra were easily obtained and were identical to previously reported spectra. Ionization thresholds for a number of molecules have been studied from which accurate IPs have been determined [e.g., iodobenzene (IP = $8.75^{+0.01}_{-0.04}$ eV) o-xylene (IP = $8.54^{+0.01}_{-0.04}$ ev)].

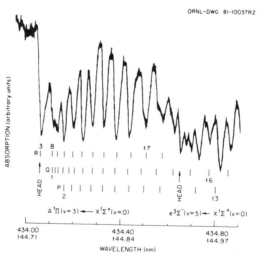

ORNL-DWG 81-10037R2

Fig. 4. A portion of the vacuum ultraviolet absorption spectrum of CO recorded using third harmonic generation in xenon. The upper wavelength scale is the laser wavelength in air and the lower scale is the vacuum ultraviolet wavelength corrected for the refractive index of air.

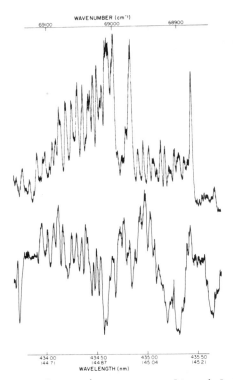

WAVENUMBER (cm⁻¹)

Fig. 5. The two-photon (one vacuum ultraviolet plus one blue) ionization spectrum (trace) and one-photon absorption spectrum (lower) of ^{14}NO in the region of the $F^2\Delta$ (v = 3) ← $X^2\pi_{3/2,1/2}$ transition. The upper wavelength scale and the wave number scale have been corrected for the refractive index of air.

Third harmonic generation results quite naturally in a colinear beam of VUV and blue light. Therefore, by removing the VUV bandpass filter separating chambers I and II (i.e., same configuration as Fig. 1), it is possible to study two-photon, two-color ionization experiments. Figure 5 displays the two-photon (one VUV plus one blue photon) ionization spectrum (upper trace) and one-photon VUV absorption spectrum (lower trace) of NO in the region of the $F^2\Delta$ (v = 3) ← $X^2\pi$ transition. The transitions involved are shown in Fig. 6. The two leading peaks of each bandhead represent the transitions from fine structure $^2\pi_{3/2,1/2}$ splitting of the ground state. An indication of the spectral simplification provided by this technique can be seen by comparing the sharp, rotationally resolved two-photon ionization spectrum with the absorption spectrum. Although there is one-to-one correspondence in some areas of the spectrum, in most regions the absorption spectrum is vastly more complicated.

None of the $^2\Sigma$ and $^2\pi$ Rydberg states known to be in this wavelength region are observed nor are any valence states. Valence

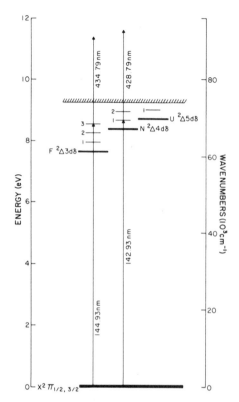

Fig. 6. The partial energy level diagram of NO showing the F$^2\Delta$ state and the two-photon transition responsible for the ioniza-tion signal shown in Fig. 5.

states are observed in MPI spectra of NO only when they are mixed with nearby Rydberg states.[16] The absence of the $^2\Sigma$ and $^2\pi$ states is easily explained. All states in this energy region are crossed by a repulsive $^2\Sigma^+$ state which dissociates to ground state atoms. Due to the $\Delta\Lambda$ = 0, ±1 selection rule for perturbation and predis-sociation, the $^2\Sigma$ and $^2\pi$ states can predissociate and thus have a shortened lifetime. In these cases the predissociation is faster than the ionization step and these states are not seen in our experiments. The $^2\Delta$ states, however, cannot predissociate via the $^2\Sigma^+$ state and thus have lifetimes long enough for ionization to occur. Higher vibrational levels of the $^2\Delta$ states do, however, show some signs of predissociation. It should be noted that these same kinetic considerations occur for four-photon MPI experiments resonant with these Δ and π states, yet they are observed in the studies of Zakheim and Johnson.[16] However, the detection of four-photon MPI requires a focused laser due to the higher order process and the one-photon ionization steps occurs with much higher power density (this step is probably saturated). Thus ionization *can* compete with predissociation. In contrast, in the present experi-ments, only an unfocused beam is used and predissociation dominates.

Clearly, increased laser power density obtained either by a higher powered dye laser or by focusing the emerging THG and blue beam would result in more states being observed. Thus, besides simplifying overlapping bands, an idea of relative lifetimes of states can be obtained. If the ionizing beam could be delayed in time relative to the VUV beam, exact lifetimes could be obtained.

CONCLUSIONS

In this paper, we have demonstrated the utility of performing spectroscopic studies in the VUV region with a laser-based light source. Furthermore, the simplicity of bulk gas experiments using THG VUV has been amply shown. The apparatus can be extremely simple, based on proportional counters and photon detectors and the N_2 laser pumped dye laser used here is already a relatively common tool in spectroscopy laboratories. We are currently extending this technique to a molecular beam apparatus[2,5,17] which will allow for very versatile, high resolution photoionization for mass spectroscopy and photoelectron spectroscopy.

REFERENCES

1. See, for instance, R. Hilbig and R. Wallenstein, IEEE J. Quantum Electron QE17, 1566 (1981); and references therein.
2. C. D. Cooper, A. D. Williamson, J. C. Miller, and R. N. Compton, J. Chem. Phys. 73, 1527 (1980).
3. J. C. Miller, R. N. Compton, M. G. Payne, and W. R. Garrett, Phys. Rev. Lett. 45, 114 (1980).
4. K. Aron and P. M. Johnson, J. Chem. Phys. 67, 5099 (1977).
5. R. N. Compton, J. C. Miller, A. E. Carter, and P. Kruit, Chem. Phys. Lett. 71, 87 (1980).
6. J. C. Miller and R. N. Compton, Phys. Rev. A 25, 2056 (1982).
7. M. G. Payne, W. R. Garrett, and H. C. Baker, Chem. Phys. Lett. 75, 468 (1980).
8. J. H. Glownia and R. K. Sander, Private Communication.
9. R. Wallenstein, Opt. Comm. 33, 119 (1980).
10. H. Egger, R. T. Hawkins, J. Bokor, H. Pummer, M. Rothschild, and C. K. Rhodes, Opt. Lett. 5, 282 (1980).
11. M. Rothschild, H. Egger, R. T. Hawkins, H. Pummer, and C. K. Rhodes, Chem. Phys. Lett. 72, 404 (1980).
12. M. Rothschild, H. Egger, R. T. Hawkins, J. Bokor, H. Pummer, and C. K. Rhodes, Phys. Rev. A 23, 206 (1981).
13. H. Zacharias, H. Rottke, and K. W. Welge, Opt. Comm. 35, 185 (1980).
14. H. Zacharias, H. Rottke, J. Danon, and K. W. Welge, Opt. Comm. 37, 15 (1981).
15. J. C. Miller, R. N. Compton, and C. D. Cooper, J. Chem. Phys. 76, xxxx (1982).
16. D. Zakheim and P. M. Johnson, J. Chem. Phys. 68, 3644 (1978).
17. J. C. Miller and R. N. Compton, J. Chem. Phys. 75, 22 (1981); J. Chem. Phys. 75, 2020 (1981).

LASER-INDUCED CONTINUUM STRUCTURE IN MULTIPHOTON IONISATION SPECTROSCOPY

P.E. Coleman and P.L. Knight

Blackett Laboratory, Imperial College, London SW7 2BZ, U.K.

and

K. Burnett

JILA and Physics Department, University of Colorado, Boulder, Co 80309

ABSTRACT

We report non-perturbative results on multiphoton transition and up-conversion enhancement by dressing photoionisation continua using radiative interaction analogous to configuration interaction in autoionisation.

INTRODUCTION

Multiphoton processes with significant continuum contributions are enhanced by exploiting autoionising resonances embedded in the continuum by configuration interaction [1]. Continuum resonances can be induced by radiative coupling from an initially empty discrete state; such "pseudo-autoionising" states [2] can be shifted and broadened at will by altering the frequency/intensity of the inducing laser field.

Transitions from state a to the dressed continuum F are described by matrix elements

$$V_{aF} = V_{aE} \frac{(q_a + \varepsilon)}{(\varepsilon - i)} \tag{1}$$

where q_a is the Fano asymmetry parameter and ε is the detuning of E from the embedded state in units of the width of the induced resonance (I.R.). For an I.R. q_a is the ratio of the real part of the (two-photon) Rabi frequency from a to the embedded state via the continuum to the imaginary part of this frequency. Recent measurements [3] report $q \sim 7$ for Cs transitions.

We have investigated the use of laser-induced resonances in several multiphoton processes [4]. Figure 1 is a level scheme for two-photon resonant three-photon ionisation where V embeds the empty discrete state k in the continuum E. In lowest order the matrix elements in the Golden Rule expression involving E are replaced by those involving dressed continuum states F as in eq. (1). The enhancement near resonance $\sim q_2^2$, with a zero minimum at $\varepsilon = -q_2$

ISSN:0094-243X/82/900378-06$3.00 Copyright 1982 American Institute of Physics

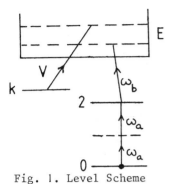

Fig. 1. Level Scheme

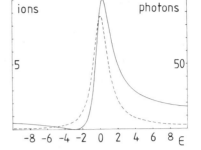

Fig. 2. Enhancement of ion rate and up-conversion rate for $q_2 = 2.5$, $q_o = 3$.

(fig. 2). Also plotted is the enhancement in the up-conversion process $2\omega_a + \omega_b \rightarrow \omega_{vuv}$ calculated from the lowest order $|\chi^{(3)}|^2$, with continuum-enhanced matrix elements given by eq. (1). By enhancement we mean the ratio of rates with and without I.R. embedding. The up-conversion has peak enhancement $\sim(q_2^2 \, q_o^2)$ since matrix elements from 0 and from 2 to F are enhanced.

INTENSE FIELD RESONANCES

Intense-field excitation of structured continuum resonances will lead to power-broadening of Fano-like lineshapes, Rabi oscillations and AC-Stark splittings [5,6]. We describe these [4] by diagonalising the entire Hamiltonian including dressing V and

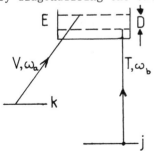

Fig. 3. Level Scheme

and probe T interactions (Fig. 3.) with laser fields of frequency ω_a, ω_b exciting continuum states E detuned D as shown. State k is initially empty and embedded into E. The diagonalized states e are obtained using two-channel Fano theory [7], and are stationary superpositions of j, k, E. The wavefunction at time t, $|\phi(t)\rangle$ is expanded in e with time-independent coefficients determined by normalisation and the initial condition of starting in j. The probability of being in the

undressed state ℓ is $P_\ell(t) = |\langle \ell|\phi(t)\rangle|^2$, $\ell = k, j, E$. We find that Rabi oscillations between j and the structured continuum can occur if the Rabi frequency Ω exceeds ionization widths. Ω depends on q: $\Omega = E(+) - E(-)$ where $E(\pm)$ are roots of

$$(e - E_k' + \frac{i}{2}\gamma_k)(e - E_j' + \frac{i}{2}\gamma_j) - (\gamma_j\gamma_k/4)(q - i)^2 = 0 \quad (2)$$

Here E_k', E_j' are AC-Stark shifted detunings of the k-E and j-E excitation, with $E_k' - E_j' = D$ the dynamic detuning including light shifts; γ_ℓ are the one-photon ionization widths of $\ell(= k, j)$ to E; q is

$$q = P \int_0^\infty dE \frac{T_{jE}V_{Ek}}{e - E} \Big/ \frac{1}{2}(\gamma_k\gamma_j)^{\frac{1}{2}} \quad (3)$$

The roots E(±) are complex and describe the energies and widths of the strong-coupled superposition of j and k (the Autler-Townes dressed states). In fig. 4 we plot Re E(±) versus detuning D, indicating the width ImE(±) by shading. For resonant excitation, (D = 0) the imaginary part of E(-) vanishes when the embedding equals the probe ionization interaction $\gamma_k = \gamma_j$, implying that the lifetime of this Autler-Townes superposition state goes to infinity. Any population in this superposition is trapped.

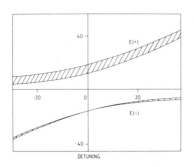

Fig. 4. Autler-Townes energies for $\gamma_j = \gamma_k = 2\pi$, q = 5.

In fig. 5, we plot populations in j and k and ion population $P_I = 1 - P_j - P_k$ as a function of time for q = 5, D = 0, $\gamma_j = \gamma_k = 2\pi$. Population oscillates between j and k due to combined V and T, damps because of ionization losses but half the population is trapped as t→∞ in a coherent superposition of j and k immune to photoionisation [8]. For unequal coupling strengths the Rabi oscillations persist in P_j and P_k and also in a "stair case" of plateaus in P_I; as t→∞, $P_I(\infty) = 1$ and all population is ionized.

The ion population $P_I(t)$ as a function of the detuning D of the probe field from the structured continuum exhibits the asymmetric Fano profile shown in fig. 2 for weak excitation ($\gamma_j << \gamma_k$). As the probe T increases the Fano lineshape power-broadens and distorts. In the wings P_I becomes unity (complete ionization) after a short time but an interference minimum persists which moves in time (fig. 6)

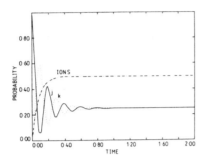

Fig. 5. time- dependence of populations

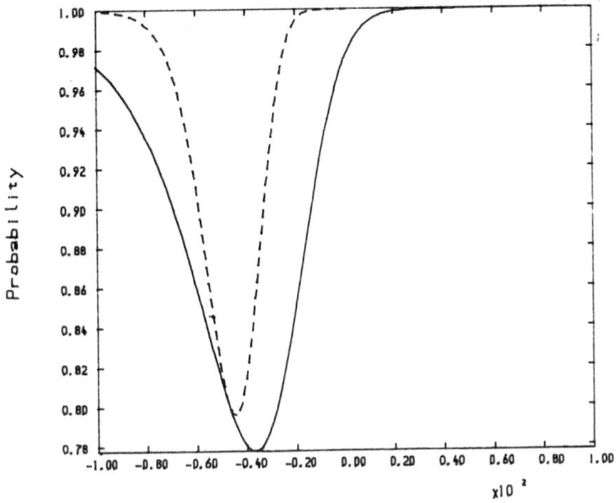

Fig. 6. Ion probability versus detuning for q = 5, $\gamma_j = 4\gamma_k$
$\gamma_k = 2\pi$ for t = 1 (solid) and t = 5 (dashed) in units of γ_k

The photoelectron spectra $P_E(t)$ directly exhibit the two quasi-energies $E(\pm)$. For intense fields such that many Rabi oscillations occur in an ionization lifetime, an AC-Stark split doublet is produced with energies $ReE(\pm)$ and width $ImE(\pm)$. Again multi-channel interference produces asymmetries; for D = 0 and $\gamma_j = \gamma_k$, trapping manifests itself in the disappearance of one of the components of the doublet. Such split structure could be observed by probe

382

absorption to a further high-lying resonance [5] or to a low-lying discrete state.

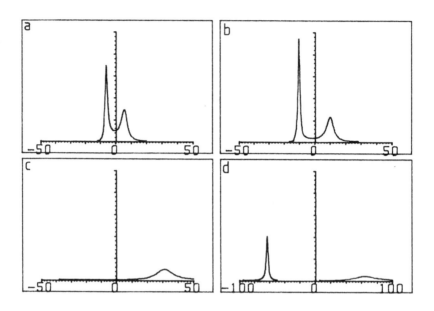

Fig. 7. Photoelectron energy spectrum for $q = 5$, $\gamma_k = 2\pi$, $D = 0$ and (a) $\gamma_j/2\pi = 0.04$, (b) $\gamma_j/2\pi = 1/9$, (c) $\gamma_j/2\pi = 1$, (d) $\gamma_j/2\pi = 4$.

In fig. 7 we show the long-time photoelectron energy spectrum P_E ($t\to\infty$) for various "probe" intensities.

Continua in multiphoton processes can be structured by laser excitation. Such induced structure is not merely a skewed Lorentzian. Multichannel interferences are essential in the description of this system. Merely by turning on V, population can not only <u>return</u> periodically from the continuum but may never be excited at all.

REFERENCES

1. J.A. Armstrong and J.J. Wynne, in "Nonlinear Spectroscopy" (North Holland, Amsterdam) 1977, Ed. N. Bloembergen.
2. L. Armstrong Jr., B.L. Beers and S. Feneuille, Phys. Rev., A12, 1903, (1975).
3. Y.I. Heller, V.F. Lukinykh, A.K. Popov and V.V. Slabko, Phys. Lett., 82A, 4, (1981).
4. P.E. Coleman, J.N. Elgin, P.L. Knight and K. Burnett, in "Quantum Electronics and Electro-Optics", (Wiley, Chichester) Ed. P.L. Knight; P.E. Coleman, P.L. Knight and K. Burnett, to be published.
5. P. Lambropoulos, Appl. Opt., 19, 3926, (1980); P. Lambropoulos and P. Zoller, Phys. Rev., A24, 379, (1981).
6. K. Rzążewski and J.H. Eberly, Phys. Rev. Lett., 47, 408, (1981).
7. U. Fano, Phys. Rev., 124, 1866, (1961).
8. P.E. Coleman and P.L. Knight, J. Phys. B., (in press), 1982.

TWO-PHOTON EXCITATION OF ARGON

Michael S. Pindzola
Department of Physics, Auburn University, AL 36849

Marvin G. Payne
Health and Safety Research Division
Oak Ridge National Laboratory, TN 37830

ABSTRACT

We calculate two photon excitation parameters for various excited states of argon assuming the absorption of near resonance broadbandwidth laser radiation. Results are given for the case of two photons absorbed from the same laser beam as well as the case of absorbing photons of different frequency from each of two laser beams. We use multiconfiguration Hartree-Fock wave functions to evaluate the second-order sums over matrix elements. Various experimental laser schemes are suggested for the efficient excitation and subsequent ionization of argon.

THEORY

Multiphoton excitation is describable in terms of rate equations[1,2] when the generalized n-photon Rabi frequency and the excited-state photoionization rate are much less than the laser bandwidth. This is generally the case when powerful excimer lasers are used to excite atomic levels in rare gases. For the case of two photons absorbed from the same laser beam, the two-photon excitation rate may be written as

$$W = \alpha I^2 \, G(\omega) \, , \tag{1}$$

where I is the laser intensity, $G(\omega)$ is the bandwidth-dependent laser line shape function, and α is the two-photon excitation parameter. For the case of absorbing photons of different frequency from each of two laser beams, the two photon excitation rate may be written as

$$W = \beta \, I_1 I_2 \, G(\omega_1, \omega_2) \, , \tag{2}$$

where β is a function of the possible laser frequency combinations, $\omega_1 + \omega_2$, needed to achieve near resonance absorption.

In the non-relativistic dipole approximation the two-photon excitation parameters are given by

$$\alpha = \frac{8\pi^3}{c^2} \left| \oint_n \langle \psi_f | \Delta | \psi_n \rangle \langle \psi_n | \Delta | \psi_i \rangle (E_i - E_n + \omega)^{-1} \right|^2 \, , \tag{3}$$

ISSN:0094-243X/82/900384-06$3.00 Copyright 1982 American Institute of Physics

and
$$\beta = \frac{8\pi^3}{c^2} \left| \sum_{n} <\psi_f|\Delta|\psi_n><\psi_n|\Delta|\psi_i>[(E_i - E_n + \omega_1)^{-1} \right.$$

$$\left. + (E_i - E_n + \omega_2)^{-1}] \right|^2 , \qquad (4)$$

where c is the speed of light, $\sum_{n}$ represents a sum over bound states and an integration over the continuum, and atomic units are used (1 a.u. = 27.2 eV). The dipole operator is given by $\Delta = \sum_{i=1}^{N} \hat{\epsilon} \cdot \vec{r}_i$ in the "length" form and $\Delta = \sum_{i=1}^{N} \hat{\epsilon} \cdot \vec{\nabla}_i/\omega$ in the "velocity" form, where N is the number of atomic electrons and $\hat{\epsilon}$ is the direction of the radiation-field polarization.

CALCULATION FOR ARGON

We calculated two-photon excitation parameters for various ground to excited state transitions in argon. We generated wavefunctions for the ground and excited bound states of argon in the Hartree-Fock approximation.[3] Total energies for various single and multiple configuration calculations are given in Table I.

Table I: Argon Bound State Energies

Configuration	Energy (a.u.)
$3p^6 \, ^1S$	-526.8175
$(3p^6 + 3p^4 \, 3d^2) \, ^1S$	-526.9335
$3p^5 4p \, ^1S$	-526.3642
$(3p^5 4p + 3p^3 3d^2 4p) \, ^1S$	-526.4516

The 4p ^1S excitation energy determined from the multiconfiguration results is 13.11 eV which agrees reasonably well with the experimental value of 13.27 eV.[4] A further intermediate coupling calculation shows that the 4p $^1S_0$ state is 83% pure, having a 17% mixture of the 4p $^3P_0$ state. The calculated fine structure splitting is 0.211 eV compared to 0.206 eV from experiment.[4]

In LS coupling two photon transitions are allowed from the ground state to the $4p^1S$, $4p^1D$ and $4f^1D$ excited states in argon. The dipole matrix elements found in Eqs. (3) and (4) were reduced to radial quadrature following standard algebraic methods. We assumed linear polarization of the radiation field. The resulting sums over radial intermediate states were determined using the inhomogeneous differential equation technique.[5] We used an iterative Hartree-Fock method to generate the perturbed bound states. Our results for β are presented in Figs. (1)-(3).

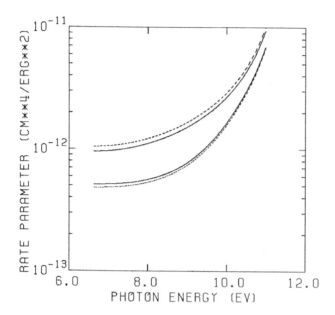

Fig. 1   Argon Two-Photon Rate Parameter β for 4p^{1}S.

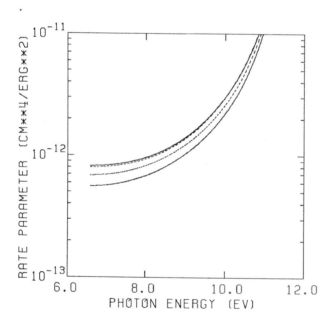

Fig. 2   Argon Two-Photon Rate Parameter β for 4p^{1}D.

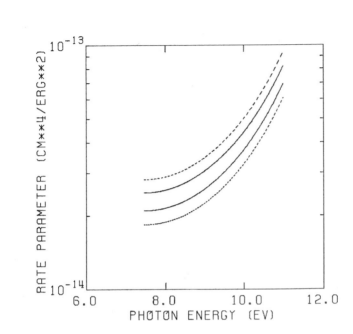

Fig. 3 Argon Two-Photon Rate Parameter β for 4f^{1}D.

In all three figures the upper dashed curve is a single configuration Hartree-Fock (SCHF) length calculation while the lower dashed curve is a SCHF velocity result. The photon energies run from threshold for two photon absorption to 11.0 eV. The curves in all three figures are climbing to single photon resonances with various 4s and 3d levels located above 11.0 eV. The upper solid curve in all three figures is a multiconfiguration Hartree-Fock (MCHF) length calculation while the lower solid curve is a MCHF velocity result. Only ground state 3p→3d correlations are included in our MCHF results. Calculational details are more fully explained in a previous work on the two-photon excitation of oxygen.[6]

The excited states of rare gases are strongly influenced by the spin-orbit interaction of the core electrons. We thus made the transformation from LS to jK coupling in keeping with the level schemes used in spectroscopic tables.[4] In calculating the rate parameters the sums over radial intermediate states remained the same, we only changed the angular weighting factors. Our MCHF results for α are given in Table II.

Table II: Argon Two-Photon Rate Parameters

Pair Coupled State	Wavelength (Å)	$\alpha_L(cm^4/erg^2)$	$\alpha_V(cm^4/erg^2)$
3/2 4p[5/2]$_2$	1894	1.23×10^{-13}	7.69×10^{-14}
3/2 4p[3/2]$_2$	1883	2.56×10^{-14}	8.61×10^{-15}
3/2 4p[1/2]$_0$	1868	1.69×10^{-13}	9.05×10^{-14}
1/2 4p[3/2]$_2$	1864	6.44×10^{-14}	7.15×10^{-14}
1/2 4p[1/2]$_0$	1840	8.60×10^{-14}	4.53×10^{-14}

In a previous work[7] we calculated the two photon excitation parameter for the xenon 3/2 6p[1/2]$_0$ state at 2496 Å. A SCHF length calculation gave 1.04×10^{-12} cm^4/erg^2 while the SCHF velocity result was 4.61×10^{-13} cm^4/erg^2. Relativistic effects further increased the xenon excitation parameter. Thus the argon two-photon transition rates are around an order of magnitude smaller than those found in xenon.

Various experimental laser schemes can be used for the efficient excitation and subsequent ionization of argon.[2] We report the excitation parameter for two methods. The first method involves the use of an ArF excimer laser at 1933 Å and a parallel beam obtained by anti-Stokes shifting 1933 Å radiation in HD. The second method uses two beams, one from F$_2$ at 1576 Å and a second is obtained at 2293 Å by frequency summing ND-YAG light with light from a frequency doubled dye laser. Our MCHF results for β are given in Table III.

Table III: Argon Two-Photon Laser Schemes

Pair Coupled State	Wavelength (Å)	$\beta_L(cm^4/erg^2)$	$\beta_V(cm^4/erg^2)$
3/2 4p[1/2]$_0$	1807 1933	6.77×10^{-13}	3.62×10^{-13}
3/2 4p[1/2]$_0$	1576 2293	7.67×10^{-13}	3.89×10^{-13}

We hope that accurate measurements of α and β as a function of laser bandwidth for argon will soon be made. The results will provide stringent tests for atomic structure calculations as well as stochastic theories of the laser light.

REFERENCES

1. P. Zoller, Phys. Rev. A19, 1151 (1979).
2. M. G. Payne, C. H. Chen, G. S. Hurst, S. D. Kramer, W. R. Garrett and M. S. Pindzola, Chem. Phys. Letters 79, 142 (1981).
3. C. F. Fisher, Comp. Phys. Commun. 1, 151 (1969); 14, 145 (1978).
4. C. E. Moore, "Atomic Energy Levels", NBS Circular No. 35 (U.S.G.P.O., Washington, D.C. 1971).
5. A. Dalgarno and J. T. Lewis, Proc. R. Soc. (London) A233, 70 (1955).
6. M. S. Pindzola, Phys. Rev. A17, 1021 (1978).
7. M. S. Pindzola, M. G. Payne and W. R. Garrett, Phys. Rev. A24, 3115 (1981).

390

2 π ELECTRON SPECTROMETER FOR MULTI-PHOTON IONIZATION STUDIES

P. Kruit, M.J. van der Wiel
FOM-Institute for Atomic and Molecular Physics, Kruislaan 407,
1098 SJ Amsterdam, The Netherlands.

F. H. Read
Schuster Lab., Department of Physics, The University of
Manchester, Manchester M13 9PL, UK.

ABSTRACT

A novel type time-of-flight electron spectrometer is described. It features full collection over 2 π sr. at a resolution of 20 meV. The instrument is based on the principle that a diverging magnetic field has the property of aligning the trajectories of electrons that travel from the strong field region to the weak field region. Some experimental results from the multiphoton ionization of Xenon are given. Limitations concerning the rate of change of the magnetic field and the permissible electron energies are considered.

INTRODUCTION

In experiments on multiphoton ionization of atoms or molecules it may be desired to measure the energies of the resulting photoelectons in order to get information on additional photon absorption, shifts of the continuum level, vibrational excitation of the molecular ion or other details of the process. The number of such electrons created per laserpulse has to be limited severely in order to prevent the electrons from being decelerated by the space-charge of the ions. In practice the number of electrons may have to be as small as about 100 per pulse, distributed over all directions and over a range of energies between 0 and 10 eV. So it is important to accept a large solid angle and to detect all energies simultaneously.

In this paper we describe a spectrometer that has these features, using the properties of an inhomogeneous magnetic field, as described earlier by Beamson et al[1,2]. in a design of a photoelectron spectromicroscope. More details on theoretical, computational and experimental results will be given in a future publication[3].

PRINCIPLE OF OPERATION

An electron is initially emitted at an angle θ; with respect to the direction of a magnetic field B; , and with a velocity v. The angular momentum of the circular motion is

$$1_i = \frac{m^2v^2 \sin^2\theta_i}{eB_i}i , \tag{1}$$

e and m being the electron charge and mass. If the variation of the magnetic field is adiabatic, i.e. the field experienced by an electron changes negligibly in the course of one revolution of the helical motion, then the angular momentum is a conserved quantity. This implies that the angle of the helical motion in the region of low field (B_f) is given by

$$\frac{\sin\theta_f}{\sin\theta_i} = \left(\frac{B_f}{B_i}\right)^{\frac{1}{2}} . \tag{2}$$

The transverse component of the velocity is therefore reduced. Since the total velocity is unchanged the longitudinal component increases from $v \cos \theta_i$ to

$$v_{z,f} = v \left(1 - \frac{B_f}{B_i} \sin^2\theta_i\right)^{\frac{1}{2}} .$$

By using a magnetic field that diverges from 1 Tesla at the source region to 0.001 Tesla at the detector region, electrons initially emitted over 2π steradians are formed into a beam of half-angle 2°. If this 'parallelization' is achieved in the first few millimeters of a 0.5 m flight tube, the measured forward velocity is equal to the total velocity within 1%.

A further consequence of the adiabatic invariance is that the magnetic flux linked by the electron orbit is a constant of the motion, so the electron in always bounded by the same field lines. Since the lateral displacement of a given field line from the axis of cilindrical symmetry is proportional to $B^{-\frac{1}{2}}$, we see that the instrument functions as an electron image magnifier with magnification

$$M = (B_i/B_f)^{\frac{1}{2}}. \tag{3}$$

This aspect hardly bears any importance for the multiphoton ionization studies and is more extensively described in reference 3.

ADIABATIC PARAMETERS

If the magnetic field, experienced by an electron changes too much in the course of one revolution, the electron will leave the field line to which it was bound. In an analytical treatment of the problem[3], not given here, we find that a characteristic parameter governing the non adiabatic behavior is:

$$\chi_2 = \frac{mv}{4e} \int_{z_i}^{z_f} \frac{(dB/dz)^2}{B^3} \, dz. \tag{4}$$

This parameter is connected to the fractional change (χ_1) in B_z during one pitch of the orbit through the formula

$$\chi_2 = \frac{1}{8\pi} \int_{z_i}^{z_f} \chi_1^2 \frac{dz}{\Delta z_{pitch}} \quad . \tag{5}$$

Although the first-order effect of non-adiabatic behaviour is only a displacement of the image at the detector, the second-order effect is an incomplete parallelization of the trajectories, thus influencing the energy resolution.

A detailed computer simulation of the electron motion in a diverging field shows that the adiabatic parameter χ_2 should be smaller than about 0.2 in order to have the ideal properties for a source region of a few tenths of a millimeter, being the typical dimension of a laser focus in multiphoton ionization studies.

DESIGN OF THE APPARATUS

The configuration of the electromagnet, and also of other parts of the apparatus, is shown in figure 1. The water-cooled coil gives 7000 Ampere-turns, creating a field between the pole pieces of 1T. A field of 10^{-3}T is produced in the drift tube (of length 0.5m) by means of a simple coil. The high magnetic permeability of iron causes the field to drop to almost zero inside the hole in the pole pieces. This would give a high contribution to the adiabatic parameter χ_2 at this point. To provide a smooth change in field the cross-sectional area of the iron circuit in the pole-pieces is reduced giving saturation of the iron at this point. From the measured axial magnetic field we find that the adiabatic parameter for the spectrometer is:

$$\chi_2 = 0.063 \ E^{\frac{1}{2}}. \tag{6}$$

where E is the electron energy in eV.

Insulated cylindrical grids are mounted inside the flight tube to enable the electrons to be either accelerated or retarded. Electrons arriving at the end of the flight tube strike a double channel plate detector, operated as an analogue signal amplifier. The detected signal is fed to a Philips PM 3310 digital oscilloscope which records the signal in 256 channels of a charge coupled device, with each channel representing a time interval of 20 ns. We modified the scope in order to be able to transfer the digitized signal to an on-line computer system at a

20 Hz repetition rate. Conversion from time to energy is performed only for the high-resolution part of the time spectrum. To obtain the information of all possible electron energies, a retarding voltage in the flight tube is scanned.

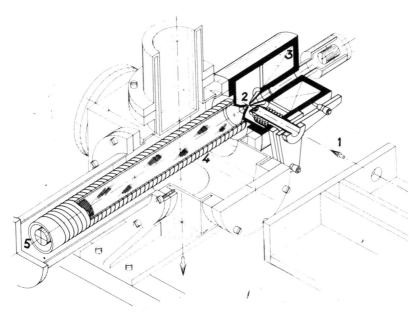

Figure 1: Magnetic field parallelizer and flight tube. 1) laser beam 2) laser focus - source of electrons 3) coil and iron circuit to give 1 T field 4) coil to give 0.001 T field 5) double channel plate detector.

EXPERIMENTAL RESULTS

In figure 2 we present an energy spectrum for the multiphoton ionization of Xenon. Three photons are resonant with the 6s state, while two more photons are needed for ionization. The spectrum shows the possibility for both energy resolution better than 20 meV and detection of low probability (1%) events, namely the absorption of additional photons in the continuum.

More experimental results are presented in references 4 and 5.

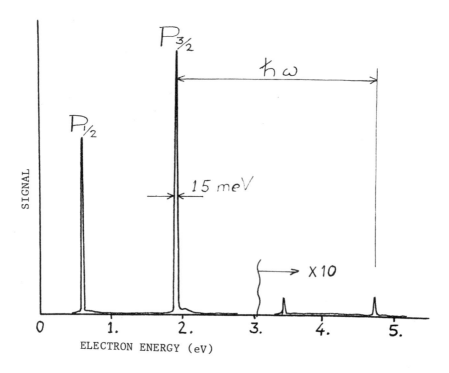

Figure 2: Electron energy spectrum for the five photon ionization of Xenon to the $P_{1/2}$ and $P_{3/2}$ continua.

This work was subsidized by the Foundation FOM and ZWO.

REFERENCES

1. Beamson G, Porter H.Q. and Turner D.W., J. Phys. E., Sci. Instrum 13, 64 - 66 (1980).
2. Beamson G., Porter H.Q., and Turner D. W., Nature 290, 556-561 (1981).
3. Kruit, P. and Read, F.H., to be published (1982).
4. Kruit, P., Kimman, J. and van der Wiel, M.J., J. Phys. B., At. Mol. Phys. 14 L 597-L 602 (1981).
5. Kimman, Kruit, P. and van der Wiel, M.J., to be published in Chem. Phys. letters (1982).

Two-Photon Excitation of the Lowest g-States of Xe$_2^*$

W. Gornik, E. Matthias, and D. Schmidt

Institut für Atom- und Festkörperphysik, Freie Universität Berlin

D-1000 Berlin 33 (West) Germany

Applying two-photon excitation techniques we have studied atomic and molecular states in the energy region of the $5p^5 6s$ configuration of Xe I. In this way, states can be selectively populated which cannot be reached by dipole transitions from the ground state.

A detailed description of the experimental set-up is given in Ref. |1|. Frequency doubled laser light of 285-300 nm was focussed into a xenon gas vessel with typical power densities of 20 MW/cm^2 in the focus. The VUV fluorescence was detected using a solarblind photomultiplier and photon counting techniques. In Fig. 1, an example of the excitation spectrum is shown. The quadratic dependence of the fluorescence intensity on incident laser power and gas pressure prove that the observed structures arise from two-photon excitation of molecular states. In addition, the $^3P_2$ state was weakly excited, providing evidence for parity changing two-photon transitions.

From the results of polarization studies the symmetries of the excited molecular states were determined |2|. In view of theoretical potential energy curves |3| we arrive at an assignment for the molecular states as indicated in Fig. 1. Full consistency between theoretical and experimental data, however, could not be reached.

Information about the relaxation scheme of excited g-states can be obtained by spectral and time-resolved fluorescence studies. Fig. 2 summarizes first results. Discriminating the second continuum against the total fluorescence by means of a cutoff filter we observe different fluorescence behavior for g-states belonging to the $^3P_1$ or $^3P_2$ dissociation limits, indicating different

ISSN:0094-243X/82/900395-02$3.00 Copyright 1982 American Institute of Physics

decay schemes. Further investigations, especially time-resolved studies of the fluorescence are in progress.

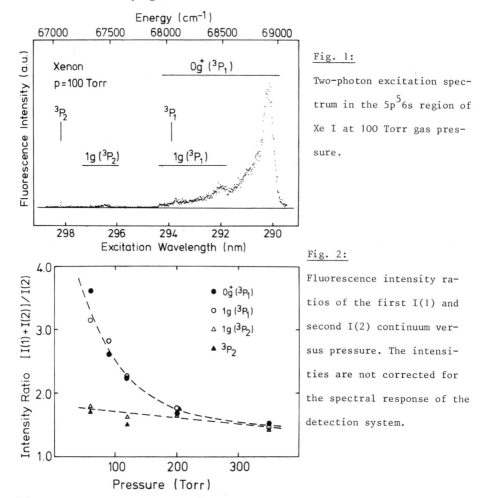

Fig. 1:

Two-photon excitation spectrum in the $5p^5 6s$ region of Xe I at 100 Torr gas pressure.

Fig. 2:

Fluorescence intensity ratios of the first I(1) and second I(2) continuum versus pressure. The intensities are not corrected for the spectral response of the detection system.

|1| W. Gornik, S. Kindt, E. Matthias, and D. Schmidt, J.Chem.Phys. 75, 68 (1981)

|2| D.L. Andrews and W.A. Ghoul, J.Chem.Phys. 75, 530 (1981)

|3| W.C. Ermler, Y.S. Lee, K.S. Pitzer, and N.W. Winter, J.Chem.Phys. 69, 976 (1978)

"Resonance Enhanced Multiphoton Ionization
of the Trifluoromethyl Radical"

Michael T. Duignan, Jeffrey W. Hudgens
and Jeffrey R. Wyatt

Laser Chemistry Group
Code 6110
Naval Research Laboratory
Washington, D. C. 20375
Telephone (202) 767-2519

ABSTRACT

The multiphoton ionization spectrum of CF_3 radicals

between 415 and 490 nm, proceeding through a three-photon

resonance, is presented. Several previously unreported

transitions are observed.

The technique of mass selective resonance enhanced

multiphoton ionization (REMPI) has been successfully applied

to the trifluoromethyl radical in the gas phase. CF_3 does

not flouresce and absorbs only in the VUV making its detec-

tion and study experimentally difficult. The radicals were

generated by either thermal pyrolysis of CF_3I or by infrared

multiphoton dissociation (IRMPD) of CF_3I, CF_3Br or $(CF_3)_2$ CO.

A composite REMPI spectrum, recorded in the 415-490 nm

region using four different laser dyes, is presented in

Figure 1. The spectral features are consistent with a

three-photon resonant transition of the lone electron to a

3p Rydberg orbital. A single additional photon ionizes the

ISSN:0094-243X/82/900397-03$3.00 Copyright 1982 American Institute of Physics

radical. The spectrum is marked by extensive excitation of an 820 cm^{-1} progression identified as the umbrella mode frequency (ν_2') of the ion, and expected by the large conformational change (pyrimdal $\rightarrow$ planar) associated with the transition. REMPI band heads agree well with those reported in one-photon VUV absorption (caption Fig. 1). Several transitions at frequencies >69000 cm^{-1} previously unobserved are reported.

IRMPD of a variety of substrates was employed for collisionless production of CF_3. The IRMPD process, however, tends to leave the radicals with excess internal energy - complicating an already congested spectrum.

REMPI is proposed as a sensitive technique for real-time monitoring of transient chemical species and as a valuable addition to absorption, fluorescence and photoelectron spectroscopies.

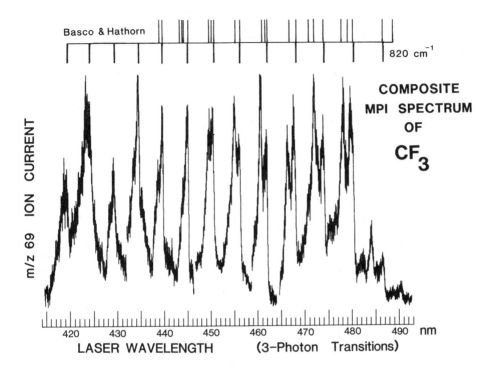

Figure 1. Composite REMPI spectrum of CF_3, using four different laser dyes (Stilbene 420, Coumarins 440, 450, and 480) and not normalized to laser power. Uppermost lines denote band heads observed by Basco & Hathorn (Chem. Phys. Lett. 8, 291(1971)) in VUV absorption at $1/3^{rd}$ the indicated wavelength. This suggests MPI spectrum consistent with a three-photon resonance, ionized by a single additional photon. The next set of lines show the 820 cm^{-1} progression interpreted as ν_2', the umbrella mode of the upper state.

400

LASER-INDUCED PHOTOIONIZATION OF MOLECULAR HYDROGEN:
A TECHNIQUE TO MEASURE ROVIBRATIONAL GROUND STATE POPULATIONS

E. E. Marinero, C. T. Rettner, and R. N. Zare
Department of Chemistry, Stanford University, Stanford, CA 94305

ABSTRACT

Using tunable anti-Stokes orders of a frequency-doubled dye laser, rotationally-selective excited-state populations in H_2 are prepared by two-photon excitation. The ensuing photoionization allows direct determination of (v'',J'') populations.

INTRODUCTION

Detection of molecular hydrogen in a quantum-specific manner is of fundamental importance in chemical physics. The lowest-lying electronic states in H_2 are only accessible using radiation below 110 nm and its infrared activity is weak (only quadrupole allowed). Thus standard methods such as laser-induced fluorescence and infrared chemiluminescence are inapplicable especially at low H_2 densities.

An alternative approach based on three-photon ionization is here described. Two aspects of the technique are emphasized: collisionally unrelaxed distributions are measured and an XUV frequency range of 673 cm^{-1} is spanned by simply scanning the rhodamine dye laser over just 82 cm^{-1}.

EXPERIMENTAL

The experimental scheme is shown in Figure 1. A commercial

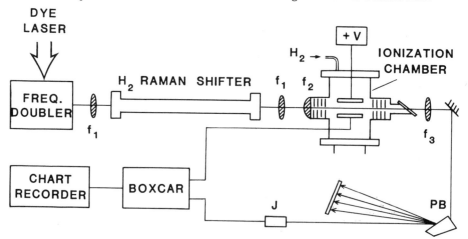

Fig. 1. Experimental scheme for state-specific detection of H_2.
Key: f_1, f_2, f_3 = 50, 10 and 40 cm focal length lenses, +V = repeller voltage, P.B.: Pellin Broca Prism, J = Joulemeter.

ISSN:0094-243X/82/900400-02$3.00 Copyright 1982 American Institute of Physi

Nd-YAG-pumped frequency-doubled dye laser is Raman-shifted in H_2.
Enhanced 4th anti-Stoke conversion is achieved by sending collinearly
through the shifter both the fundamental and the doubled-dye laser
radiation. The undispersed anti-Stokes orders are then focussed into
a low-pressure (15 mtorr) H_2 cell. Combinations of different anti-
Stokes orders lead to two-photon excitation of single rotational
levels of the E,F $^1\Sigma_g^+$ state over a wide range of (v', J') levels. As
first pointed out by Kligler et al.[1] at power densities >10^9 Watts/
cm^2 the most efficient process following the two-photon process is
the subsequent ionization of the excited state. In our experiments,
we are indeed able to ionize 100% of the excited-state population.
A spectrum is generated by collecting the photoions as a function of
wavelength. Figure 2 shows a portion of a typical ion-spectrum.

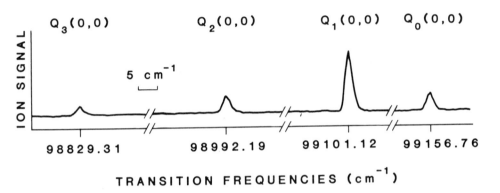

Fig. 2. Ion spectrum of room temperature H_2. Excitation of the
(0,0) band is achieved by the combined photon energies of the 3rd
and 4th anti-Stokes of the Raman-shifted doubled dye laser. The
signal amplitudes follow the Boltzmann distribution.

Rotational populations can be directly measured after correcting for
laser power and rotational line strengths.

REFERENCES

1. D. J. Kligler and C. K. Rhodes, Phys. Rev. Lett. 40, 309 (1978);
   D. J. Kligler, J. Bokor, and C. K. Rhodes, Phys. Rev. A21, 607
   (1980).

Two-photon excitation and ionization of H-atoms with tunable
VUV at Lyman-α

by H. Rottke, H. Zacharias and K. H. Welge
Department of Physics, University of Bielefeld
Bielefeld, Germany

A b s t r a c t

Two kinds of experiments with tunable VUV laser light have been carried

out with H atoms: (1) High resolution resonant two-photon ionization,

$$H(1\ ^2S_{1/2}) + h\nu_\alpha \rightarrow H(2\ ^2P_{1/2,3/2}) + h\nu_{uv} \rightarrow H^+ + e$$

with tunable Lyman-α light and (2) resonant two photon excitation of hig

Rydberg states,

$$H(1\ ^2S_{1/2}) + h\nu_\alpha \rightarrow H(2\ ^2P) + h\nu_{uv} \rightarrow H(n\ ^2S,\ n\ ^2D) \rightarrow H^+ + e$$

followed by ionization through collisions. In both cases an atomic beam

configuration has been employed and H^+ ions were detected with a mass

filter. The smallest bandwidth of tunable Lyman-α radiation achieved was

~ 1.0 GHz, enough for partial resolution of the hyperfine structure

doublet of the $1\ ^2S_{1/2}$ ground state. The application to produce polarize

protons by high resolution two photon ionization is discussed. Rydberg

states have been excited up to $n \sim 48$ and detected by collisional ioniza

tion under the given experimental conditions.
ISSN:0094-243X/82/900402-20$3.00 Copyright American Institute of Physics

INTRODUCTION

Considerable progress has been made recently in generating tunable VUV
laser light by four wave mixing in the wavelength region down to the
LiF cut off ($\sim$ 105 nm) [1-5]. This progress, made possible by the development of powerful, narrowband tunable pulsed dye laser has opened
of course a wide range of laserspectroscopic applications in the VUV
[6-9]. In this paper we report two kinds of experiments done with atomic
hydrogen employing tunable VUV laser light to excite the atoms at the
Lyman-α line ($\sim$ 121.6 nm) to the 2 ^{2}P state, followed by ionization
with a second UV photon or by further excitation to high Rydberg states,
and their ionization.

One-photon excitation of hydrogen atoms to the 2 ^{2}P state has been performed previously with tunable VUV laser light by observing the absorption in a gas discharge through hydrogen at pressures of about 50 mTorr
[10]. Two-step photoionization, H(1 ^{2}S) + $h\nu_\alpha \rightarrow$ H(2 ^{2}P) + $h\nu_{uv} \rightarrow$ H$^+$ + e
with tunable VUV radiation has been achieved recently at low densities
under atomic beam conditions with mass selective ion detection [11].
The objectives of the present experiments have been two-fold: (1) High-
resolution two-photon ionization of atomic hydrogen, that is investigation of the possibility to increase the VUV laser resolution at the
Lyman-α wavelength to a degree where the hyperfine splitting of the
H(1 ^{2}S) ground state of 1.42040 GHz could be resolved. In this case
the two-step photoionization would produce polarized protons. (2) Investigation of the possibility to produce atomic hydrogen in high Rydberg
states by two-step excitation, followed by ionization of the H-Rydberg
atoms: H(1 ^{2}S) + $h\nu_\alpha \rightarrow$ H(2 ^{2}P) + $h\nu_{uv} \rightarrow$ H(n ^{2}S,n ^{2}D) $\rightarrow$ H$^+$ + e. This
would offer the possibility for investigations of Rydberg hydrogen atoms
under low density conditions, for instance in external fields.

Experimental

Resonant two-photon ionization

The two-photon ionization experiments have been carried out with an
experimental set-up similar to that employed in a previous work [11].
Tunable VUV light was produced around the Lyman-α line (λ_α = 121,6 nm)
by frequency tripling of UV light (λ_{uv} = 364,8 nm) in Krypton. The light
beam, containing both wavelengths together was directed perpendicularly
onto a beam of atomic hydrogen, and ionized the atoms by the two-step
absorption mechnism:

$$H(1\ ^2S) + h\nu_\alpha \rightarrow H(2\ ^2P) + h\nu_{uv} \rightarrow H^+ + e^- \qquad (1)$$

Ions were detected by a secondary electron multiplier after passage
through a quadrupol mass filter. A peculiarity of the excitation procedure is that both frequencies are tuned simultaneously and that the sum
of the energy of the two photons is very close to the ionization limit.

The UV light was generated by mixing the fundamental of a Nd:YAG laser
(Quanta Ray) at 1.064 μm with the tunable light from a pulsed dye laser
in the visible around 554.6 nm in a KD*P crystal. The Nd:YAG laser serve
to simultaneously pump the dye laser. Operating the dye laser oscillator
with an external confocal spherical interferometer (FSR : 8 GHz, finesse
: 200), the bandwidth of the visible light was $\Delta\nu_{vis}^{ext.}$ ~ 120 MHz, that
is nearly Fourier transform limited corresponding to the duration of
5 - 6 nsec of the laser pulses [12]. After amplification in a three stag
amplifier chain the system yielded pulses of 15 mJ energy in the visible
Without the external confocal interferometer the bandwidth was $\Delta\nu_{vis}^0$ ~

600 MHz and the pulse energy about 25 mJ/pulse in the visible. Dyes used were Rhodamin 590 in the oscillator and longitudinally pumped third aplifier stage and Rhodamin 560 in the two transversally pumped preamplifier stages.

Without any frequency selective element inserted into the Nd:YAG laser cavity, the bandwidth of the Nd:YAG fundamental at 1.064 μm was $\Delta\nu_{IR}^{o} \sim 30$ GHz. Insertion of one etalon (FSR : 43 GHz, finesse : 15) resulted in a bandwidth $\Delta\nu_{IR}^{1} \sim 3$ GHz which could be further reduced to $\Delta\nu_{IR}^{1ELN} \sim 700$ MHz by additionally operating the oscillator in the electronic line narrowing mode (Quanta Ray) [13].

The smallest bandwidth at 1.064 μm was obtained by inserting a second etalon (FSR : 7.9 GHz, finesse : 5.2) into the oscillator and employing the electronic line narrowing mode of operation. Single longitudinal mode operation of the oscillator should occur yielding Fourier transform limited pulses with $\Delta\nu \sim 65$ MHz, corresponding to the pulse duration of about 9 nsec. However, in practical operation the system did not oscillate on one single longitudinal mode in every shot, but the frequency rather jumped from pulse to pulse between two adjacent longitudinal cavity modes of ~ 250 MHz separation or two adjacent modes oscillated simultaneously during one pulse, resulting in an effective bandwidth $\Delta\nu_{IR}^{2ELN} \sim 250$ MHz when averaged over many pulses.

The sum frequency mixing generated UV light ($\lambda_{uv} \cong 364,8$ nm) with a bandwidth $\Delta\nu_{uv}^{1} \sim 3.5$ GHz for $\Delta\nu_{vis}^{o} \sim 600$ MHz and $\Delta\nu_{Ir}^{1} \sim 3$ GHz or $\Delta\nu_{uv}^{1ELN} \sim 800$ MHz and $\Delta\nu_{uv}^{2ELN} \sim 330$ MHz according to a dye laser bandwidth $\Delta\nu_{vis}^{ext} \sim$

120 MHz and Nd:YAG laser bandwidths $\Delta\nu_{IR}^{1ELN} \sim 700$ MHz and $\Delta\nu_{IR}^{2ELN} \sim 250$ MHz respectively. The pulse energies of the 364.8 nm light were about 6 mJ/pulse for $\Delta\nu_{uv}^{1} \sim 3.5$ GHz and ~ 3 mJ/pulse for the bandwidths $\Delta\nu_{uv}^{1ELN} \sim 800$ MHz and $\Delta\nu_{uv}^{2ELN} \sim 330$ MHz.

For third harmonic generation the UV light was focused by a 70 mm focal length lens into a cell containing Krypton at 130 mbar pressure [1-3]. The VUV together with the UV light, emerging from the Krypton cell was collimated by a MgF$_2$ lens with a focal length of 60 mm at 121.6 nm (100 mm at 364.8 nm) and directed into the ionization region of the quadrupol mass filter. Corresponding to the bandwidths at 364.7 nm given above the bandwidths of the 121.6 nm light were $\Delta\nu_{vuv}^{1} \sim 9$ GHz, $\Delta\nu_{vuv}^{1ELN} \sim 2.5$ GHz and $\Delta\nu_{vuv}^{2ELN} \sim 1$ GHz. At the smallest bandwidth of ~ 1 GHz and 3 mJ/pulse UV input energy the VUV pulse energy was ~ 5 nJ/pulse (or $3 \cdot 10^9$ photons/pulse) corresponding to a conversion efficiency of about $1.5 \cdot 10^{-6}$ in the frequency tripling process. In order to convert the linearly polarized radiation at Lyman-α to a circular polarization a quarter wave plate of MgF$_2$ with (35 ± 5) % transmission at 121.6 nm could be inserted into the light beam between the MgF$_2$ lens and the entrance into the mass filter.

Hydrogen atoms were produced by a microwave discharge through flowing molecular hydrogen in a glass tube of 12 mm diameter. The gas was expanded through a capillary of 10 mm length and 1 mm inner diameter at the end of the discharge tube into the first stage of a differentially pumped vacuum system. The second stage of the system was separated from the first one by a scimmer with a hole of 0.25 mm diameter placed at 49 mm distance from the orifice of the capillary. The quadrupol mass filter in the second stage of the vacuum system was placed with its axis in the molcular beam direction. Within the ionization region of the mass filter, given by a cylindrical cage structure of 8 mm diameter and 12 mm length, and separated by 30 mm from the scimmer the molecular beam was crossed perpendicularly by the VUV and UV laser beams. Determined by the geometry, the divergence of the hydrogen beam was about 1.5° corresponding to a residual Doppler width $\Delta\nu_D \sim 550$ MHz at the Lyman-α wavelength, perpendicular to the beam axis.

The microwave resonator was located about 40 cm up-stream from the inlet capillary. The discharge was operated at 0.6 Torr hydrogen pressure and 100 Watt microwave power. Under these conditions the $[H]/[H_2]$ ratio was measured to be about 0.3 in the beam behind the scimmer resulting in a H atom density of $\sim 1 \cdot 10^{10}$ cm^{-3} in the beam within the ionization region.

Rydberg H-atom excitation

In order to selectively excite high Rydberg levels by resonant two-photon absorption,

$$H(1\ ^2S) + h\nu_\alpha \rightarrow H(2\ ^2P) + h\nu_{uv} \rightarrow H(n\ ^2S,\ n\ ^2D) \rightarrow H^+ + e^-$$

first the fundamental for the Lyman-α radiation production ($\sim$ 364.8 nm) had to be separated from the VUV light beam, and secondly UV light $h\nu_{uv}$, independently tunable, had to be used. The experimental arrangement is illustrated schematically in Fig. 2. The light beams at $\sim$ 121.6 nm and $\sim$ 364.8 nm emerging collinearly from the Krypton frequency tripling cell were spectrally separated by a MgF$_2$ prism placed at a distance of 60 mm from the MgF$_2$ collimating lens of the Krypton cell. The VUV beam passed through the ionization region of the mass filter crossing the hydrogen beam at right angles as described in the preceeding section. The independently tunable UV light was obtained from a second laser, crossing both the VUV light and hydrogen beam perpendicularly as indicated in Fig. 2.

The radiation at the Lyman-α wavelength was generated as described above. In this experiment the bandwidth of the Nd:YAG laser fundamental at 1.064 μm was $\sim$ 3 GHz, using one etalon (FSR: 43 GHz, finesse: 15) in the Nd:YAG oscillator. The bandwidth of the visible light from the dye laser was $\sim$ 600 MHz yielding about 3.5 GHz bandwidth at 364.8 nm. At 121.6 nm a pulse energy of about 10 nJ/pulse ($\hat{=}$ 6 $\cdot$ 10^9 photons/pulse) was achieved by frequency tripling at an input energy of about 6 mJ/pulse at 364.8 nm.

The tunable UV light for excitation from $2\ ^2P$ to the Rydberg states $n\ ^2S$, $n\ ^2D$ was also generated by sum frequency mixing the fundamental of a second Nd:YAG laser with the tunable visible light from a dye laser (Quanta Ray) pumped by the second harmonic of this Nd:YAG laser. The tuning range applied in this experiment was from ~365 nm to ~ 367 nm, corresponding to a tuning range from ~ 555.6 nm to ~ 560.3 nm in the visible dye laser wavelength. At an input pulse energy of about 30 mJ/pulse from the dye laser and ~ 120 mJ/pulse from the Nd:YAG laser fundamental to the KD^*P crystal about 5 mJ/pulse UV light was generated with a bandwidth of ~ 15 GHz.

The VUV and tunable UV laser pulses (pulse duration: ~ 5 nsec) were synchronized electronically such that they arrived simultaneously in the interaction region with the hydrogen beam except for a jitter of less than 1 nsec.

Atomic hydrogen was produced as described above. In this experiment two kinds of molecular beam arrangements have been employed. In one case the gas stream from the capillary (∅ 0.5 mm, length 5 mm) was not collimated by a scimmer but it was crossed by the two laser beams in the ionization region at a distance of ~ 20 mm from the orifice of the capillary. In this one stage arrangement the average pressure in the atomic beam within the ionization region was ~ $2.8 \cdot 10^{-5}$ Torr. At a dissociation efficiency $[H]/[H_2]$ of about 30 % this resulted in an average atomic hydrogen density of ~ $1 \cdot 10^{11}$ cm^{-3} in the atomic beam. The second arrangement was with the scimmer in its place, as used in the two photon ionization experiments, described in the preceeding section.

Results

Resonant two-photon ionization of H atoms

Fig. 4 shows a result of the resonant two-photon ionization experiment
where the dye laser and the Nd:YAG laser fundamental bandwidths were
$\sim$ 120 MHz and $\sim$ 700 MHz, respectively, resulting in $\sim$ 2.5 GHz total
bandwidth at Lyman-α. The spectrum represents the H^+ ion signal obtained
by scanning the VUV-laser wavelength through the Lyman-α transition.
For comparison, Fig. 3 shows the analogous previously published result
[11], obtained with about 9 GHz total bandwidth at 121.6 nm. Both the
UV and VUV light beams, tuned simultaneously by their direct connection,
were linearly polarized in these experiments.

In Fig. 4 the two fine structure components $2\ ^2P_{1/2}$ and $2\ ^2P_{2/3}$, with
10.96913 GHz splitting [14], are clearly resolved. From the spectrum one
reads a linewidth of $\sim$ 2.5 GHz which is practically equal to the total
bandwidth of the VUV laser radiation $\Delta\nu_\alpha$. This was to be expected because
the residual Doppler width $\Delta\nu_D$, resulting from the atomic beam divergence
was relatively small. Assuming Gaussian line profiles one has for the
total linewidth $\Delta\nu_{tot}$: $(\Delta\nu_{tot})^2 = (\Delta\nu_d)^2 + (\Delta\nu_D)^2$ showing that the
Doppler contribution to $\Delta\nu_{tot}$ will only be a few percent.

The intensity ratio of the two lines is $\sim$ 1:2, what is to be expected
from the statistical weight of the transitions $1\ ^2S_{1/2} \rightarrow 2\ ^2P_{1/2}$ and
$1\ ^2S_{1/2} \rightarrow 2\ ^2P_{3/2}$. This intensity distribution is, however, by no means
self evident because in tuning the laser wavelength from the $1\ ^2S_{1/2} \rightarrow$
$2\ ^2P_{1/2}$ to the $1\ ^2S_{1/2} \rightarrow 2\ ^2P_{3/2}$ resonance, the two-photon energy,

$h\nu_{vuv} + h\nu_{uv}$, is varied through the ionization threshold of the isolated H-atom (IP = 109678.7737 cm^{-1} [15]). The sum energy of the VUV and UV photons for excitation through the 2 $^2P_{1/2}$ and 2 $^2P_{3/2}$ states are 109678.5607 cm^{-1} and 109679.0486 cm^{-1} respectively, which is in the first case 0.213 cm^{-1} less and in the second case 0.2749 cm^{-1} more than the ionization potential.

The absorption through the 2 $^2P_{1/2}$ level thus leads to excitation of very high Rydberg states, which are evidently readily ionized by perturbations, for instance small electric fields or collisions with the residual gas. This observation agrees with the previous experiments at lower resolution [11].

Fig. 5 shows the result obtained with the smallest laser bandwidth achieved in this work when the Nd:YAG laser oscillator was operated with two etalons and the electronic line narrowing device and the dye laser with external confocal interferometer. Different from the experiments at wider bandwidth, where linearly polarized VUV and UV light was used, this measurement was done with circularly polarized VUV light produced by the MgF$_2$ quarter-wave plate in the laser beams.

As can be seen in Fig. 5, the 1.420406 GHz hyperfine structure splitting [16] of the hydrogen ground state into 1 $^2S_{1/2}$(F = 0,1) is partially resolved. Fitting the individual lines by Gaussian line profiles one obtaines an effective linewidth $\Delta\nu_{eff}$ = 1.15 GHz, slightly smaller than the hyperfine structure splitting. Taking the residual Doppler width of the atomic beam ($\Delta\nu_D \sim 550$ MHz) into account we obtain from $(\Delta\nu_{eff})^2 = (\Delta\nu_D)^2 + (\Delta\nu_\alpha)^2$, a bandwidth $\Delta\nu_\alpha \sim 1.0$ GHz for the Lyman-α radiation. This is consistent with the 1.1 GHz width estimated for the VUV laser light from the widths of the Nd:YAG laser fundamental (~ 250 MHz) and the dye laser (~ 120 MHz).

Further improvement of the resolution would be principally possible of course by stabilizing the Nd:YAG laser frequency to a single longitudinal cavity mode, yielding a Fourier transform limited bandwidth of about 65 MHz according to the pulse duration of about 9 nsec. With this improvement full resolution of the hyperfine structure doublet could be achieved.

Resolving the hyperfine structure splitting of the hydrogen ground state leads to a potentially interesting application of the resonant two-photon ionization with $2\,^2P$ as intermediate state, because this process is capable of producing spin polarized protons, as was shown theoretically by Parzynski [17] and Fedorov [18]. The way a polarization of the proton spin shows up in the ionization process of unpolarized H atoms is readily demonstrated for the ionization of atoms in the hyperfine structure state $1\,^2S_{1/2}(F = 0)$ via the intermediate $2\,^2P_{1/2}(F = 0,1)$ state using circularly polarized light for the excitation process $1\,^2S_{1/2}(F = 0) \rightarrow 2\,^2P_{1/2}$. For right hand circularly polarized light the selection rule is $\Delta m_F = +1$ implying that only transitions from $1\,^2S_{1/2}(F = 0, m_F = 0)$ to $2\,^2P_{1/2}(F = 1, m_F = 1)$ are possible. Here it is understood that the quantization axis of the atomic system is chosen along the propagation direction of the light beam.

Only the nuclear spin state $I = 1/2$, $m_I = + 1/2$ contributes to the hyperfine structure state $|2\ ^2P_{1/2}, F = 1, m_F = 1\rangle$, so that the state can be written as:

$$|2\ ^2P_{1/2}, F = 1, m_F = 1\rangle = |2\ ^2P_{1/2}, m_y = + 1/2\rangle |I = 1/2, m_I = + 1/2\rangle$$

The following ionization process from $2\ ^2P_{1/2}$ does not affect the nuclear spin state. The nuclear spin component along the propagation direction of the light beam is thus $m_I = + 1/2$ for all protons produced in this way, that is the resonant two photon ionization process:

$$H[1\ ^2S_{1/2}(F = 0)] + h\nu_\alpha(\sigma_+) \rightarrow H[2\ ^2P_{1/2}] + h\nu_{uv} \rightarrow p + e$$

yields 100 % spin polarized protons.

The degree of proton polarization P for spin 1/2 particles is defined by the relation:

$$P = (n_{1/2} - n_{-1/2})/(n_{1/2} + n_{-1/2}) \qquad [18]$$

n_{m_I} being the number of particles with nuclear spin component m_I along a suitably chosen quantization axis, which is here the common propagation direction of the light beams.

The two photon ionization mechanism

$$1\ ^2S_{1/2}(F) + h\nu \rightarrow 2\ ^2P_J + h\nu_{uv} \rightarrow p + e$$

with right circularly polarized exciting ($h\nu_\alpha$) and ionizing ($h\nu_{uv}$) light results in the polarization degrees P(F, J):

$$P(0,1/2) = 1, \quad P(0,3/2) = -7/11 \qquad [18]$$

$$P(1, J) = -\frac{1}{3} \cdot P(0, J) \qquad [17]$$

In deriving these results validity of second order time dependent perturbation theory and the dipol approximation for the interaction of the light beams with the atomic system was assumed. For ionization via the $2\,^2P_{1/2}$ state the discussion above, together with the relation $P(1, J) = -\frac{1}{3} P(0, J)$ shows that the final proton polarization is independent of the polarization of the ionizing light.

As the experiment (Fig. 5) was carried out with circularly polarized Lyman-α light the protons corresponding to the peaks of the $1\,^2S_{1/2}$ $(F = 0) \rightarrow 2\,^2P_{1/2}$ transition in Fig. 5 should be 100 % polarized, and those of the $1\,^2S_{1/2}(F = 1) \rightarrow 2\,^2P_{1/2}$ transition should have 33 1/3 % polarization with the polarization direction reversed. Because the resolution was not yet complete in these experiments, the actual degrees of polarization were accordingly smaller.

As to the number of polarized protons produced in this experiment, we estimate for the peak of the $1\,^2S_{1/2}(F = 0) \rightarrow 2\,^2P_{1/2}$ line a yield of the order of 10^5 protons/pulse. Possibilities to increase the proton yield considerably are to increase the light intensity at the Lyman-α wavelength and the atomic beam flux.

Rydberg atoms

Fig. 6 shows the spectral dependence of the H^+ production obtained by excitation of $H(1\ ^2S_{1/2})$ atoms to $H(2\ ^2P)$ with fixed frequency Lyman-α radiation and further excitation to high Rydberg states $H(n\ ^2S,\ n\ ^2D)$ by tunable UV light. From the Rydberg states the H atoms are ionized in some way discussed below.

This experiment was carried out without the scimmer, that is the two laser beams intersected each other in the freely expanding gas about 20 mm behind the orifice of the capillary. In this mode of operation the effective Doppler width perpendicular to the axis of the gas stream and determined by the cross section of the laser beams was roughly speaking about 30 GHz at the Lyman-α wavelength and correspondingly about 10 GHz in the UV at about 366 nm. On the other hand the bandwidths of the Lyman-α and UV radiation were $\sim$ 10 GHz and $\sim$ 15 GHz respectively. Because of the linewidth in the spectrum of $\sim$ 45 GHz the highest Rydberg state resolved is n = 48 as indicated in Fig. 6.

The question arises which of the various possibilities, i. e. collisions, external electric fields, thermal radiation, UV light was responsible for ionization of the Rydberg atoms. Under the conditions of the present experiment by far the major cause for ionization of atoms in all resolved states were collisions. This was proven by an experiment in the differentially pumped vacuum system, with the scimmer placed between the capillary and the excitation region, thus reducing the effective pressure in the interaction region from about $2.8 \cdot 10^{-5}$ Torr to $\sim 3 \cdot 10^{-6}$ Torr. All other experimental conditions remained essential-

ly the same. The spectrum in Fig. 7 shows the result of this experiment.
It was taken with the gain of the detection electronic increased by a
factor of $2 \cdot 10^3$ with respect to the spectrum in Fig. 6. Resolved lines
up to about n = 43 nearly disappeared, i. e. the atoms excited to these
states were nearly no more ionized. Atoms in higher levels were still
ionized effectively by some mechanism, as indicated by the steep increase
of the signal for n > 43. The present experiments have shown the feasibility of producing hydrogen atoms in high Rydberg states under low pressure
and atomic beam conditions by resonant two photon excitation. In a recent
development we have produced highly excited H-atoms up to about n = 60
by this two step excitation in cryogenetic environment (T ~ 15 K) and
ionized the atoms by an electric field. This indicates the feasibility
of experiments also in magnetic fields.

References

[1]   R. Mahon, T. J. McIlrath, D. W. Koopman; Appl. Phys. Lett. 33, 305 (1978)

       R. Mahon, T. J. McIlrath, V. P. Myerscough, D. W. Koopman; IEEE J. Quantum
       Electron, QE-15, 444 (1979)

[2]   D. Cotter; Opt. Comm. 31, 397 (1979)

[3]   R. Hilbig, R. Wallenstein; IEEE J. Quantum Electron., QE-17, 1566 (1981)

[4]   H. Langer, H. Puell, H. Röhr; Opt. Comm. 34, 137 (1980)

[5]   F. S. Tomkins, R. Mahon; Opt. Lett. 6, 179 (1981)

[6]   J. R. Banic, R. H. Lipson, T. Efthimiopoulos, B. P. Stoicheff; Opt. Lett.
       6, 461 (1981)

[7]   J. C. Miller, R. N. Compton; J. Chem. Phys., to be published

[8]   M. Rothschild, H. Egger, R. T. Hawkins, H. Pummer, C. K. Rhodes; Chem.
       Phys. Lett., 72, 404 (1980)

       M. Rothschild, H. Egger, R. T. Hawkins, J. Bokor, H. Pummer, C. K. Rhodes;
       Phys. Rev., A23, 206 (1981)

[9]   H. Zacharias, H. Rottke, K. H. Welge; Opt. Comm. 35, 185 (1980)

[10]   R. Wallenstein; Opt. Comm., 33, 119 (1980)

[11]   H. Zacharias, H. Rottke, J. Danon, K. H. Welge; Opt. Comm., 37, 15 (1981)

[12]   H. Zacharias, R. Schmiedl, R. Wallenstein, K. H. Welge; in: Laser 79,
       Opto. Electronics, ed. W. Waidelich (IPC Science and Technology Press,
       Guildford, 1979) pp. 74-80

[13]   Y. K. Park, R. L. Byer ; Opt. Comm., 37, 411 (1981)

[14]   J. C. Baird, J. Brandenberger, K.-I. Kondaira, H. Metcalf; Phys. Rev.,
       A5, 564 (1972)

[15]   G. W. Erickson; J. Phys. Chem. Ref. Data, 6, 831 (1977)

[16]   S. Bashkin, J. O. Stoner, Jr.; Atomic energy-level and Grotrian diagrams
       (North Holland, Amsterdam 1978) p. 3

[17]   R. Parzynski; Opt. Comm. 34, 361 (1980)

[18]   M. F. Fedorov; Opt. Comm. 26, 183 (1978)

Figure captions

Fig. 1   Experimental set-up. Two-stage pumping with scimmer used for two-photon high resolution ionization and one-stage pumping without scimmer used for two-photon high Rydberg state excitation experiments. VUV light beam passed ionization cage (2) perpendicularly to plan of drawing. (1): quadrupol mass filter, (5) gas inlet for hydrogen gas from microwave discharge.

Fig. 2   Scheme of laser beam arrangement in two-photon high Rydberg state excitation/ionization experiment.

Fig. 3 H^+ yield as function of frequency tuned through $1\,^2S \rightarrow 2\,^2P$ transition (From previous publication [ ], reproduced for comparison)

Fig. 4 H^+ yield as function of frequency tuned through $1\,^2S \rightarrow 2\,^2P$ transition. Effective resolution ~ 2.5 GHz.

Fig. 5 H^+ yield as function of frequency tuned through $1\,^2S \rightarrow 2\,^2P$ transition. Effective resolution ~ 1.0 GHz.

Fig. 6   Rydberg state excitation/ionization spectrum. VUV wavelength fixed at Lyman-α. UV wavelength tuned. Background pressure $\sim 3 \cdot 10^{-5}$ Torr; One-stage pumping arrangement.

Fig. 7   Same as Fig. 5. Background pressure $\sim 3 \cdot 10^{-6}$ Torr; two-stage pumping system.

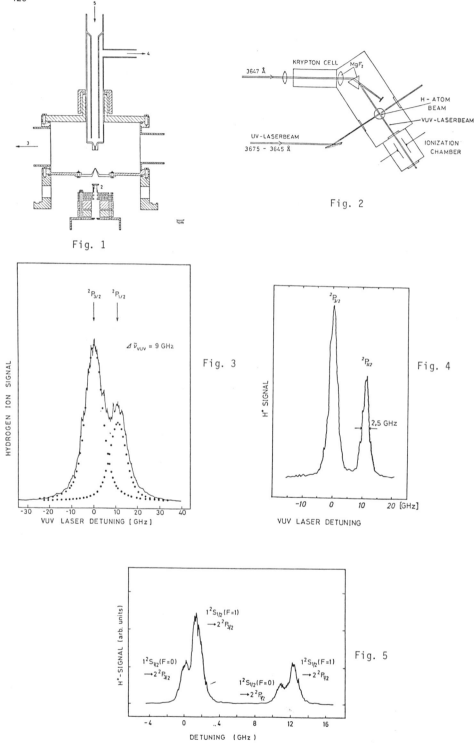

Fig. 1

KRYPTON CELL

3647 Å

MgF₂

H – ATOM BEAM

VUV-LASERBEAM

UV-LASERBEAM
3675 – 3645 Å

IONIZATION CHAMBER

Fig. 2

HYDROGEN ION SIGNAL

$^2P_{3/2}$ $^2P_{1/2}$

$\Delta\tilde{\nu}_{VUV} = 9$ GHz

Fig. 3

VUV LASER DETUNING [GHz]

H⁺ SIGNAL

$^2P_{3/2}$

$^2P_{1/2}$

2,5 GHz

Fig. 4

VUV LASER DETUNING

H⁺-SIGNAL (arb. units)

$1^2S_{1/2}(F=1)$
$\rightarrow 2^2P_{3/2}$

$1^2S_{1/2}(F=0)$
$\rightarrow 2^2P_{3/2}$

$1^2S_{1/2}(F=0)$
$\rightarrow 2^2P_{1/2}$

$1^2S_{1/2}(F=1)$
$\rightarrow 2^2P_{1/2}$

Fig. 5

DETUNING (GHz)

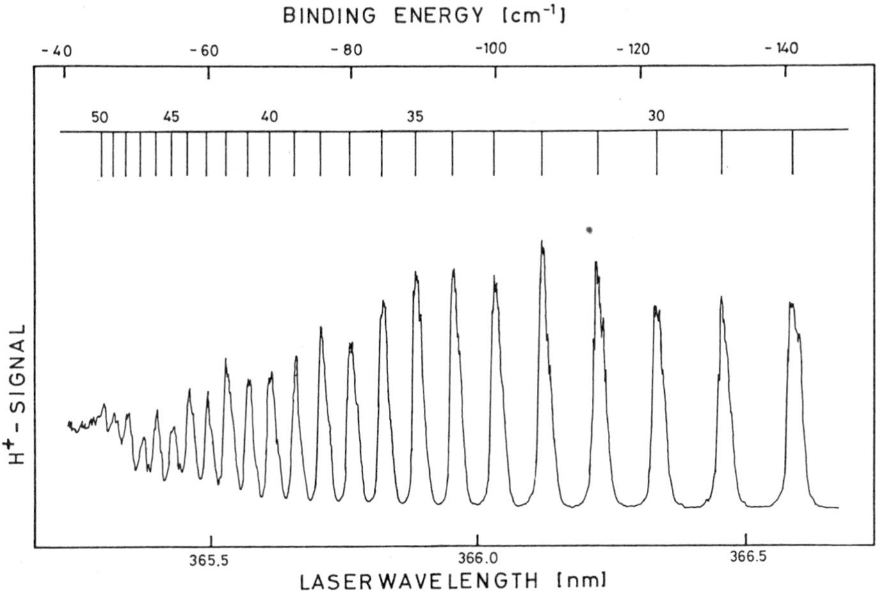

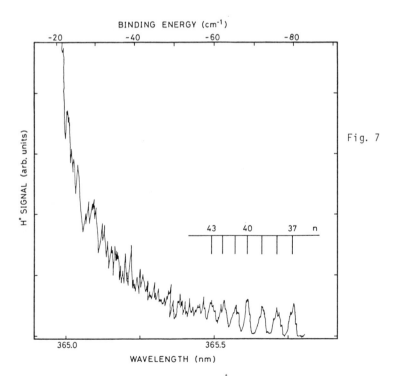

BINDING ENERGY [cm⁻¹]
-40
-60
-80
-100
-120
-140
50
45
40
35
30
H⁺ - SIGNAL
365.5
366.0
366.5
LASERWAVELENGTH [nm]

Fig. 6

BINDING ENERGY (cm⁻¹)
-20
-40
-60
-80
H⁺ SIGNAL (arb. units)
43
40
37
n
365.0
365.5
WAVELENGTH (nm)

Fig. 7

Generation of Light Below 100nm in Hg Vapor

by

R. R. Freeman and R. M. Jopson
Bell Laboratories, Murray Hill, NJ 07974

J. Bokor
Bell Laboratories, Holmdel, NJ 07733

ABSTRACT

We have used mercury as a medium in which to generate useful quantities of both coherent and incoherent light for wavelengths shorter than 100nm.

The uses for a pulsed, laboratory scale source of light for wavelengths shorter than 100nm fall naturally into two categories: narrow band, high spectral brightness sources for spectroscopy of gas phase species, and widely tunable sources of somewhat larger bandwidth for studies of condense matter.[1] It has been the goal of many investigators for sometime to produce tunable, narrow band, coherent radiation via four-wave mixing below 100nm. The use of laser generated plasmas to produce a broad continuum of incoherent light which is subsequently filtered by a monochromator has been recently reported.[2] In this paper, we report on the use of mercury not only as an efficient medium for four-wave mixing for wavelengths shorter than 100nm, but as a convenient material as a target for laser produced plasmas as well.

The immediate problem that presents itself in all four-wave mixing schemes for which the output is shorter than 100nm is that all window materials are opaque. This problem is usually overcome by employing some form of differential pumping through restricted slits.[3] In this work we assembled a windowless apparatus based on a heat pipe design. The apparatus is shown in Fig. 1.

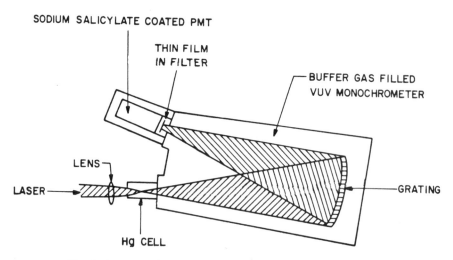

SODIUM SALICYLATE COATED PMT

THIN FILM IN FILTER

BUFFER GAS FILLED VUV MONOCHROMETER

LENS

LASER

GRATING

Hg CELL

Fig. 1 Apparatus for tripling and four-wave mixing in Hg vapor.

A mercury cell, with a quartz input window, is mounted on the entrance port of a one meter, normal incidence monochromator. There is no window on the exit of the mercury cell; instead, the monochromator is filled with approximately 1 Torr of pure He. The He is transparent at the wavelengths of interest ($\lambda > 80$nm) and serves to constrain the Hg vapor within the confines of the cell. The monochromator is necessary, not to determine the bandwidth of the generated radiation (indeed, the slits of the monochromator are set fully open), but to disperse the fundamental laser beams out of the optical path of the generated light. A thin film (0.1μ) indium filter is placed in front of the sodium salicylate coated PMT to shield against scattered laser light in the monochromator.

We began our investigations into the usefulness of mercury for four-wave mixing in this region of the spectrum by tripling a single input laser. Fig. 2 shows a schematic diagram of the excitation scheme: a variable energy (0-20 mJ/pulse), 10pps, 5 nsec pulse duration laser pulse was focused into the mercury cell with a spot size of roughly 10^{-2}cm.

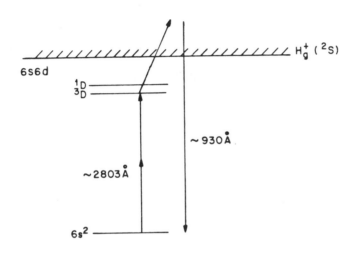

Fig. 2. Energy schematic of tripling process in Hg vapor.

In the vicinity of 280.0nm, there exists two possible two-photon resonances in Hg: $6s^2 \rightarrow$ 6s6d, where the 6s6d state is split into nominally ^1D and ^3D. The tripled output of this process is expected to be tunable, and resonantly enhanced around 930A.

Fig. 3 shows a plot of the relative tripled output power as a function of the wavelength of the fundamental.

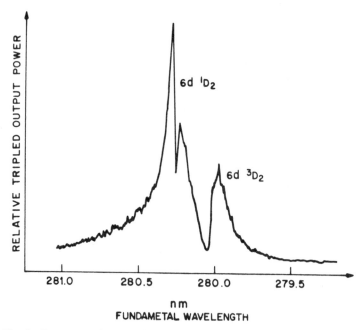

Fig. 3. Spectrum of output in the vicinity of 93nm as a function of the fundamental laser frequency.

This data was taken with roughly 5×10^{10} W/cm^2 at the focus, with a Hg density of approximately 2×10^{16} cm^{-3} and a He pressure of 5 Torr. There are four major features of this spectrum: 1) the resonance enhancement at both the $^1D_2$ and $^3D_2$ is broad; 2) there is a *dip* in the relative output when the fundamental is tuned to exactly the $6s^2 \rightarrow 6s6d$ ($^1D_2$) two-photon resonance; 3) there is a pronounced asymmetry in the tripled output around the $^1D_2$ resonance; and 4) there is a deep interference minimum between the two resonance enhancements. We will discuss each of these features in more detail below. At the peak of the output we estimate the efficiency of the process at approximately 10^{-6}, although exact figures are difficult to extract due to uncertainities in reflectance of the grating and transmission of the thin metal filter.

The width of the resonance enhancement and the depth of the dip on exact two-photon resonance depends upon both the intensity of the light and the density of the Hg. There have been detailed predictions of such behavior given in the literature,[4] but to our knowledge no previous experimental observations of this saturation have been reported. Fig. 4 shows how the dip on exact resonance develops as the laser intensity is increased from 7×10^9 W/cm^2 to approximately 4×10^{10} W/cm^2, with the Hg density *wig* 5×10^{15} cm^{-3}.

Fig. 5 shows the development of the spectrum with increasing laser intensity with a constant Hg density of *wig* 5×10^{16} cm^{-3}. Note that at this increased density the asymmetry in the relative output around the dip becomes quite pronounced.

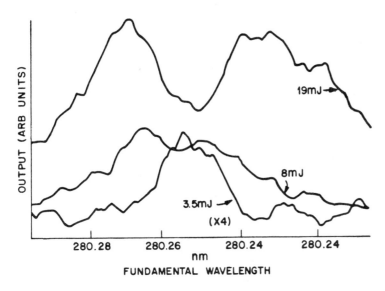

Fig. 4. Evolution of the "dip" at exact two-photon resonance as a function of increasing laser intensity.

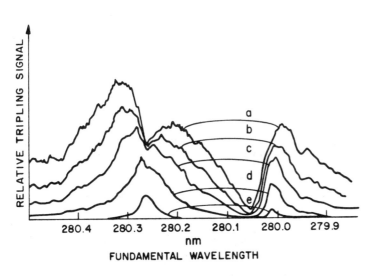

Fig. 5 Spectrum of output in the vicinity of 93nm as a function of the fundamental laser frequency for various intensities of light: a) 3×10^{10} W/cm^2; b) 2×10^{10} W/cm^2; c) 1.5×10^{10} W/cm^2; d) 8×10^9 W/cm^2; e) 2×10^9 W/cm^2.

In Fig. 6 the laser intensity is kept constant at approximately 4×10^{10}W/cm^2, while the Hg density is increased from $2\times10^{+15}$cm^3 to greater than 2×10^{16}cm^{-3}.

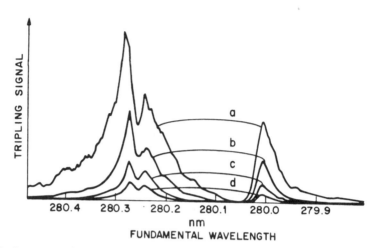

Fig. 6. Spectrum of output in the vicinity of 93nm as a function of the fundamental laser frequency for various densities of Hg vapor: a) 2×10^{16}cm^{-3}; b) 8×10^{15}cm^{-3}; c) 5×10^{15}cm^{-3}; d) 2×10^{15}cm^{-3}.

We attribute the breadth of the resonance, the dip on resonance, and the asymmetry around the dip to saturation due to depletion of the fundamental. As the fundamental beam propagates through the material it experiences loss due to two photon absorption. This absorption peaks on two-photon resonance; the strength of tripling also peaks on two-photon resonance so that the observed output is a competition between the two effects. Because the tripling is proportional to I^3, when the density is high enough the two-photon loss on resonance (proportional to $I^2 \cdot N$) wins out and actually reduces the overall tripling output from the cell. Thus the breadth of the resonance enhancement is due to a relative reduction of the efficiency (as compared to a medium with no two-photon absorption) as two-photon resonance is approached. The asymmetry is a consequence of the resonantly enhanced Kerr effect: because the fundamental beam experiences a strong absorption resonantly enhanced at the two-photon frequency, the index of refraction is also changing in a dispersive sense about the two-photon frequency. On the low frequency side of two-photon resonance this index change is such that it helps compensate the linear phase mismatch inherent in the material. On the high frequency side the phase mismatch is made worse and the output is less.

The deep minimum between the two resonance enhancements is due to a simple interference: the total nonlinearity of the medium is due to the sum of two terms, each with a resonance denominator centered about its respective resonance frequency. The tripled output is proportional to the square of the sum of the two terms. If the two-photon resonance expression is dominated by one path[5], the laser frequency from which the minimum occurs is given approximately by $(\sqrt{f_1}\omega_2 + \sqrt{f_2}\omega_1)/2 \cdot (\sqrt{f_1} + \sqrt{f_2})$ where ω_i is the atomic frequency of state i and f_i is the oscillator strength of the transition between the intermediate state and state i.

We have had some initial success in modeling the output spectrum by employing a simple set of coupled differential equations.

$$\frac{d}{dz} E_\omega - X_A |E_\omega|^2 = 0 \tag{1}$$

$$\frac{d}{dz} E_{3\omega} - X_G E_\omega^3 \, exp\{i(3k_\omega - k_{3\omega})z\} = 0 \tag{2}$$

Here X_A is the (complex) frequency dependent two-photon absorption coefficient and X_G is the (complex) frequency dependent third harmonic coefficient. We've assumed that two-photon absorption induced by $E_{3\omega}$ is negligible, that two-photon absorption induced by $E_{3\omega}$ on E_ω is negligible and that pump depletion due to third harmonic production is negligible. Using these two equations, we have been able to qualitatively reproduce all of the major features of the spectrum shown in Fig. 3. To produce a detailed quantitative fit to the spectrum will require a treatment of considerably greater sophistication: focused beam equations, atom by atom saturation, and Raman type couplings between $E_{3\omega}$ and E_ω. Nevertheless, we feel that the basic physical phenomena are well understood here, and remarkably well described by these two simple equations.[4]

There have recently been reported several experiments which demonstrate how useful Hg vapor is for generating tunable, coherent light as short as 105nm.[5] We have extended the demonstrated range of four-wave mixing in Hg by the addition of another laser to our original tripling apparatus. In this work we wanted to make the output tunable, and demonstrate the tunability by sweeping through an autoionizing resonance in Hg. In this case, we also made use of an atomic beam apparatus in order locate the autoionizing state of interest and to have an independent measure of its width and shape.

A beam of Hg atoms was produced in an oven and made to cross the two laser beams between the plates of an ion detector. The first laser was tuned to exact two-photon resonance ($6s^2 \rightarrow 6s6d \, (^1D_2)$) and the second laser was scanned in the vicinity of 290nm. The same laser beams could be directed into the apparatus shown in Fig. 1 so that the four-wave mixing output could be measured. (The ion signal and four-wave mixing output were recorded sequentially.) Fig. 7 shows the relative outputs for both the ion and four-wave mixing as a function of the second laser frequency.

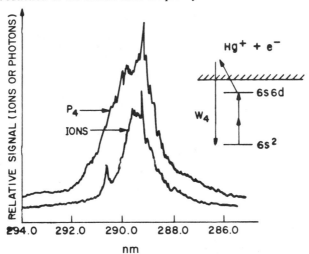

Fig. 7 Relative output of four-wave mixing (P_4) near 94.5nm and relative output of ions measured in an atomic beam for the same laser frequencies.

The autoionizing state is a J=1 of nominal configuration $5d^9 6s^2 7p$; the extra peak on the low frequency side of the ion signal is J≥2 state since these are allowed in three-photon ionization, but not in four-wave mixing.

We measured an output of approximately 0.01 μJ/pulse at the peak of the autoionizing state, roughly equal to the maximum *observed tripling output*. At high densities of Hg (>10^{16}cm^3) the four-wave sum frequency was substantial even off the autoionizing peak, yielding approximately 0.005 μJ/pulse (although the measurement of the exact output energy was subject to the same limitations as in the tripling case). In this case, the tunability of the process would be limited solely by the dye range of the second laser. Assuming an effective fundamental laser range of 1000nm to 200nm (making use of standard IR and UV nonlinear optics to extend the dye laser range), the four-wave mixing output, using the 6s6d two-photon resonance, covers 123nm to 82.4nm. Tomkins and Mahon[6] have recently reported high conversion efficiencies in Hg vapor over an extended range in the VUV using lower lying two-photon resonances; thus, Hg vapor has been shown to be a highly efficient, convenient material for the generation of coherent light for wavelengths longer than 150nm to shorter than 83nm.

In the course of performing the four-wave mixing experiments, there were several occasions when the input beams were accidently focused onto the metal mercury surface. When this happened, we observed in the visible a bright plasma ball at the laser focus, and a substantial, broad band output in XUV through the monochromator. We rearranged the mercury cell to allow the full laser beam to impinge on the mercury surface within a few centimeters of the input slits. The configuration used to measure the plasma spectrum is shown in Fig. 8 (see Ref. 7).

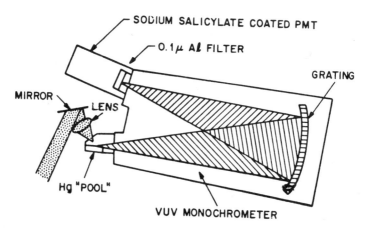

Fig. 8. Apparatus used to measure the spectrum of the mercury plasma.

We became interested in this plasma as a source of broadband radiation in the XUV not only because the conversion of laser light to radiation shorter than 80nm proved to be substantial (greater than 1%) but because unlike solid metal targets used in the past[2] the mercury surface "self-healed" after each laser shot. Thus, we did not need to reposition the target material after each shot making a repetition rate of 10pps possible.

When the monochromator was filled with 5 Torr of pure He, we observed a continuum emission below 80nm which cutoff at approximately 50.5nm at the onset of the He continuum absorption. Fig. 9 shows the relative signal strength between 50 and 55nm, showing the absorption lines arising from the principle series in He.

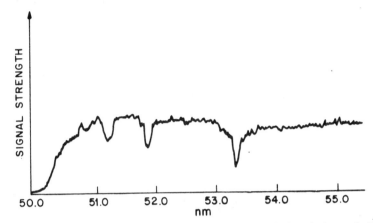

Fig. 9. Spectrum of continuum emission from a laser induced plasma in Hg. The dips are the absorptions arising from the five Torr of He in the monochromator. The abrupt cutoff at 50.5nm is due to the onset of continuum absorption.

For wavelengths longer than 80nm, we observed a background continuum with overlaying strong emission lines due to recombination in HgII and HgIII. When the monochromator is pumped out (for short times to avoid serious Hg contamination of the monochromator) the recorded output continues to wavelengths shorter than 40nm where we observe the expected cutoff due to loss of reflectivity of the normal incidence grating. This behavior is shown in Fig. 10.

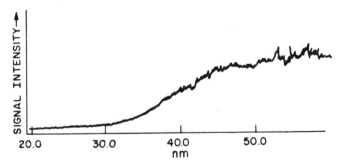

Fig. 10 The recorded signal from a laser induced plasma in Hg from 20 to 60nm when no He is used in the monochromator. The fall off below 40nm is due to loss of reflectivity of the normal incidence monochromator.

We are currently instrumenting a grazing incidence monochromator in order to measure the true short wavelength cutoff of the plasma spectrum.

We made preliminary estimates of the intensity of the output and found only 10^6 photons/sec/10% bandwidth at 50nm. This is low, and we suspect it is due to the poor grating reflectivity (probably less than 1%) and severely restricted solid angle (10^{-4}) of the 1 meter monochromator. We are currently setting up to measure the through-put with a 1/3 meter normal incidence machine and expect to be able to increase the useful output by many orders of magnitude.

In conclusion, we have demonstrated that mercury is a material with properties that make it quite useful for four-wave mixing to wavelengths as short as 83nm and for a target for a laser-induced plasma source with continuum output to wavelengths shorter than 40nm.

REFERENCES

[1] T. J. McIlrath and R. R. Freeman, "Laser Techniques for Extreme Ultraviolet Spectroscopy", this volume.

[2] P. K. Carroll, E. T. Kennedy and G. O'Sullivan, Appl. Optics 19, 1454 (1980); G. O'Sullivan, P. K. Carroll, T. J. McIlrath and M. L. Ginter, Appl. Optics 20, 3243 (1981).

[3] J. A. R. Samson, "Techniques of Vacuum Ultraviolet Spectroscopy", Pied Publications, Lincoln Nebraska, 1980.

[4] H. Scheingraber and C. R. Vidal, Optics Comm. 38, 75 (1981); H. Puell, H. Scheingraber and C. R. Vidal, Phys. Rev. A 22, 1165 (1980); H. Kidal and S. Brueck, JQE, QE-16 566 (1980); E. Stappaerts, JQE, QE-15, 110 (1979).

[5] D. C. Hanna, M. A. Yuratich, D. Cotter, "Nonlinear Optics of Free Atoms and Molecules", Springer Verlag, NY 1979.

[6] F. S. Tomkins and R. Mahon, "Generation of Narrow-Band Tunable Radiation in the 1200A Spectral Region", this volume; F. S. Tomkins and R. Mahon, Optics Lett. 6, 179, 1981.

[7] "Laser Produce Plasmas in Hg: A Source of Pulsed XUV Radiation", R. M. Jopson, R. R. Freeman and J. Bokor, post deadline paper, this volume.

SPECTROSCOPY OF CO AND NO USING COHERENTLY GENERATED VUV

C.R. Vidal

Max-Planck-Institut für Physik und Astrophysik
Institut für extraterrestrische Physik
8046 Garching, W.-Germany

ABSTRACT

Generation of coherent vuv radiation using the methods of non-linear optics is briefly reviewed. It is followed by a discussion of the presently existing methods of vacuum uv laser spectroscopy which so far have been mostly applied to the CO- and NO-molecule.

INTRODUCTION

In recent years the technology of generating coherent vacuum uv radiation has reached the point where these new light sources become attractive for high resolution laser spectroscopy. For generating coherent radiation in the vuv one has essentially two different possibilities. The first method employs inversion systems such as, for example, the widely used noble gas halogen excimer lasers[1] which are capable of stimulated emission in the vuv. The second technique uses the methods of nonlinear optics where coherent vuv radiation is generated by means of sum frequency upconversion or harmonic generation of high power lasers operating in the visible part of the spectrum[2]. The first group of systems can generate peak powers as high as 1 GW with a linewidth of typically 10 cm^{-1} and larger, whereas the second group gives significantly lower powers with a linewidth which can easily be as small as 0.1 cm^{-1} and smaller. Peak powers of the latter systems range from several 10 W for tunable, resonantly enhanced sum frequency mixing[3,4,5] up to 1 MW for nonresonant harmonic generation using fixed frequency high power solid state lasers[6,7]. For the purposes of high resolution spectroscopy in the vuv the small linewidth and the beam quality are the quantities of paramount importance. For this reason, mostly the second group of systems using the methods of nonlinear optics is of interest to high resolution spectroscopy in the vuv.

Recently, the two techniques mentioned have been combined. The harmonic signal of tunable dye lasers was amplified by means of high power noble gas halogen excimer systems. In this manner, the linewidth and beam quality which can be achieved with the methods of nonlinear optics, could be combined with the large output power of the excimer systems[8,9,10]. With a Fourier transform limited linewidth a spectral resolution $\nu/\Delta\nu$ of about 10^7 was achieved over a limited tuning range which is given by the bandwidth of the excimer system[11]. This has to be seen in contrast to a spectral resolution of 10^5 which can be achieved with conventional technology using large vuv spectrometers. In addition, all of the coherent vuv sources mentioned above have a spectral brightness which is many orders of magnitude above the spectral brightness of all existing synchrotrons. The aim of our work at Garching has been to perform high resolution laser spectroscopy on small molecules of astrophysical interest such as H_2, CO and others where the main absorption bands are in the vuv spectral region. For this purpose we have pursued three major areas of research:

(1) development of coherent vuv sources using the methods of nonlinear optics
(2) high resolution laser spectroscopy in the vuv
(3) development of new methods for analyzing molecular spectra which are particularly well suited for laser spectroscopic measurements.

In the first area we have carried out a series of experiments which were primarily designed to provide a detailed quantitative understanding of the different physical processes which are responsible for the conversion efficiency in resonant and non-resonant frequency mixing experiments. The aim was to find the optimum conditions for operating these systems.

In the second area of laser spectroscopy in the vuv we are essentially still in an exploratory phase where we try to find out how the numerous methods of laser spectroscopy in the visible and infrared part of the spectrum can be transferred into the vuv. The methods are tested on simple molecules of astropyhsical interest.

In the following a brief review of the first two areas of research will be given. The third area covering our methods for analyzing molecular spectra will not be described since it has recently been reviewed[12].

COHERENT VUV SOURCES USING THE METHODS OF NONLINEAR OPTICS

In general, an electromagnetic wave traveling through a gaseous medium tends to polarize the atoms and molecules of the gas. For small electric fields this polarization P is proportional to the field amplitude E where the proportionality constant is given by the linear susceptibility $\chi^{(1)}(\omega)$. At large field amplitudes this linear relationship is no longer correct and higher order terms become important giving rise to optical harmonics which are of interest here.

In trying to describe the nonlinear properties of any medium one has to look for a selfconsistent solution which obeys the Schrödinger equation as well as Maxwell's equations. The Schrödinger equation describes the response of the medium to a local applied electric field and provides the material properties such as the dielectric constant as a function of the electric field amplitude. Maxwell's equations, on the other hand, describe the propagation of the electromagnetic fields as a function of the local properties of the medium. A general solution of this kind exists only for the limiting case of a two-level system, a solution known as the Bloch equations. For our purposes, however, one has to consider a multilevel system where the Schrödinger equation can be treated by means of a perturbation approach[13,14]. For a homogeneous and isotropic medium the field dependent material properties can be expressed by the linear susceptibility $\chi^{(1)}$ and the lowest order nonlinear susceptibilities $\chi^{(3)}$. The Maxwell equations, on the other hand, lead to the fundamental equation of nonlinear optics where the polarization is generally separated into different contributions which are linear and nonlinear in the electric field amplitudes.

In order to illustrate the dominant physical processes which characterize the nonlinear properties of a gaseous medium, the following discussion is restricted to the limiting case of the third harmonic generation. The sum frequency mixing and the generation of higher harmonics basically do not reveal any new physics. For the case of the third harmonic generation the nonlinear polarizations can be written as[2,7]

$$P_1^{NL}(\omega) = \frac{N}{4}\left\{ 3\chi_T^{(3)}(\omega)E_3 E_1^* E_1^* + \chi_s^{(3)}(\omega)E_1 |E_1|^2 + \chi_s^{(3)}(\omega,3\omega)E_1 |E_3|^2 \right\} \quad (1)$$

$$P_3^{NL}(3\omega) = \frac{N}{4}\left\{ \chi_T^{(3)}(3\omega)E_1^3 + \chi_s^{(3)}(3\omega)E_3 |E_3|^2 + \chi_s^{(3)}(3\omega,\omega)E_3 |E_1|^2 \right\} \quad (2)$$

The different contributions to the nonlinear polarization at the fundamental wave, P_1^{NL}, and at the third harmonic wave, P_3^{NL}, have the following physical significance. The first term in Eq.(2) containing the nonlinear susceptibility $\chi_T^{(3)}(3\omega)$ is the desired term giving rise to the third harmonic generation, whereas the first term in Eq.(1) describes the inverse process by which some of the third harmonic radiation is converted back into the fundamental wave. Besides the desirable terms containing $\chi_T^{(3)}$ there are additional terms containing $\chi_s^{(3)}$ which originate from the same order of the perturbation approach to the Schrödinger equation and which are important for the ultimate conversion efficiency of any nonlinear medium. The real parts of $\chi_s^{(3)}$ represent intensity dependent changes of the refractive index where, for example, the last term of Eq.(2) describes the change of the refractive index at the harmonic frequency due to the fundamental intensity proportional to $|E_1|^2$. The imaginary parts of $\chi_s^{(3)}(\omega)$ and of $\chi_s^{(3)}(3\omega)$ are responsible for the two-photon absorption of the fundamental and of the harmonic wave, respectively, whereas the imaginary part of the mixed terms $\chi_s^{(3)}(\omega,3\omega)=\chi_s^{(3)}(3\omega,\omega)$ gives rise to Raman-type gain or losses.

For practical applications any nonlinear medium has to meet two major requirements:

(1) On a microscopic scale the nonlinear medium has to provide a sufficiently large nonlinear susceptibility for every volume element exposed to the fundamental wave. The nonlinear susceptibility is essentially given by

$$\chi_T^{(3)}(3\omega) = \hbar^{-3} \sum_{a,b,c} \frac{\mu_{ga}\mu_{ab}\mu_{bc}\mu_{cg}}{(\omega_{ag}-\omega)(\omega_{bg}-2\omega)(\omega_{cg}-3\omega)} \quad (3)$$

and can be optimized by a suitable choice of the medium with the corresponding complex transition freuquencies ω_{mn} and a proper choice of the frequency ω [14a].

(2) On a macroscopic scale all the different contributions originating from different volume elements of the nonlinear medium have to be added correctly in phase giving rise to a constructive interference. This requirement is taken care of by the phase matching condition. For the third harmonic generation

this yields the simple relation

$$n_1 = n_3 \tag{4}$$

where the refractive index at the fundamental frequency has to be equal to the refractive index at the harmonic frequency providing the same phase velocity for both waves.

As first demonstrated by Harris and coworkers[15,16] the preceding two requirements can be met by a gaseous two-component system which has a sufficiently small absorption cross section for the fundamental and the harmonic waves. The first component is selected according to its nonlinear susceptibility for a given frequency ω, whereas the second component has to provide the phase matching of the nonlinear system.

For the generation of gaseous two-component systems metal vapor noble gas mixtures which are contained in a heat pipe oven[17,18], and mixtures of noble gases[19,20,5] have been used.

In the small signal limit the third harmonic intensity Φ_3 is given by

$$\Phi_3 = \frac{n_1}{n_3} \left\{ \frac{12\pi^2\omega}{c^2 n_1^2} \; NL \; |\chi_T^{(3)}(3\omega)| \right\}^2 \Phi_1^3 F(\tau, \Delta kL) \tag{5}$$

where NL is the column density of the gaseous component which dominates the nonlinear susceptibility $\chi_T^{(3)}$ and Φ_1 is the intensity of the fundamental wave. F is the phase matching factor which assumes the maximum value of unity, if the optical depth τ and ΔkL vanish. Δk is the wave vector mismatch. Depending on τ and ΔkL, F is generally smaller than unity.

In trying to achieve the highest possible harmonic intensity Φ_3 different methods can be pursued. First of all, one can optimize $\chi_T^{(3)}$ by a proper choice of the nonlinear medium and by exploiting the resonant denominators in Eq.(3). However, the one- and three-photon resonances should be avoided because they lead to a strong absorption of the fundamental or the harmonic wave. On the other hand, the two-photon resonance which is associated with an electric dipole forbidden transition, can be very useful at moderate input intensities as long as the two-photon absorption due to $Im\{\chi_s^{(3)}(\omega)\}$ in Eq.(1) is sufficiently small.

As a further means of raising the harmonic intensity Φ_3 one can raise the fundamental intensity Φ_1. However, limitations are imposed by saturation phenomena not contained in Eq.(5) and eventually higher order nonlinear susceptibilities become important which originate from higher order perturbations of the Schrödinger equation. These terms give rise to higher harmonic waves which have been investigated by Reintjes and coworkers[21] and which for practical applications in laser spectroscopy have to be separated from the other harmonic waves. This can be achieved to some extent already be enhancing a particular harmonic wave through a proper choice of the nonlinear material with a bad three-photon resonance and a very good five-photon resonance and by suitable phase matching. Furthermore, the higher order terms lead to additional terms in Eqs.(1) and (2) which give rise to an effective field dependence of the third order terms.

As a final and very effective method for optimizing the harmonic intensity one can increase the column density NL in Eq.(5). This should not exceed a value where the optical depth at the harmonic frequency is equal to about unity. For larger values of the column density the conversion efficiency eventually becomes independent of the column density because an equilibrium is reached inside the non-linear medium where the harmonic intensity generated per unit length is balanced by the absorption losses per unit length[22]. It should be noted that with increasing column density the requirements for phase matching become more stringent in order to achieve a phase matching factor F close to unity. However, one can relax the requirements for phase matching somewhat by focusing the incident beam because the phase matching curve looses its oscillatory structure which it has for a parallel beam. In focusing the beam the confocal parameter should not be smaller than the length L of the nonlinear medium. Any stronger focusing does not increase the conversion efficiency of a phase matched system and gives rise only to undesirable higher order terms. It should also be noted that the density N together with the density of the phase matching medium affects also the pressure broadening of the nonlinear susceptibility. As a result for resonant frequency mixing, an increase in density may give rise to a lower differential nonlinear susceptibility per wavelength interval as shown for the MgKr system[4].

In trying to raise the harmonic intensity Φ_3 by increasing the fundamental intensity Φ_1 one eventually has no longer a third power law and long before energy conservation leads to a depletion of the fundamental wave saturation takes place. The dominant process for this saturation is due to a field dependence of the phase matching condition

$$n_1(E) = n_3(E) \tag{6}$$

which is caused by the nonlinear susceptibilities $\chi_s^{(3)}$ which can be approximated by

$$\chi_s^{(3)}(\omega) = \hbar^{-3} \sum_{a,b,c} \frac{\mu_{ga}\mu_{ab}\mu_{bc}\mu_{cg}}{(\omega_{ag}-\omega)(\omega_{bg}-2\omega)(\omega_{cg}-\omega)} \tag{7}$$

$$\chi_s^{(3)}(3\omega,\omega) = \hbar^{-3} \sum_{a,b,c} \frac{\mu_{ga}\mu_{ab}\mu_{bc}\mu_{cg}}{(\omega_{ag}-3\omega)(\omega_{bg}-2\omega)(\omega_{cg}-3\omega)} \tag{8}$$

and which have the same two-photon resonance as the nonlinear susceptibility $\chi_T^{(3)}$. Directly around the two-photon resonance the summation over all states b is dominated by one term and all nonlinear susceptibilities have the same line profile. The complex transition frequency ω_{mn} is given by $\omega_{mn} = \Omega_{mn} - i\Gamma_{mn}$. Right on resonance the nonlinear susceptibilities are therefore dominated by the imaginary part, whereas off-resonance by the real part.

As a result, saturation occurs in all frequency mixing experiments due to a field dependent destruction of the phase matching condition. For non-resonant frequency mixing $Re\{\chi_s^{(3)}\}$ gives rise to a field dependent change of the refractive index[7]. For resonant fre-

436

quency mixing we have again a field dependence of the refractive index, however, because of a different reason. In this case, $\text{Im}\{\chi_s^{(3)}\}$ is responsible for a strong two-photon absorption and the corresponding change in population densities modifies the effective refractive index of the nonlinear medium[23,4].

In trying to optimize $\chi_T^{(3)}$ with respect to $\chi_s^{(3)}$ for obtaining the highest possible conversion efficiency one has only little freedom because of the similarity of the relations for $\chi_T^{(3)}$ and $\chi_s^{(3)}$. Any change in one of the nonlinear susceptibilities immediately affects all the others. Furthermore, for a Gaussian laser pulse in space and time the field dependence can only be partially compensated by an initial mismatch in the phase matching condition.

More detailed calculations of the saturation for two-photon resonant sum frequency mixing have recently been carried out by Scheingraber et al.[24]. As shown in Fig. 1 the resonant enhancement in the small signal limit is actually turned into a minimum of the conversion efficiency at high input intensities. Figure 1 also shows that the two-photon resonance is shifted due to the AC Stark effect and that the line profile for the third harmonic intensity is broadened due to power broadening.

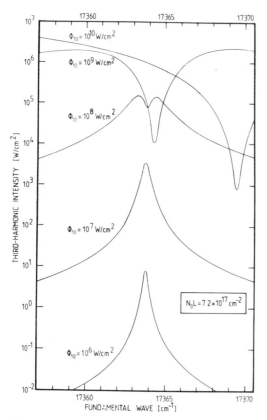

Fig. 1. Maximum achievable third harmonic intensity versus frequency of the fundamental wave around the $5s^2-5s5d$ two-photon resonance in strontium. The example is calculated for a column density $NL=7.2\times10^{17}\text{cm}^{-1}$ with an optimized initial mismatch for input intensities ranging from 10^6 to 10^{10} W/cm^2.

An investigation of the linewidth dependence in the small signal limit considering the laser line width, Doppler- and pressure broadening was recently carried out by Leubner et al.[25] where an expression first given by Stappaerts et al.[26] was Fourier transformed into an expression which is easily accessible to standard asymptotic techniques. Closed form expressions of the line profiles were derived for all situations of practical interest.

As a result of our work which has now been essentially completed, we have obtained a detailed quantitative understanding

of resonant and non-resonant frequency mixing, of its spectral properties and of the optimum conditions[27] in practical applications.

VUV LASER SPECTROSCOPY

This area of research is at this stage still in its infancy and
several groups in laboratories such as those in Toronto, Chicago,
Bielefeld and Garching try to explore the different possibilities of
transfering the numerous existing methods of laser spectroscopy[28]
into the vuv spectral region. In any work of this kind one is con-
fronted with a few basic considerations:

(1) Presently, the most attractive coherent vuv sources are all
    pulsed. Freeman et al.[29] have demonstrated, however, cw fre-
    quency mixing around 170 nm in a Sr vapor cell installed in-
    side a super-conducting magnet. Due to its low conversion
    efficiency this source does not yet appear to be very useful
    for vuv laser spectroscopy. So far, there are also no cw in-
    version systems in the vuv.

(2) The linewidth of the pulsed systems are at best Fourier trans-
    form limited. In this case, a pulse duration of 1 nsec cor-
    responds to a spectral resolution of typically 2×10^6 in the
    near vuv.

(3) Radiation in the vuv will ionize a lot of atoms and photo-
    dissociates or photoionizes almost all molecules. This
    seriously affects detection schemes of laser spectroscopy in
    the vuv.

(4) Experiments in the vuv suffer from severe technical con-
    straints with respect to the available optical components.

(5) Calibration of wavelengths can be carried out with existing,
    secondary standards in the vuv[30]. However, nonlinear optics
    provides easy access to the more accurate standards in the
    visible which eventually will replace existing standards in
    the vuv.

In order to discuss the difficulties in vuv laser spectroscopy
and to present the work presently going on, an admittedly superficial
summary of the different methods in laser spectroscopy is given in
Table I. The outline follows essentially the recent book of Demtröder[28] and indicates by means of a small circle all methods which
have already been successfully realized in the vuv.

In going through the list of Table I it is worth making a few
comments concerning the feasibility and the value of the different
methods in the vuv.

Starting with the methods of absorption spectroscopy it should
be noted that Doppler limited absorption spectroscopy and excitation
spectroscopy on the CO-molecule[31,20] and NO-molecule[32] have already
been performed. Optoacoustic spectroscopy is generally very attractive for systems with a low quantum efficiency and should be interesting in particular in conjunction with the investigation of photodissociation of molecules where most of the photon energy is turned
into potential energy and not into kinetic energy.

Table I   Methods of laser spectroscopy

A circle indicates all methods already performed in the vacuum uv

I.   Absorption spectroscopy

1. Doppler limited absorption spectroscopy 0
2. Excitation spectroscopy 0
3. Optoacoustic spectroscopy
4. Optogalvanic spectroscopy
5. Ionization spectroscopy 0
6. Intracavity absorption spectroscopy

II. Optical pumping and fluorescence

1. Fluorescence spectroscopy 0
   life time measurements, Franck Condon factors,
   branching ratios, collisional cross sections
2. Stepwise excitation
3. Double resonance spectroscopy

III. High resolution, sub-Doppler spectroscopy

1. Spectroscopy on beams
2. Saturation and polarization spectroscopy
3. Heterodyne spectroscopy
4. Multiphoton spectroscopy (0)
5. Level crossing spectroscopy

IV. Time resolved spectroscopy

1. Life time measurements 0
2. Coherent spectroscopy
   quantum beat spectroscopy, photon echos, optical nutation,
   pulse Fourier transform spectroscopy

V.   Nonlinear optics

1. Multiphoton spectroscopy 0
2. Harmonic generation and frequency mixing 0
3. Raman spectroscopy

Optogalvanic spectroscopy appears to be less attractive in the vuv because the method is easier to use and more sensitive with cw lasers in the visible and infrared especially since highly excited states which are accessible to vuv radiation, can easily be excited in a discharge by means of collisions. Work on state selective optogalvanic spectroscopy[33] is going on in our laboratory making the advantage of state selectivity using vuv sources less important.

Ionization spectroscopy has already been done on the CO- and NO-molecule[34]. It is a method with a high intrinsic sensitivity.

Intracavity absorption should be particularly interesting with broadband excimer lasers. It requires, however, a high quality spectrometer.

In the area of optical pumping in the vuv we have just succeded to obtain for the first time spectrally resolved fluorescence of the A-X and C-X system of the NO-molecule from individual, laser excited rotational vibrational levels[35]. Franck Condon factors, their J-dependence and branching ratios have been measured. Life time measurements which do not require a spectral resolution of the fluorescence, have been carried out on the CO-molecule[31] and the NO-molecule[32].

Stepwise excitation and double resonance spectroscopy should be particularly attractive if only one of the transitions is in the vuv, whereas the other transition is in the more convenient visible or infrared spectral region. In this manner, one can carry out 'vuv laser spectroscopy in the visible' where the system is prepared in the vuv and is interrogated and analyzed in a more convenient spectral region.

In the area of high resolution, sub-Doppler spectroscopy work is in progress in our laboratory on pulsed beams which generate molecules of low rotational temperature. In addition, polarization spectroscopy on the CO-molecule is in progress using the technique of two-step polarization labeling[36,37] where the CO-molecule is pumped on the A-X transition and is interrogated on the B-A transition in the visible.

Heterodyne spectroscopy will probably be less attractive with existing coherent vuv sources. Multiphoton spectroscopy, however, has already been done, for example, on the CO-molecule[38,39], the H_2-molecule[11] and several atomic species. In the vuv, however, they have not yet been done with sub-Doppler resolution.

Finally, level crossing spectroscopy should certainly be feasible and valuable because of its simplicity requiring vuv sources of moderate linewidth.

Except for life time measurements the whole area of time resolved spectroscopy such as coherent spectroscopy has not been touched at all in the vuv. However, it could eventually turn into a particularly valuable area of spectroscopy because in view of existing pulsed coherent vuv sources it allows one to carry sub-Doppler, high resolution experiments from the frequency domain into the time domain making the technique of injection locking especially interesting.

The last area of spectroscopy in Table I is probably the most extensively investigated one. Multiphoton experiments on CO have been mentioned already. The technique of frequency mixing and harmonic generation was described in the previous section and was also applied to the NO-molecule[40] and the H_2-molecule[9]. Raman spectroscopy will for most cases be less attractive in the vuv than in the visible where experiments are easier to carry out.

The preceding, rather superficial summary of the state of the art shows that a lot of exiting vuv laser spectroscopy is still waiting to be done. It should also be noted that nowhere in this paper a comparison with the techniques using synchrotrons was presented. For this the reader is referred to a very useful discussion by Jortner and Leach[41] who have looked into this problem more from the point of view of the synchrotron discussing possible applications to molecular physics.

REFERENCES

1. Ch.A. Brau, 'Rare gas halogen excimers', in: Excimer Lasers, edited by C.K. Rhodes, Topics in Applied Physics, Vol. 30, p. 87, Springer Berlin, 1979.
2. C.R. Vidal, Appl. Optics 19, 3897 (1980).
3. S.C. Wallace and G. Zdasiuk, Appl. Phys. Letters 28, 449 (1976).
4. H. Junginger, H.B. Puell, H. Scheingraber and C.R. Vidal, IEEE J. Quant. Electr. QE-16, 1132 (1980); QE-17, 557 (1981).
5. R. Hilbig and R. Wallenstein, Appl. Optics 21, 913 (1982).
6. D.M. Bloom, G.W. Bekkers, J.F. Young and S.E. Harris, Appl. Phys. Letters 26, 687 (1975).
7. H. Puell, K. Spanner, W. Falkenstein, W. Kaiser and C.R. Vidal, Phys. Rev. A 14, 2240 (1976).
8. H. Egger, T. Srinivasan, K. Hohla, H. Scheingraber, C.R. Vidal, H. Pummer and C.K. Rhodes, Appl. Phys. Letters 39, 37 (1981).
9. H. Pummer, T. Srinivasan, H. Egger, K. Boyer, T.S. Luk and C.K. Rhodes, Opt. Letters 7, 93 (1982).
10. J.C. White, J. Bokor, R.R. Freeman and D. Henderson, Opt. Letters 6, 293 (1981).
11. M. Rothschild, H. Egger, R.T. Hawkins, J. Bokor, H. Pummer and C.K. Rhodes, Phys. Rev. A 23, 206 (1981).
12. C.R. Vidal, 'Precise determination of potential energy curves from spectroscopic data' in LASER 81, New Orleans, Dec. 1981.
13. J.A. Armstrong, N. Bloembergen, J. Ducuing and P.S. Pershan, Phys. Rev. 127, 1918 (1962).
14. H. Puell and C.R. Vidal, Phys. Rev. A 14, 2225 (1976).
14a. R.T. Hodgson, P.P. Sorokin, and J.J. Wynne, Phys. Rev. Letters 32, 343 (1974).
15. S.E. Harris and R.B. Miles, Appl. Phys. Letters 19, 385 (1971).
16. J.F. Young, G.C. Bjorklund, A.H. Kung, R.B. Miles and S.E. Harris, Phys. Rev. Letters 27, 1551 (1971).
17. C.R. Vidal and F.B. Haller, Rev. Scient. Instrum. 42, 1779 (1971).
18. H. Scheingraber and C.R. Vidal, Rev. Scient. Instrum. 52, 1010 (1981).
19. H. Langer, H. Puell and H. Röhr, Opt. Commun. 34, 137 (1980).
20. R. Hilbig and R. Wallenstein, IEEE J. Quant. Electr. QE-17, 1566 (1981).
21. J. Reintjes, C.Y. She and R.C. Eckardt, IEEE J. Quant. Electr. QE-14, 581 (1978).
22. H. Scheingraber, H. Puell and C.R. Vidal, Phys. Rev. A 18, 2585 (1978).
23. H. Puell, H. Scheingraber and C.R. Vidal, Phys. Rev. A 22, 1165 (1980).
24. H. Scheingraber and C.R. Vidal, Opt. Commun. 38, 75 (1981).

25. C. Leubner, H. Scheingraber and C.R. Vidal, Opt. Commun. 36, 205 (1981).
26. E.A. Stappaerts, G.W. Bekkers, J.F. Young and S.E. Harris, IEEE J. Quant. Electr. QE-12, 330 (1976).
27. H.B. Puell and C.R. Vidal, IEEE J. Quant. Electr. QE-14, 364 (1978).
28. W. Demtröder, 'Laser Spectroscopy', Springer Series in Chemical Physics, Vol. 5, Springer Berlin, 1981.
29. R.R. Freeman, G.C. Bjorklund, N.P. Economou, P.F. Liao and J.E. Bjorkholm, Appl. Phys. Letters 33, 739 (1981).
30. W. Kaufman and B. Edlén, Phys. Chem. Ref. Data 3, 825 (1974).
31. A.C. Provorov, B.P. Stoicheff and S. Wallace, J. Chem. Phys. 67, 5393 (1977).
32. J.R. Banic, R.H. Lipson, T. Efthimiopoulos and B.P. Stoicheff, Opt. Letters 6, 461 (1981).
33. C.R. Vidal, Opt. Letters 5, 158 (1980).
34. H. Zacharias, H. Rottke and K.H. Welge, Opt. Commun. 35, 185 (1980).
35. H. Scheingraber and C.R. Vidal, paper ME 13 of this volume.
36. N.W. Carlson, F.V. Kowalski, R.E. Teets and A.L. Schawlow, Opt. Commun. 29, 302 (1979).
37. N.W. Carlson, A.J. Taylor and A.L. Schawlow, Phys. Rev. Letters 45, 18 (1980).
38. F.H.M. Faisal, R. Wallenstein and H. Zacharias, Phys. Rev. Letters 39, 1138 (1977).
39. S.V. Filseth, R. Wallenstein and H. Zacharias, Opt. Commun. 23, 231 (1977).
40. S.C. Wallace and K.K. Innes, J. Chem. Phys. 72, 4805 (1980).
41. J. Jortner and S. Leach, J. de chim. physique 77, 7 (1980).

Generation of Narrowband Tunable VUV Radiation

R. Hilbig and R. Wallenstein

Fakultät für Physik, Universität Bielefeld

48 Bielefeld, West Germany

Nonresonant sum- and difference-frequency mixing of the fundamental (ω_L) and the second harmonic (ω_{uv}) output of a powerful narrowband pulsed dye laser excited by a Nd-YAG laser has been investigated in Xe and Kr. The sum-frequency $\omega_{vuv} = 2\omega_{uv} + \omega_L$ is tunable in spectral regions of negative dispersion between 110 nm, and 130 nm. The maximum VUV pulse power exeeds 20 W ($5 \cdot 10^{10}$ photons/pulse). The difference frequency $\omega_{vuv} = 2\omega_{uv} - \omega_L$ provides VUV light pulses with up to 60 W ($2.3 \cdot 10^{11}$ photons/pulse) at wavelengths between 185 nm and 207 nm. Coherent VUV light of shorter wavelength (159.5 nm to 186.6 nm) is obtained by mixing the UV dye laser radiation with the infrared output (ω_{IR}) of the Nd-YAG laser ($\omega_{vuv} = 2\omega_{uv} - \omega_{IR}$). With UV light of shorter wavelength ($\omega'_{uv} = \omega_{uv} + \omega_{IR}$) the difference frequency conversions $\omega_{vuv} = 2\omega'_{uv} - \omega_L$ and $\omega_{vuv} = 2\omega'_{uv} - \omega_{IR}$ allow the generation of VUV light with wavelengths between 122.6 nm and 160 nm. Thus with a single dye laser which is operated in the most efficient operating range of Nd-YAG laser pumped dye lasers (ω_L = 550 - 670 nm) the investigated conversion schemes generate intense coherent VUV light which is continuously tunable between 110 nm and 210 nm.

Besides the nonresonant frequency conversion two-photon resonant frequency mixing has been studied in detail in Xe and Hg. In these experiments the resonant enhancement of the nonlinear processes increased the VUV pulse power to values in the range of 0.5 to 10 KW.

In Xe the two-photon transitions 5p - 6p, 5p - 7p and 5p - 4f have been used for a resonantly enhanced generation of VUV light at $\omega_{vuv} = 2\omega_1 \pm \omega_L$, where ω_1 is the two-photon transition frequency and ω_L is tunable in the spectral range of λ_L = 220 - 760 nm. While the sum-frequency generates radiation in the XUV (λ_{vuv} = 73 nm to 101 nm) the difference frequency is continuously tunable between 129 nm and 220 nm. The experimental results provide, for example, detailed information

ISSN:0094-243X/82/900442-03$3.00 Copyright 1982 American Institute of Physics

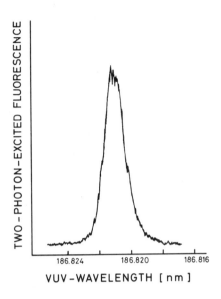

Fig. 1 Two-photon excitation of the 4p[1/2]0 level of Argon. The narrowband VUV
radiation (with a pulse power of more than 2KW) is generated by difference-frequency mixing in Hg vapor. The excitation is observed via the
transition (4p[1/2]0 → 4s[3/2]1) at λ = 751.5 nm

on conversion limiting saturation phenomena and their dependence on the input

power and the detuning of ω_1 from the two-photon resonant value.

In Hg the resonant frequency conversion is investigated for the resonances 6^1S_0

- 7^1S_0 and 6^1S_0 - 6^1D_2. Below the ionization limit intense VUV is observed at

the sum-frequency in the wavelength range of λ_{vuv} = 119 nm - 127 nm in the vicin-

ity of excited states 6s np (n$\geq$9). Continuously tunable radiation is provided by

the sum-frequency above the ionization limit (λ_{vuv} = 111.7 - 116 nm) and by the

difference frequency at λ_{vuv} = 177 - 188 nm.

While the intensity of the VUV light generated by the nonresonant frequency conversion is sufficient for precise linear spectroscopy, the high pulse power obtained by the resonant frequency mixing allows the application for nonlinear spectroscopy in the XUV (see Fig. 1). Since very intense narrowband VUV is generated at wavelengths as short as 120 nm the two-photon excitation will be advantageous for doppler-free investigations in the XUV at λ = 60 to 100 nm.

GENERATION OF TUNABLE, COHERENT 79 nm RADIATION BY FREQUENCY MIXING

H. Egger, T. Srinivasan, K. Boyer, H. Pummer and C. K. Rhodes
Department of Physics, University of Illinois at Chicago Circle
Post Office Box 4348, Chicago, Illinois 60680 USA

ABSTRACT

Tunable coherent radiation in the 79 nm range (200 mW) has been generated by sum frequency mixing of two ArF* photons and one visible dye laser photon.

SUMMARY

Excimer laser systems provide highly attractive fundamental sources for frequency conversion to the XUV because of their demonstrated ability to deliver tunable, very high spectral brightness radiation in the ultraviolet region. Coherent, tunable radiation at practically any wavelength in the 20 nm - 100 nm region can be obtained by either harmonic generation or frequency mixing using excimer lasers and dye lasers as fundamental sources. Fig. 1 shows the wavelengths that can be achieved by these processes. Here, we report the generation of tunable coherent radiation at $\sim$ 79 nm by sum frequency mixing 2 ArF* photons and one visible dye laser photon.

The visible dye photon was obtained from a XeF* pumped, tunable dye laser (FL 2002, Lambda Physik), operated with stilbene 420. The radiation is tunable from 416 nm to 458 nm with an output energy of $\sim$ 5 mJ, and pulse duration $\sim$ 15 ns. The ArF* laser used in this experiment is a modified version of a system described previously[1]. The output of an excimer pumped pulsed dye laser (FL 2000, Lambda Physik) operating with Rhodamine 6G at $\sim$ 580.5 nm is passed through a grating - pinhole arrangement to reduce the amplified spontaneous emission and is amplified in a single stage dye amplifier. The absolute wavelength calibration was done using a standard source and monochromator. The output of the amplifier is focussed (50 cm focal length lens) into a Sr heat pipe to generate the third harmonic of the dye laser pulse. This UV radiation is amplified in a double pass and two subsequent single pass ArF* amplifiers, using spatial filters and a grating - pinhole arrangement to suppress the amplified spontaneous emission. The output characteristics of this ArF* system are: energy $\sim$ 200 mJ, pulse duration $\sim$ 7 ns and wavelength 193.5 nm. For frequency mixing, the dye laser and ArF* radiations are focussed through a 1 mm pinhole, by a 1 m focal length lens into the nonlinear medium. The spatial overlap of the two foci is achieved by compensating for the dispersion of the focussing lens with a telescope in the dye laser beam.

Earlier experiments[2,3] indicate that the absorption of the generated radiation by the nonlinear medium limits the conversion efficiency. Fig. 2 shows the schematic of a cell designed to reduce this limitation. The nonlinear medium, H_2, flows through the central tube of 6 mm inner diameter. A buffer gas, which exhibits a low absorption cross section at the mixed wavelength, in this case Ne, flows through the outer tube. By keeping the Ne pressure slightly higher than the H_2 pressure, the H_2 is confined to the inner tube and the maximum absorption length is limited to 6 mm. By shifting the focus to the exit port of the H_2 cell, the absorption length can be reduced essentially to the confocal parameter of the focussing lens. With this configuration, 200 mW of tunable coherent radiation in the 79 nm range has been observed. An increase of 2 orders of magnitude in the energy of the usable radiation has been achieved by this method of minimizing the absorption of the XUV radiation.

To illustrate the spectroscopic utility of this radiation, the source was tuned across the $^2P_{1/2}$ 9d autoionizing resonance in argon. This was done by introducing $\sim$ 200 mtorr of Ar in the differential pumping stage and tuning the dye laser. Fig. 3 shows the measured absorption profile normalized to the data from Hudson and Carter[4]. The linewidth of this source can be estimated from scans across the absorption lines of D_2 in this region. Fig. 4 shows such a scan. The measured absorption curve agrees well with the data taken by Dehmer et al.[5] with 2.5 cm^{-1}. resolution. The narrowest feature in this scan has a width $\sim$ 2 cm^{-1}. If this line is assumed to be Doppler broadened (0.7 cm^{-1}), a source width of $\sim$ 2 cm^{-1} corresponding to a bandwidth of $\sim$ 0.7 cm^{-1} for the dye lasers is obtained. These bandwidths can be reduced by either introducing intracavity etalons in the pulsed dye lasers or using a pulse amplified single frequency CW dye laser.

With the same experimental arrangement, tripled radiation with peak powers up to 30 W has been obtained with H_2, D_2, Kr, Ar and CO as the nonlinear medium, Ne as the buffer gas and the ArF* radiation as the fundamental source[2]. This coherent radiation at $\sim$ 64 nm is tunable over the gain profile of the ArF* amplifier. In recent experiments with a similar setup, preliminary measurements indicate that 20 kW radiation at $\sim$ 64 nm corresponding to a conversion efficiency of $\sim$ 2 x 10^{-5} can be obtained with 10 mJ ArF* radiation of pulse duration $\sim$ 10 ps.

Other wavelength regions that have been reached so far are indicated in Fig. 1. Radiation at 103.3 nm and 61.6 nm has been obtained by third and fifth harmonic conversion of XeCl radiation with Ar as the nonlinear medium[6,7]. Third harmonic radiation of Xe$_2$* at 57 nm has been generated by Hutchinson et al.[8] By sum frequency mixing of one KrF* or XeCl* laser photon with 2 dye laser photons, Caro et al.[9] have observed radiation at 115 nm and 127 nm, respectively. Generation of tunable, coherent, narrowband ($\Delta\nu$ = 0.01 cm^{-1}) radiation at $\sim$ 83 nm by two photon near resonant frequency tripling in Xe gas, with a high spectral brightness KrF* source ($\sim$ 248 nm) has also been reported previously[10]. The experimental apparatus is shown schematically in Fig. 5. The high spectral brightness ultraviolet radiation is focussed by a 10 cm focal length lens into a 350 μm pinhole, through which Xe flows from the tripling cell to the differentially pumped chamber. With 10 torr of Xe, third harmonic radiation with peak power $\sim$ 40 mW corresponding to a conversion efficiency $\sim$ 10^{-8} is observed. It must be noted that the effect of absorption of the third harmonic signal by the nonlinear medium has not been minimized in this experiment. An increase of $\sim$ 10^2

in the usable 83 nm radiation can be expected by using the flow geometry shown in Fig. 2.

Another wavelength region that had been reached using the high spectral brightness KrF* radiation is in the vicinity of Lyman β (102.6 nm). Sum frequency mixing of 2 KrF* photons (pulse energy $\sim$ 60 mJ and pulse duration $\sim$ 10 ns) with 1 visible dye photon (Rhodamine 6G, pulse energy $\sim$ 6 mJ and pulse duration $\sim$ 10 ns) has yielded a few milliwatts of tunable coherent radiation at $\sim$ 102.6 nm. Once again, no attempt was made to reduce the absorption of the radiation.

It has been shown[11] that for optimum conversion efficiency, $b\Delta k = -2$ ($\Delta k = k_3 - 3k_1$) for the above processes. Estimates indicate that in our experiments, $b\Delta k$ is much smaller than the optimum value. Hence, by using a phase matching gas to increase the mismatch to $-2/_b$, the conversion efficiency can be improved substantially. With optimum phase matching, further increase in the generated signal can be expected with the increase in the pressure of the nonlinear medium. The usable upconverted radiation can further be increased by modifying the detection system as shown in Fig. 6. With such an arrangement, absorption cross sections $\sim 10^{-20} cm^2$ can be easily measured with good discrimination by suitably adjusting the length and pressure of the absorbing medium.

In conclusion, coherent, tunable, narrowband radiation in the XUV has been generated by frequency tripling of ArF* and KrF* radiation and by sum frequency mixing 2 excimer photons and one visible dye photon. With improved conversion efficiency and suggested modifications in the detection system, this radiation will prove to be a significant tool for high resolution spectroscopic applications.

The authors wish to acknowledge the expert technical assistance of J. Wright, S. Vendetta and M. Scaggs. This research was supported by the Air Force Office of Scientific Research, the Office of Naval Research, the National Science Foundation under grant no. PHY78-27610 and the Department of Energy under contract nos. DE-AC02-79ER10350, DE-AC02-80ET33065, and DE-AC08-81DP40142.

REFERENCES

1. H. Egger, T. Srinivasan, K. Hohla, H. Scheingraber, C. R. Vidal, H. Pummer and C. K. Rhodes, Appl. Phys. Lett. <u>39</u>, 37 (1981).

2. H. Pummer, T. Srinivasan, H. Egger, K. Boyer, T. S. Luk, and C. K. Rhodes, Opt. Lett. March 1982 (to be published).

3. R. Mahon, T. J. McIlrath, V. P. Myerscough, and D. W. Koopman, IEEE J. Quantum Electron. QE-<u>15</u>, 444 (1979).

4. R. D. Hudson and V. L. Carter, J. Opt. Soc. Am. <u>58</u>, 227 (1968).

5. P. M. Dehmer, P. S. Dardi and W. A. Chupka, Argonne National Laboratory, Radiological and Environmental Research Division Annual report, Oct. '78 - Sept. '79 ANL - 79-65, Part I, P-7.

6. J. Reintjes, Opt. Lett. $\underline{4}$, 242 (1979).

7. J. Reintjes, R. Christensen and L. L. Tankersley in Digest of Conference on Lasers and Electro-optics (Optical Society of America, Washington D. C. 1981) Paper WB4.

8. M. H. R. Hutchinson, C. C. Ling and D. J. Bradley, Opt. Commun. $\underline{18}$, 203 (1976).

9. R. G. Caro, A. Costela and C. E. Webb, Opt. Lett. $\underline{6}$, 464 (1981).

10. H. Egger, R. T. Hawkins, J. Boker, H. Pummer, M. Rothschild and C. K. Rhodes, Opt. Lett. $\underline{5}$, 282 (1980).

11. G. C. Bjorklund IEEE J. Quantum Electron. QE-$\underline{11}$, 287 (1975).

449

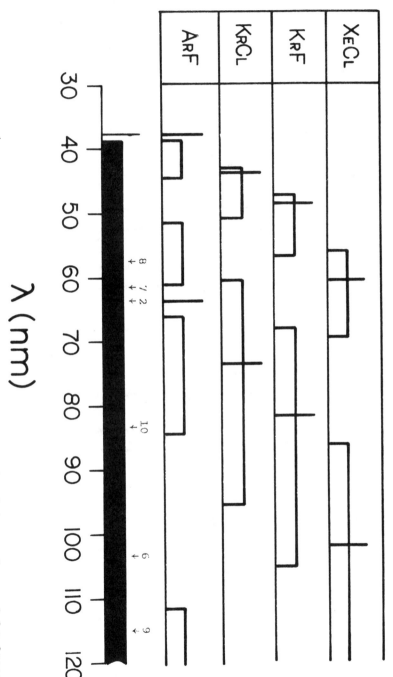

Fig. 1 Wavelength regions that can be covered by nonlinear processes involving excimer and dye lasers. Observed wavelengths are indicated by the vertical arrows; the associated number denotes the corresponding reference.

450

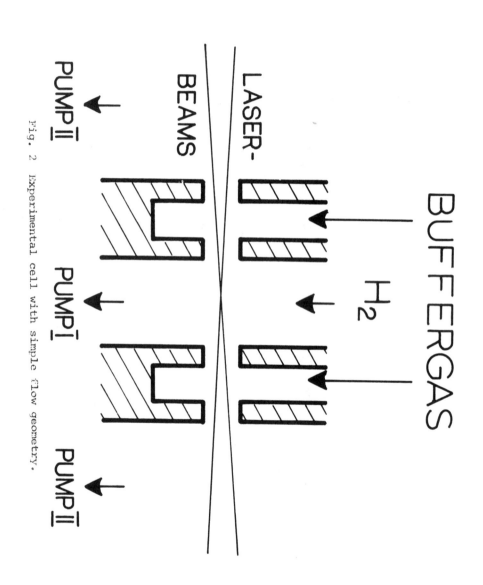

Fig. 2 Experimental cell with simple flow geometry.

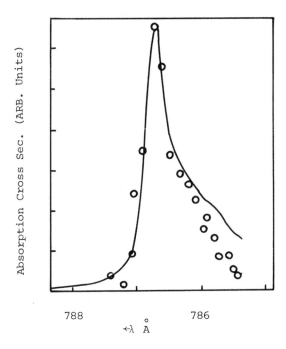

Fig. 3 Measured absorption profile of $2_{P_{1/2}}$ 9d' autoionizing resonance of argon normalized to data from Hudson and Carter[4].

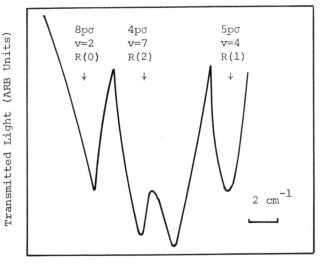

Fig. 4 Relative absorption of D_2 around 79.3 nm.

Absorption Cross Sec. (ARB. Units)

788   786

←λ Å

Transmitted Light (ARB Units)

8pσ   4pσ   5pσ
v=2   v=7   v=4
R(0) R(2) R(1)
↓   ↓   ↓

2 cm^{-1}

λ →

452

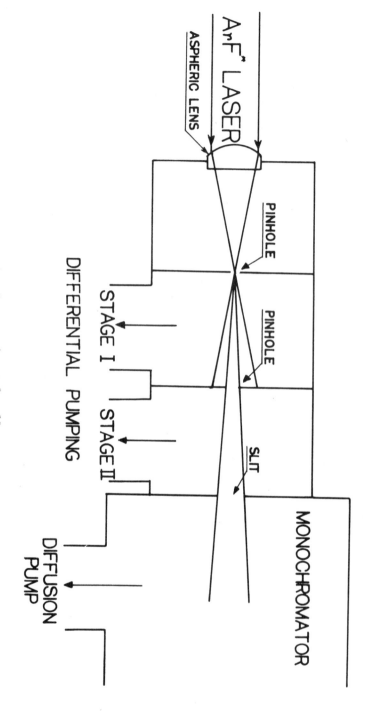

Fig. 5 Experimental cell.

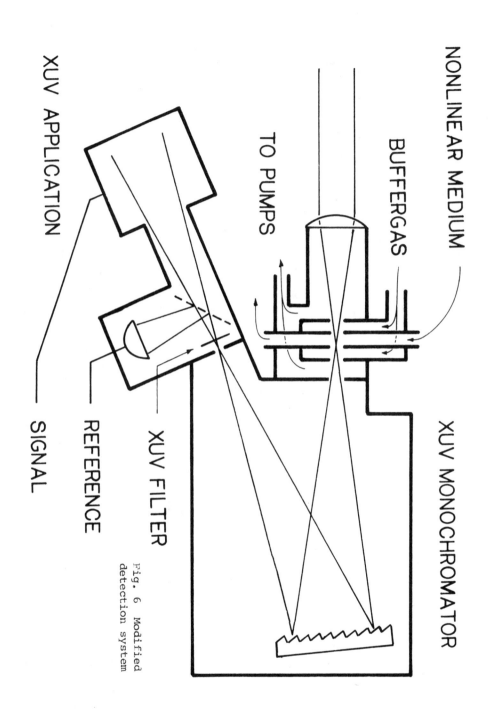

NONLINEAR MEDIUM

BUFFERGAS

TO PUMPS

XUV MONOCHROMATOR

XUV APPLICATION

SIGNAL

REFERENCE

XUV FILTER

Fig. 6 Modified
detection system

UPCONVERSION OF LASER RADIATION
TO γ-RAY ENERGIES

C. B. Collins
Center for Quantum Electronics
University of Texas at Dallas
P.O. Box 688, Richardson, TX 75080

ABSTRACT

Both the coherent and incoherent pumping with intense optical radiation of electromagnetic transitions in nuclei were reviewed in this work. In the first case the anti-Stokes upconversion of conventional laser radiation was shown to be a potentially viable means for pumping a γ-ray laser. With the most favorable possible arrangement of nuclear energy levels, the threshold for stimulated output at 10 KeV might be reached at a few tens of Joules for isomeric media with lifetimes of the order of tens of years. Whether such an arrangement of excited states actually exists will require additional experimentation.

In the case of the incoherent pumping of nuclear material, the resonant excitation of energetic states of nuclei by X-radiation from laser plasmas or exploding wires was modeled with results similar to those for the case of coherent pumping. Thresholds for stimulated output could be as low as a few hundred Joules into the primary laser pulse producing the X-ray flash. Again the possibilities were found to be critically dependent upon unknown parameters, particularly upon the efficiency for the conversion of input energy into X-rays. The overall possibilities for the excitation of a γ-ray laser with optical pumping were concluded to be reasonably encouraging and the need for additional spectroscopic data was clearly defined.

INTRODUCTION

The objective of this study has been to estimate the feasibility of an optically pumped, γ-ray laser. Surprisingly, it is the paucity of spectroscopic data for potential laser media that makes a feasibility demonstration into a formidable problem, rather than any known obstacle to be overcome. It may even be easy to stimulate nuclear radiation. The extensive data base, essential to quantitative modeling, does not yet exist nor do the facilities for building it. Based upon existing knowledge, the prospects seem encouraging enough that the initiation of a modest experimental effort to provide the critical data seems warranted.

The most promising means for realizing a γ-ray laser have evolved from a theoretical breakthrough[1] in which it was realized that optical resonances might exist in the nucleus and so provide the key for the use of high power, but conventional laser radiation, to stimulate the release of stored nuclear energy. Several specific mechanisms were subsequently described,[2-6] all involving optical pumping of nuclear material. Some processes[6] are very analogous to the optical pumping techniques used to produce the ruby laser in 1960, but transformed in a manner suitable for nuclear transitions. Whether or not any of these approaches can reach threshold depends upon the resolution of basic issues lying in an interdisciplinary region between quantum electronics and nuclear physics that have not been previously addressed. Because of this interdisciplinary context requiring the fusion of concepts taken from previously unrelated fields of physics, it is useful to review the technical background both from the viewpoint of nuclear physics and from laser physics. Then a synthesis into a new perspective, that of nuclear quantum electronics can be considered.

TECHNICAL CONSIDERATIONS

Nuclear Concepts

At the nuclear level the storage of energy can approach tera-Joules (10^{12}J) per liter, thus presenting radically improved possibilities for all processes depending upon the use of concentrated energy. As generally perceived[7] this stored energy results from the motion of the constituents of a nucleus that is quantized into a series of stationary states characterized by their energy, angular momentum, and parity, much as in the case of the motion of atomic electrons. If the nucleus is in one of its excited states, it generally undergoes a transition to a lower state with a corresponding transfer of energy to an electromagnetic radiation field, to atomic shell electrons, or to newly created particles. However, it has been recently shown[1-6] that the channel in which the transition energy and angular momentum is coupled into the radiation field has the potential to support the stimulated emission of γ-radiation. Such a stimulated release of the energy stored in nuclear transitions would occur at the rate at which resonant electromagnetic radiation passed through the laser medium and could lead to output powers as great as 3×10^{21} Watts/liter.[6] This is an astronomical level of intensity and has not been achieved on earth by any means previously.

Viewed from the perspective of nuclear physics, the release of this energy into electromagnetic fields would represent a third source of nuclear power, distinct from either fission or fusion. It would have particular advantages denied to the extant sources of nuclear power that accrue from the fact that the nuclear reaction is propagated by photons rather than by material particles. In the scheme originally suggested for coherent optical pumping[1-6] it had been proposed that the energy would be extracted from the population of a metastable, isomeric state of nuclear excitation, completely inverted with respect to some lower level of energy. Since the lifetimes of such states can reach hundreds of years, such initial conditions of energy storage could be readily attained. Moreover, if the fuel isomer is produced, as many isomers are, by the capture of thermal neutrons in another isotope, there is virtually no energy loss to the fission reactor in which it is bred. After separation of the irradiated material, even the pure isomeric fuel has no potential for spontaneous explosion in any size of accumulation as the critical mass for single photon gain is infinite because of the spurious broadening of the transition from the metastable state.[8] Apparently, the pure fuel isomer could be ultimately utilized only upon excitation by the pump laser or maser acting upon the optical resonances described in the following section.

From the perspective of nuclear physics the clearest advantage of this photon-propagated laser process is in the "critical mass" needed to support the extraction of energies. The cross sections per nucleus for electromagnetic processes are larger than for comparable processes mediated by particle collisions. In general, the cross section for a fully-allowed process pumped by the optical input as discussed below should be of the order of the Breit-Wigner value, $\lambda^2/2\pi$. For typical transition energies around 10 KeV, the resulting cross section reaches 10^{-17} cm^2 or 10^7 barns! This is to be contrasted to the 570 barn cross section per nucleus for the fission of ^{235}U induced by collisions with thermal neutrons.[9] Masses of the order of <u>milligrams</u> should lie at the threshold for critical size in the laser process, in contrast to critical masses of kilograms for fission. While the energy storage of the photon-propagated reactions appears to be limited to the order of 1 MeV per nucleus in contrast to the 170 MeV available from fission, the absence of a need for a moderator compensates this somewhat. Moreover, the response time of the photon propagated process greatly exceeds that of either fission or fusion, another advantage accruing from the propagation of the reaction at the speed of light.

In some cases not only the power but even the released energy has remarkable properties. For example the full stimulation of a typical isomer such as ^{180}Hf, because of cascading would release of the order of 600 MJ/gram, much of it as 400 KeV γ-rays. If emitted isotropically this would correspond to 1000 Roentgens/gram of fuel on a spherical surface 1 Km in diameter in air at sea level pressure. However, with properly directed pumping so that lasing occurred, the same output would be achieved in a highly directed beam. Since only one input photon is lost for each output photon produced the total energy requirements from the pump are small, even for full consumption of the fuel. For example, in this case an integrated input energy of the order of only 10 J/gram would be required to effect the release of the 600 megaJoules/gram.

Finally it should be recognized that this photon-propagated process tends to consume the radioactivity of its fuel leaving either stable wastes or isotopes with short lifetimes to spend before reaching stability. Unlike conventional nuclear power release, since neither fission nor induced radioactivity is to be expected, the neutron-rich fragments leading to long-lived waste materials[9] would not be produced.

Unfortunately, if analyzed strictly from the perspective of nuclear physics, the feasibility for a γ-ray laser capable of

achieving these remarkable benefits is completely negative. A
recent review along such lines[8] dealt extensively with concepts
involving the use of a neutron flux for pumping the laser medium,
either in real-time or as a preparatory step to be followed by a
rapid separation of isotopes within their relatively short natural
lifetimes. All those proposals were concluded to require infeasibly
high levels of particle fluxes to pump the inversions, exceeding
even those available from nuclear explosions, and to require neutron
moderators having virtually infinite thermal capacities. It was
generally concluded that the relative difficulty in configuring a
γ-ray laser lies not in the smallness of the stimulated emission
cross section, but rather in the requirement for remarkably high
levels of pumping fluence needed to compensate the spontaneous
power radiated at threshold. Although the theoretical breakthrough
which might resolve these problems had been published[1] two years
prior to the review of conventional approaches, it was not recognized
that the fluences typically available from conventional, high power
lasers, exceed those values available from particle fluxes and it
is conceivable that threshold levels of nuclear excitation might be
reached through some type of conversion of incident laser radiation.
While it is probable that such levels of optical input might lie
above the damage threshold of the host material, even in the most
extreme cases, the disassembly of the laser material can proceed
only at the velocity at which a shock can propagate through it
while the development of the γ-ray output can occur at the speed of
light, so that single shot operation could be usefully considered.
A full realization of the possibilities required the perspective of
quantum electronics be employed in addition to that of nuclear
physics.

Nuclear Quantum Electronics

While the storage of energy in nuclear media can be treated
adequately within the traditional framework of nuclear physics, its
release through conventional processes is prevented for the reasons
described above. However, several nonlinear optical processes of
quantum electronics have analogs at the nuclear level[1-6] that may
provide for the realization of the potential performance discussed
above. The particular process described in the first works[1-5] was
a type of anti-Stokes Raman scattering of intense but conventional
laser radiation from nuclear isomers. The theoretical treatment
served to estimate matrix elements for a new class of two-photon
Mössbauer transitions[4] making possible, in principle, the frequency
upconversion of optical laser photons to gamma ray energies. Such
a process represents the means for accelerating the natural decay
of radioactive elements.

Even more recently it has been realized,[6] that particular
process is only a special, coherent case of optical pumping of
nuclear transitions. There is an incoherent technique which is
almost as promising.

Perhaps surprisingly, the conclusions to be reached from
modeling either process are not particularly negative. Rather, the

feasibilities of both the coherent and incoherent pumping of nuclear
transitions with "optical" radiation have been found to depend upon
various nuclear properties that are presently uncertain by several
order of magnitude. The amplitudes for upconversion depend critically upon the energies and symmetries of potentially resonant
intermediate states for the process that would lie in the "blind
spots" of conventional nuclear spectroscopy, and which must be
resolved with new technology. The laser threshold for incoherent
pumping is completely dependent upon the efficiency for the conversion to X-radiation of the laser power input to dense plasmas and
on the coefficients for coupling that radiation into the nucleus.
Reports in the literature span a range of values which would cause
threshold estimates to vary from routine to impossible. It appears
that the results of any current attempts to model these processes
must confirm the conclusions independently reached by Baldwin
et al.[8] that much additional basic study of the structure and
properties of nuclear excitation, particular near isomeric levels,
must precede any truly indicative estimate of the feasibility of a
gamma ray laser.

COHERENT OPTICAL PUMPING

The scheme originally introduced[1-5] for coherent optical
pumping can be readily visualized as a simple 3-level system if it
is assumed that a metastable, isomeric state is completely inverted
with respect to some lower level. Because of the spurious broadening of the transitions directly to lower levels, this initial
inversion is not directly useful for the production of stimulated
emission.

At sufficiently high levels of laser intensity, transitions
transfer population to virtual intermediate states, displaced in
energy from the intial state by $\pm n\hbar\omega$ to an extent necessary to
invert one or the other with respect to the lowest level. Of
course, the smaller the detuning, ΔE, between the energy of a
virtual state and the nearest real excited state of suitable symmetry the longer its lifetime will be and hence, the higher the
population that will be achieved at any fixed level of input power.
Before the stimulated intensity builds and begins to saturate the
gain in the γ-ray transition from the intermediate state to the
lowest level, the composite process can be described by a cross
section[4] which in the limit of high laser intensity becomes equal
to the Breit-Wigner cross section $\sigma^{(1)}$,

$$\sigma^{(1)} \equiv \lambda^2/2\pi \tag{1}$$

where the superscripted (1) denotes a one photon process.
A more tractable view is that at sufficiently high pumping
intensity, the forbidden transition between the isomeric state and
the lower level becomes fully allowed at the limiting cross section
given by $\sigma^{(1)}$ but at the sum (or difference) frequency. <u>The</u>

resulting multiphoton, Mossbauer width[4] is preserved down to a limit
of at least 10^5 to 10^6 Hz. Quantitatively it has been shown[4] that
if the nearest intermediate state is resonant with the pump laser
transition, the composite rate passes to the Breit-Wigner limit of
$\sigma^{(1)}$ at a power of 5.3×10^5 W/cm^2 at the apparent minimum value
for spurious broadening of 10^6 Hz, even if the transition to the
intermediate state must proceed as a magnetic dipole, M1 transition, the most common type of nuclear transition. However, the
minimum threshold fluence needed to produce this effect scales only
with $(\Delta E)^{+1}$. This is because the duration of the pumping pulse
must be increased in proportion to $(\Delta E)^{-1}$ so that the transform-
limited bandwidth of the pumping radiation should not exceed the
spurious width of the level. Simply forming the product $F = I_c \times$
$(\pi \Delta \nu)^{-1}$, where I_c is the critical intensity and $\Delta \nu$ is the spurious
width, gives the threshold fluence, F. For the case, mentioned
above the threshold fluence would be 167 mJ/cm^2 for $\Delta \nu = 10^6$ Hz.
This is quite an accessible level of input intensity and fluence
and provided that the nearly resonant intermediate state exists,
the subsequent development of stimulated output can be treated as
if the transition from the upper storage level of population to
the lower, suddenly became allowed at the Breit-Wigner rate. In
this case the coherent pumping and incoherent pumping developed
below show subsequently similar features and constraints and can be
treated in common.

INCOHERENT OPTICAL PUMPING

This scheme can be readily visualized as a 4-level system with the
energy originally being stored in a metastable isomeric level that
would be pumped further to a transfer level by the incoherent
absorption of optical radiation that had been converted to X-ray
wavelengths. The upper laser level would be pumped indirectly by
cascading from the transfer level chosen to have as short a lifetime as possible and hence, as broad an absorption bandwidth as can
be arranged. The Weisskopf estimates[10] for the widths of single
particle states of the nucleus, show that a lifetime of 1 to 10
psec would be reasonable for a level decaying through an electric
dipole, E1, cascade transition at an energy of the order of 10 KeV.
The corresponding width in the worse, 10 psec case, would be around
6.6×10^{-5} eV and this small value represents the greatest difficulty associated with optical pumping. Multiparticle resonances
do occur in various nuclei and upon occasion give lifetimes to
10^{-15} and even 10^{-18} sec. While such effects might fortuitously
enhance the pumping efficiency by orders of magnitude, they are
somewhat spurious and were not incorporated in the analysis pursued
in this work.

In the most ideal case the upper laser level can be assumed to
be longer lived so that the flux of population delivered by the
cascade through the short-lived transfer level can be integrated
for a period comparable to the duration of the external pumping
pulse. For modeling purposes it was assumed that such a matching

was effected so that the total fluence from the "optical" pump
lying within the absorption bandwidth would be usable in delivering
population to the upper laser level. Because of the spurious
broadening of the order of 10^5 to 10^6 Hz to be expected in
Mössbauer lines it was not considered reasonable to treat upper
level lifetimes, and hence pumping pulses longer than 10^{-6} sec.

If such an idealized arrangement of energy levels can be found
in practice quantum efficiencies might approach a factor of 30 to
100. Limitations arise principally from the restrictions that the
quantum of output energy must be less than 150 KeV to preserve the
Mössbauer effect and the photon of pumping light must be energetic
enough to reduce the chances of nonresonant absorption in the
medium. This gain in the photon energy can partially compensate
the great inefficiencies resulting from the mismatched bandwidths
of the pump and absorber, so that not only the spectral brightness
but also the total brightness of the final output can significantly
exceed that of the pumping flash, a minimal condition for a laser.

The most probable source of the intense level of X-ray radiation needed to pump the population of the initial level to the
transfer level is a plasma produced at the focus of a conventional
laser. The transverse pumping of a filament of nuclear medium
provides a reasonable geometry in which the mean free path for the
absorption of resonant X-ray photons is much smaller than the
typical diameter of laser plasmas, so the geometrical limit on the
pumping fluence is primarily a consequence of the extent of the
plasma. With these assumptions the resulting population in the
upper laser level in the nuclear material can be readily estimated
if the energy in the characteristic X-ray lines emitted by the
plasma is known.

STIMULATED EMISSION

As discussed by Baldwin et al.[8] and developed by Collins
et al.[6] the development of output from a γ-ray device will be
limited by several common factors, whatever the technique used to
produce the population of the upper laser level. As could be
expected, primary problems concern nonresonant, nonsaturable absorption in the medium and the geometry of the output. However, in the
cases considered here, the Fresnel number will be arranged to be
greater than 10 so that diffraction losses may be neglected.

Under small signal conditions the overall logarithmic amplification, α for a single pass down the length L, of a filament being
pumped could be approximated to be

$$\alpha = N_U \sigma_R L - N_0 \sigma_{NR} L \quad , \tag{2}$$

where σ_R and σ_{NR} denote the cross sections for the resonant emission given by Eq. (1) evaluated at the output wavelength, and for
the nonresonant absorption in the medium and diluent, respectively,

and N_U and N_0 are the upper state and initial concentrations, respectively.

In the case of <u>coherent</u> optical pumping the value of N_U to be substituted into Eq. (2) is the concentration of the isomeric state initially prepared. This results from the fact that, as explained above, the pumping flux acts mainly to switch the cross section for stimulated emission from its "forbidden" value to the Breit-Wigner value. The requirement on laser intensity is imposed by the need to develop the field strength necessary to saturate the multiphoton step and was stated earlier. If the resonant intermediate state lies energetically above the initial state, some energy will be absorbed by the switching but it is relatively a small amount equal to one pump photon per nucleus in the initial isomeric state. For a completely inverted medium with a density of 10 gr/cm^3 and a molecular weight of 200, the concentration of active nuclei could be around 3×10^{22} cm^{-3}. A pumping resonance in the violet region of the spectrum for which $\hbar\omega = 5 \times 10^{-19}$ J would, then, provide for the absorption of 15 KJ/cm^3. Because of large energy gain associated with the upconversion of the frequency, the same medium would produce an output of 48,000 KJ/cm^3 at 10 KeV.

Since σ_R/σ_{NR} is of the order of 10^{+3} to 10^{+4} under most conditions of interest, the overall gain of Eq. (2) would be positive and large in this case of coherent upconversion, even with substantial dilution which might be desirable to improve the level of control of the amplified spontaneous emission finally generated.

In the case of incoherent pumping with flash X-rays from the laser plasma the evaluation of Eq. (2) gives

$$\alpha = \sigma_R N_0 \left[\frac{\eta \sigma_x}{2\pi R} - \left(\frac{\sigma_{NR}}{\sigma_R} \right) L \right] , \qquad (3)$$

where η is the number of X-ray photons emitted by the transient laser plasma of radius, R and σ_x is the cross section given by Eq. (1) for their absorption from the metastable to the transfer level, and where the ratio of resonant to nonresonant cross sections are given by Baldwin et al. The factor $\sigma_R N_0$ is large, since N_0 is of the order of 3×10^{22} cm^{-3} in a solid. The primary condition for threshold is that Eq. (3) be positive.

As a numerical example the radius R may be assumed to be 30 μm and the plasma length, L = 0.1 cm, in order to optimize Eq. (3) somewhat. Taking $\sigma_R/\sigma_{NR} \sim 10^3$, as is usually the case, the condition for threshold is that $\eta \geq 8 \times 10^{10}$. In actual practice the active nuclei would need to be diluted by a factor of almost 50 to increase the thickness of the pumped layer so that the objective of having the Fresnel number >10 could be met.

If it is assumed that the pumping radiation is derived from an X-ray line spectrum and that it is produced from a medium chosen so that one of its strong X-ray lines is resonant with the nuclear transition to the transfer state, the ratio of the widths of nuclear

E1 transitions to typical X-ray lines[11] suggests that a relative fraction of about 10^{-5} to 10^{-4} of the energy in the X-ray line could be absorbed by the initial state population. Using the smaller value of 10^{-5} for an X-ray energy around 10 KeV the threshold condition for the case considered immediately above would require the isotropic emission from the flash of the order of 13 J in a single X-ray line.

CONCLUSION

While the viabilities of both the techniques of coherent and incoherent optical pumping of nuclear transitions depend upon the discovery of a favorable arrangement of nuclear energy levels in some real isotope, the demands upon the actual pump pulse are not particularly extreme. Synthesis of the individual elements assumed in the treatment of coherent optical pumping indicated that a threshold fluence of only 167 mJ/cm^2 per pulse would be required. The bandwidth of the pulse would need to be comparable to the width of the resonant intermediate state of the multiphoton transition which in this idealized case would correspond to a lifetime of 318 nsec, a reasonable width.

Such a surprisingly low threshold fluence is a consequence of several fundamental factors. Of primary importance is the fact that nuclear transitions usually retain their natural linewidth because of the Mössbauer effect. As a result, the stimulated emission cross sections for nuclear radiation can easily exceed those for known laser transitions at visible wavelengths. The realization[4] that there should be a multiphoton Mössbauer channel allows this benefit to be extended to non-linear wavelength conversions of the type proposed for the coherent pumping technique. A second factor that facilitates the process is also surprising from the perspective of atomic transitions. Because of the larger gyromagnetic ratios of the nucleons, the deterioration of the probability encountered when the transition must be mediated by higher multipole moments is not particularly great. The numerical estimates used in this example were based upon the model treated in detail in the literature[4] which considered that the transition to the intermediate state induced by the optical radiation was of the magnetic dipole, M1 type. However, transitions of other levels of multipolarity are even more promising, because the M1 transition limits the lifetime for energy storage in the medium to about 10 sec.[6]

The pumping of an E2 transition presents a completely different picture. Selection rules on parity and ΔL allow the possibility for the transition for the spontaneous decay of the isomeric level to be as forbidden as M4. In that case although the threshold fluence would rise to 6.3 KJ/cm^2 the lifetime of the storage level could be of the order of 10^8 to 10^9 seconds implying a storage time in the isomeric state from 3 to 30 years.

For example, if the diameter of the laser medium were chosen so that 10% of the transiting energy at optical wavelengths was absorbed on a single pass through the medium, and if a reflective enclosure were constructed so this became the principal loss in the system, then it could be considered that each input photon would make 10 passes through the material. Since the pulse duration would be much longer than the time between transits, the input intensity and fluence could be reduced by a factor of 10. In such a case the diameter would need to be of the order of the reduced input fluence divided by the energy density lost in the medium, or about 0.042 cm in the example. Then with a length of 1.0 cm, the pumping threshold would be 26 J in a pulse of about 300 nsec and the corresponding output would be 66 KJ at 10 KeV.[6]

While these system considerations are quite encouraging, as mentioned above, the implementation of this technique depends upon the discovery of appropriate intermediate states that would lie in the "blind spots" of conventional nuclear spectroscopy. Current techniques do not provide for the resolution of narrow lines, nearly degenerate with long-lived isomeric states. New techniques are needed for the estimation of the probability for the occurrence of the states that would render practicable this scheme for the coherent optical pumping of a gamma ray laser.

The case for incoherent pumping is comparable[6] with an isotropic flash of the order of 10 J of X-ray line emission being required to reach threshold in the simple case modeled. The energy cost of such a flash is very difficult to obtain since estimates of the efficiency for the conversion to X-rays of the input energy to dense plasmas seem to vary by three orders of magnitude. At optimal values of input energy and target composition conversion efficiencies up to 10% have been reported[12] with poorer yields dropping to 10^{-4} at the highest energies.[13] This implies that, at best, the threshold for γ-ray emission could be attained by pumping with the flash from a plasma produced at the focus of a 100 J laser pulse and, at worst, from the plasma produced from a 100 KJ input to an exploding wire.

Even a cursory examination of the isotope tables[14] is sufficient to reveal several promising candidates for either coherent or incoherent optical pumping. However, the degree of resolution in the measurement of transition energies and possible degeneracies is insufficient to confirm or refute the occurrence of a suitable resonance in any system that might be examined. In agreement with the observations of Baldwin et al.[8] and Collins et al.[6] it appears that the conclusions of any modeling studies attemped at this time must be that much additional basic information about the structure and properties of excited nuclear levels, particularly near isomeric states, is needed before definitive conclusions could be drawn about the feasibility of optically pumping nuclear transitions.

ACKNOWLEDGEMENT

This modeling study has included a review of both published and unpublished results obtained in collaboration in part with F. W. Lee, D. M. Shemwell, and B. D. DePaola of the University of Texas at Dallas and in part with S. Olariu and I. I. Popescu of the

Central Institute of Physics under the U.S. Romanian Cooperative Program in Atomic and Plasma Physics, partially supported by the National Science Foundation under Grant No. INT 7618982.

REFERENCES

1. C. B. Collins, S. Olariu, M. Petrascu and I. Popescu, Phys. Rev. Lett. 42, 1397 (1979).
2. C. B. Collins, S. Olariu, M. Petrascu, and I. Popescu, Phys. Rev. C20, 1942 (1979).
3. C. B. Collins, in Proceedings of the International Conference on Lasers '80, edited by C. B. Collins (STS Press, McLean, VA, 1981) pp. 524-531.
4. S. Olariu, I. Popescu and C. B. Collins, Phys. Rev. C23, 50 (1981).
5. S. Olariu, I. Popescu and C. B. Collins, Phys. Rev. C23, 1007 (1981).
6. C. B. Collins, F. W. Lee, D. M. Shemwell, B. D. DePaola, S. Olariu and I. I. Popescu, J. Appl. Phys. (pending).
7. See for example, P. J. Brussard and P. W. M. Glaudemans, Shell Model Applications in Nuclear Spectroscopy (North-Holland, Amsterdam, 1977), Chaps. 9 and 10.
8. G. C. Baldwin, J. C. Solem and V. I. Gol'danskii, Rev. Mod. Phys. 53, 687 (1981).
9. M. G. Bower, Nuclear Physics (Pergamon Press, Oxford, 1973) pp. 300-318.
10. C. M. Lederer and V. S. Shirley, Table of Isotopes, 7th Edition (Wiley, New York, 1978), Appendices 24 and 25.
11. J. H. Scofield, Phys. Rev. 179, 9 (1969).
12. D. J. Nagel, P. G. Burkhalter, C. M. Dozier, J. F. Holzrichter, B. M. Klein, J. M. McMahon, J. A. Stamper, and R. R. Whitlock, Phys. Rev. Lett. 33, 743 (1974).
13. H. Pepin, B. Grek, F. Rheault, and D. J. Nagel, J. Appl. Phys. 48, 3312 (1977).
14. C. M. Lederer and V. S. Shirley, Table of Isotopes, 7th Edition (Wiley, New York 1978).

RESONANTLY ENHANCED VACUUM ULTRAVIOLET GENERATION AND MULTIPHOTON IONIZATION IN CARBON MONOXIDE GAS*

James H. Glownia and Robert K. Sander
University of California, Los Alamos National Laboratory
Los Alamos, New Mexico 87545

ABSTRACT

Competition between three-photon resonantly enhanced vacuum ultraviolet third-harmonic generation and six-photon multiphoton ionization using the A state in gaseous carbon monoxide is observed. Excitation spectra of the third-harmonic emission exhibit increasing blue shifts and broadening with increasing pressure due to the phase matching requirements. Estimates for the efficiency and tunability show that third-harmonic generation in carbon monoxide molecules is a promising source for coherent vacuum ultraviolet light.

INTRODUCTION

The attraction of using molecular vapors as a nonlinear medium to generate vacuum ultraviolet (VUV) radiation has recently been reviewed by Wallace.[1] Despite the convenience of having a fixed gas for the nonlinear medium, single frequency pump sources, and wide tunability due to the width and number of molecular vibronic bands, only one detailed account of third-harmonic generation (THG) in molecules has been published.[2] In that study, two and three-photon resonantly enhanced THG in nitric oxide was reported. We report the observation of three-photon resonantly enhanced THG through the $A^1\Pi \leftarrow X^1\Sigma$ transition (fourth positive system) of carbon monoxide (CO) gas. In addition to detecting VUV radiation we simultaneously record signals from $A^1\Pi$-resonantly enhanced six-photon multiphoton ionization (MPI). Results of such studies in xenon[3,4] have shown that THG and MPI can be competitive processes. We report similar findings where it is observed that signals due to MPI disappear at wavelengths to the blue of the R-head in the CO A state and, concurrently, intense VUV third-harmonic radiation is detected in the forward direction.

Despite the fact that nonresonant THG in CO has failed in the past,[5] it was chosen in the present study for a number of reasons. It is a gaseous material and processes competing with third-harmonic emission are suppressed. Its high ionization potential (14.01 eV)[6] and high dissociation limit (11.11 eV)[7] make these two processes higher order and therefore unlikely in the wavelength region of interest. Furthermore, it was felt that by using the resonant enhancement due to the strong transition dipole of the A state,[8] an efficient VUV radiation source could be produced using powerful dye-laser pump sources. Finally, the extensive Franck-Condon envelope of this transition provides a wide tuning range.

*Work performed under the auspices of the US DOE.

EXPERIMENTAL

The experimental apparatus, shown in Fig. 1, is similar to that used by Miller et al.[3] It consists of a four-way cross connected to a vacuum system. The output of a frequency-tripled Nd:YAG laser (Quanta-Ray DCR-1) pumped Coumarin 440 dye laser (Quanta-Ray PDL-1) is focused by either a 75- or 100-mm focal length lens to a waist calculated to be about 10 μm. The laser flux in the focal volume was on the order of 10^{29}-10^{30} photons $cm^{-1}s^{-1}$. Photoelectrons resulting from the MPI process are monitored with a flat-plate platinum collector. Amplification occurs by an electron avalanche in the CO gas due to a negative bias of 200-300 V on the other platinum plate. Vacuum UV radiation can pass through a LiF window into a differentially pumped five-way cross filled with argon counting gas at 0.5 Torr. If VUV light strikes the negatively biased tantalum foil, the ejected photoelectrons are collected with another flat-plate platinum electrode. The tantalum foil (work function 4.2 eV) is angled, which enhances its yield for photoelectrons.[9] A VUV bandpass filter (Acton) is used to suppress background signals due to pump laser photons hitting the tantalum foil detector. The electron signals due to MPI are amplified (PAR 115) and integrated by a boxcar averager (PAR 162/164) before being recorded on an x-y plotter. Spectral grade CO (Matheson) is condensed at 77 K prior to use, suppressing intense interfering MPI signals due to iron carbonyl compounds[10,11] which have negligible vapor pressures at this temperature.

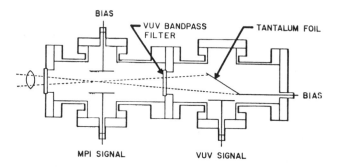

Fig. 1. Schematic diagram of the experimental apparatus used to measure the MPI and third-harmonic VUV radiation. The pump laser can also be positioned at 90° with respect to the detector.

THIRD HARMONIC GENERATION AND MULTIPHOTON IONIZATION

Figure 2 shows the MPI and THG excitation spectra for the v' = 1 and v' = 2 vibronic levels of the A state at a CO pressure of 10 Torr. The ionization signals are due to three-photon $A^1\Pi$ resonantly enhanced six-photon MPI. The signals at slower laser wavelength scan

speeds show the anticipated rotational structure.[8] To the blue of
the R-head, ionization signals can no longer be detected, and only
signals due to VUV radiation are observed. The VUV light is only
detected in the forward direction with respect to the pump laser.
The excitation spectrum exhibits increasing blue shifts and bandwidths with increasing pressure. This is shown in Fig. 3. Also
apparent in Fig. 3 are sharp dips in the VUV signal due to absorption
by e $^3\Sigma^-$ (v'=4) rovibronic levels in CO,[6] confirming identification of
the VUV light as tunable monochromatic third-harmonic emission. To
account for similar MPI-THG phenomena in xenon, a two-level model has
been used.[4] This two-level model involves coherent excitation of a
group of atoms resulting in a collection of coupled dipoles which
emit to the ground state with an enhanced oscillator strength. Qualitatively, the Rabi frequency of the three-photon pumping to the A$^1\Pi$
level and the Rabi frequency of the stimulated emission to the CO
ground state are coupled.

COMPETITION BETWEEN MPI AND THG

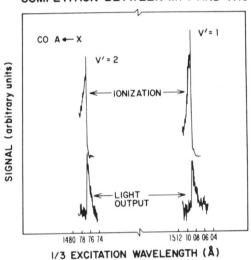

Fig. 2. The MPI and THG excitation spectra of the $^1\Pi$ v' = 1 and
v' = 2 vibronic levels in CO at a pressure of 10 Torr. At
slower dye laser wavelength scan rates the individual CO
rotational lines are apparent in the MPI spectrum.

The blue shifts and broadening of the VUV light with pressure
are consistent with the wave vector phase matching requirements.[12][13]
The phase mismatch Δk may be written as

$$\Delta k = 6\pi (n_P - n_{TH})/\lambda_P \tag{1}$$

THG IN CO

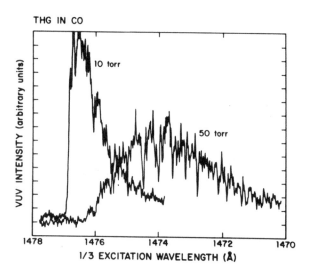

Fig. 3. Excitation spectra of the third-harmonic emission in the $^1\Pi$ v' = 2 level of CO at pressures of 10 and 50 Torr. The sharp dips in the VUV emission are due to self absorption by rovibronic levels in the e $^3\Sigma^-$ (v'=4) band of carbon monoxide.

where λ_p is the dye laser pump wavelength, and n_p and n_{TH} are the refractive indices at λ_p and $\lambda_p/3$ respectively. Bjorklund[13] has shown that the optimum phase mismatch for focused beams has a small negative value. In CO the generated VUV frequency is higher than that of the vibronic levels of the A state, and, therefore, the anomalous dispersion makes a negative Δk possible.

Measurements of the maximum THG output of CO and Xe show that they are comparable. The efficiency for THG in Xe has been previously reported.[14] It is important to note that the efficiency for THG goes as the square of the pump laser flux, thus increased THG efficiency should be possible by raising the laser flux to the limit of dielectric breakdown. We observed that this limit is noticeably higher in CO than Xe. Also, the higher $^1\Pi$ vibronic levels up to the predissociation limit exhibit third-harmonic emission, and tuning in the 1300-1500 Å wavelength region is observed as shown in Fig. 4.

Ongoing experiments are exploring the use of CO as a nonlinear medium to generate VUV radiation in the Lyman-alpha wavelength region. The v'=14 level of ^{13}CO is ideally suited for this purpose. Using an injection-locked XeCl laser as a source of 308 nm radiation, which is a near two-photon A $^1\Pi$ (v'=0) resonance, together with 578 nm Rhodamine or copper vapor laser photons, a powerful 1216 Å source appears feasible.

EXPERIMENTALLY OBSERVED THG OUTPUT
IN 5-350 TORR ^{12}CO OR ^{13}CO

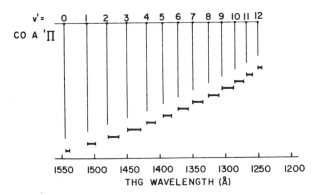

Fig. 4. Wavelengths for which third-harmonic radiation has been observed in $C^{12}O^{16}$ and for which third-harmonic should be observed in $C^{13}O^{16}$ using a single dye laser as an excitation source.

SUMMARY

In summary, it has been shown that CO vapor can be used to generate coherent tunable VUV radiation by THG using a single tunable dye laser. By simultaneously recording MPI and THG, it is found that these are competitive processes. The shifts to higher frequencies and the broadening of the third-harmonic emission with increasing CO pressure are found to be due to the phase matching requirements.

We are indebted to J. R. Ackerhalt for helpful discussions. Research is sponsored by the Department of Energy under contract W-7405-eng-36 with the University of California.

REFERENCES

1. S. C. Wallace, in Photoselective Chemistry Pt. 2, edited by J. Jortner, R. D. Levine, and S. A. Rice (Wiley, New York, 1981), pp. 169-170.

2. K. K. Innes, B. P. Stoicheff, and S. C. Wallace, Appl. Phys. Lett. 29, 715 (1976); S. C. Wallace and K. K. Innes, J. Chem. Phys. 72, 4805 (1980).

3. J. C. Miller, R. N. Compton, M. G. Payne, and W. R. Garrett, Phys. Rev. Lett. 45, 114 (1980).

4. M. G. Payne, W. R. Garrett, and H. C. Baker, Chem. Phys. Lett. 75, 468 (1980).

5. W. G. Rado, Appl. Phys. Lett. 11, 123 (1967).

6. P. H. Krupenie, The Band Spectrum of Carbon Monoxide, Nat. Bur. Stand. (US), NSRDS-5 (US GPO, Washington, D.C., 1966).

7. A. E. Douglas and C. K. Moller, Can. J. Phys. 33, 125 (1955).

8. J. D. Simmons, A. M. Bass, and S. G. Tilford, Astrophys. J. 155, 345 (1969).

9. J. A. R. Samson and R. B. Cairns, GCA Technical Report No. 66-17-N (GCA Corp., Bedford, MA, 1966).

10. M. A. Duncan, T. G. Dietz, and R. E. Smalley, Chem. Phys. 44, 415 (1979).

11. P. C. Engleking, Chem. Phys. Lett. 74, 207 (1980).

12. G. C. Bjorklund, IEEE J. Quantum Electron. QE-11, 287 (1975).

13. D. C. Hanna, M. A. Yuratich, and D. Cotter in Nonlinear Optics of Free Atoms and Molecules, edited by D. L. MacAdam (Springer-Verlag, New York, 1979), Chap. 4.

14. R. Hilbig and R. Wallenstein, IEEE J. Quantum Electron. QE-17, 1566 (1981).

MULTIRESONANT TWO-PHOTON-ABSORPTION-INDUCED FOUR WAVE MIXING
IN CRYSTALLINE RARE EARTH INSULATORS

R. L. Cone, D. A. Ender, M. S. Otteson, and Paula L. Fisher
Physics Dept, Montana State University, Bozeman, Montana 59717

and

J. M. Friedman and H. J. Guggenheim
Bell Labs, 600 Mountain Avenue, Murray Hill, NJ 07974

ABSTRACT

Coherent nonlinear optical generation of $\omega_4=\omega_1+\omega_2-\omega_3$ exhibits strong sharp intermediate (ω_1) and two-photon ($\omega_1+\omega_2$) resonances in crystalline $Tb(OH)_3$ and $LiTbF_4$, providing a novel method for high resolution coherent measurements of both excited electronic configurations and intermediate $4f^n$ states of rare earth ions. New regions of the UV and VUV are thus made accessible to existing tuneable visible and near ultraviolet lasers. Selection of sharp features from broad overlapping absorptions, line narrowing due to phase matching selectivity, and coherent transient applications are discussed.

INTRODUCTION

We report the first observation of multiresonant four wave mixing induced by two photon absorption in rare earth insulators. Coherent generation of $\omega_4=\omega_1+\omega_2-\omega_3$ exhibits strong, narrow, intermediate state resonances when ω_1 sweeps through a single photon transition to the $^5D_4$ states arising from the $4f^8$ ground configuration of Tb^{3+} while ($\omega_1+\omega_2$) is simultaneously resonant with an allowed transition to the excited $4f^75d$ configuration. Strong two photon resonances are also observed when ω_1 is fixed on resonance with the $^5D_4$ state, and ($\omega_1+\omega_2$) is scanned by tuning ω_2.

These two photon resonances demonstrate that the mixing spectra can resolve sharp features which would normally be obscured by broad overlapping uv absorptions. They thus provide a dramatic new capability for studies in this spectral region. Significant potential exists for elucidating the structure of excited configurations and for uncovering new laser transitions or materials.

The initial coherent mixing experiments have been carried out on $Tb(OH)_3$ and $LiTbF_4$ at 1.3 K. $LiTbF_4$ is similar to the laser host crystal $LiYF_4$. A number of advantages over conventional spectroscopic measurements have been demonstrated, including high resolution, frequency and directional selectivity arising from phase matching, new selection rules, and accessiblity to new and relatively unexplored regions of the UV or VUV using existing tuneable lasers. In addition, this type of multiresonant mixing provides a new mechanism for wavefront conjugation.

ISSN:0094-243X/82/900471-07$3.00 Copyright 1982 American Institute of Physics

FOUR WAVE MIXING

The wide range of phenomena arising from four wave nonlinear optical mixing have been reviewed recently by Bloembergen[1] and Levenson and Song.[2] A number of groups have begun investigations of multiresonant four wave mixing[3] and applications to condensed matter.[4] Earlier studies of multiresonant phenomena[5] and multiphoton coherent transients[6] in gases have demonstrated unique advantages for such experiments.

While Kramer and Bloembergen[7] have previously observed two photon absorption induced mixing in CuCl, and Maurani and Chemla[8] have exploited a near resonant intermediate state in their degenerate four wave mixing studies of biexcitons in CuCl and CdS, this report emphasizes direct use of the intermediate resonance. Near resonant enhancement has also been reported in polymer solutions by Shand et al.[9]

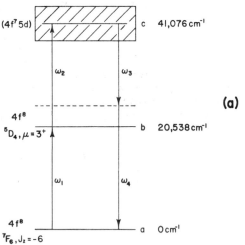

(a)

The mixing process reported here is illustrated schematically in Fig. 1. Laser beams containing ω_1, ω_2, and ω_3 are incident simultaneously at appropriate phase matching angles. The nonlinear generation of a new spatially distinct beam at frequency $\omega_4 = \omega_1 + \omega_2 - \omega_3$ exhibits multiresonant behavior described[1-3] by a third order nonlinear optical susceptibility. The specific term which is relevant to our experiments has the form:

(b)

Fig. 1. Four wave mixing mechanism, showing a) multiple resonances and b) simplest phase matching geometry.

$$\chi^{(3)}(-\omega_4,\omega_1,\omega_2,-\omega_3)$$
$$\propto \frac{\mu_{ab}\mu_{bc}\mu_{cd}\mu_{da}}{(\omega_{ba}-\omega_1+i\Gamma_{ab})(\omega_{ca}-[\omega_1+\omega_2]+i\Gamma_{ac})(\omega_{da}-[\omega_1+\omega_2-\omega_3]+i\Gamma_{ad})}$$

where the μ factors are dipole moments between the states indicated by the subscripts and the Γ factors are phase relaxation rates.

The states a and b in Fig. 1a correspond to sharp levels of Tb^{3+} arising from the ground $4f^8$ configuration.[10] The state c has been tentatively identified as a broader level arising from the $4f^7 5d$ configuration on the basis of a very strong, parity-allowed, single photon transition observed in the vicinity of 240–250 nm in the conventional ultra violet absorption spectrum[11] shown in Fig. 2. Further information on that band is given below in the discussion of results. Weak $4f^8-4f^8$ transitions extend into the $\omega_1+\omega_2$ range investigated, but the broadness of the observed resonance makes such an assignment less likely. A cooperative process involving Tb^{3+} pairs might also contribute, but it cannot explain mixing over the observed range of ω_2.

EXPERIMENTAL RESULTS

In the simplest experiments, $\omega_1=\omega_2 \neq \omega_3$. Phase matching was obtained for a crossing angle θ between the two input beams of 1–2° in a manner analogous to typical coherent anti-Stokes Raman scattering (CARS) experiments[1,2] as illustrated in Fig. 1b. In order to explore the two photon resonance in detail, three different lasers were used with a more general geometry. The phase matching adjustment was critical and in some cases required careful selection of ω_3. (This was particularly true in the triangular case depicted in Fig. 1b.) Three N_2 laser pumped dye

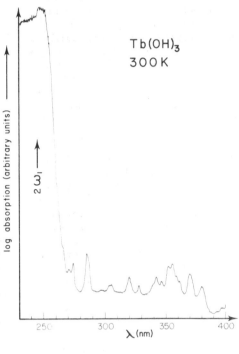

Fig. 2. Ultraviolet absorption spectrum of Tb(OH)$_3$.

lasers with peak powers of 10–50 kw were focused to 50–100 μm spots in the sample. The uniaxial crystals are typically 300 μm to 1mm thick. All beams propagated perpendicular to the c-axis and were polarized E parallel to c. The signal was isolated from the input beams by spatial and spectral filtering. Strong signals were observed, requiring reduction of photomultipler voltage to 700 volts.

The nature of the mixing process was confirmed by examination of power dependences, phase matching effects, and temporal overlap effects for the three incident laser pulses. Spectral scans with 0.1 cm^{-1} resolution showed that $\omega_4=\omega_1+\omega_2-\omega_3$. CARS was ruled out by choosing ω_3 to avoid Raman resonances. A two step second order process[12] is ruled out by the observed polarization. Thermal grating effects may be ruled out by the large differences in the ω's.

Initially,[13-16] emphasis was focused on the intermediate state resonance. Typical experimental results are shown in Fig. 3, where

474

the ω_1 resonance in the mixing experiment is compared to a conventional ω_1 absorption spectrum. (The $\omega_1+\omega_2$ resonance is unaffected over the limited ω_1 tuning range shown, due to the larger linewidths characteristic of excited configurations.) The ω_1 laser was continuously pressure scanned with a linewidth of 0.03-0.05 cm^{-1}.

It is clear that the intermediate resonance in the mixing signal gives far narrower linewidths. We have recently demonstrated[14-16] that this narrowing is a consequence of the phase matching requirement. Since the bulk index of refraction and hence $k_1=n_1\omega_1/c$ vary across the ω_1 resonance, constructive interference of the coherently generated signal will occur only over narrow regions of the inhomogeneous width corresponding to a range of k_1 values determined by the angle Θ. Dispersion of n_1 clearly arises from $\chi^{(1)}(\omega_1)$, and nonlinear contributions from terms such as $\chi^{(3)}(-\omega_1,\omega_1,\omega_1,-\omega_1)$ may be important as well since they are multiresonant.

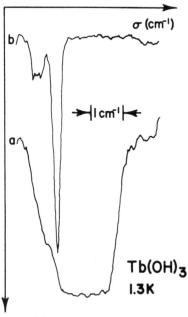

ω_1 absorption (a)
generated ω_4
intensity (b)

Fig. 3. Comparison of inhomogeneously broadened ω_1 absorption and narrowed ω_1 resonance in four wave mixing signal. ω_1=20,538cm^{-1} =ω_2 and ω_3 = 18301. Θ=1°. σ=$1/\lambda$ labels the ω_1 frequency axis.

Calculations of expected dispersion in n_1 using oscillator strengths[17] for Tb^{3+} support this model. Direct measurements of $\Delta n_1(\omega_1)$ using a triangular prism made of LiTbF$_4$ provided conclusive evidence for this interpretation.[15,16]

Results for the two photon resonance are given in Fig. 4, where the $(\omega_1+\omega_2)$ scale has been chosen for comparison with the uv absorption spectrum. The FWHM is only 230 cm^{-1}. The two photon resonance in the mixing experiment has resolved a clear feature of the Tb^{3+} ions from a featureless band in the conventional spectrum. This sharp feature is tentatively identified as a crystal field component of the $4f^7 5d$ configuration. The capability of the multiresonant mixing technique to "select" features from an overlapping absorption band provides a significant new approach to studies of this relatively unexplored spectral region in these materials. Moreover, the two-photon-absorption-induced mixing may be used to study levels beyond the absorption edge of the material.

COHERENT TRANSIENT EXPERIMENTS

The frequency selection process discussed above is a coherence effect; hence, it will ultimately be limited by the homogeneous linewidth. It is thus of fundamental importance. Since the frequency-selective mixing process can ultimately select a "homogeneous packet" within inhomogeneously broadened resonances, novel coherent transient measurements are possible. Pulse sequences appropriate for observing free induction decay of both the intermediate state and the two-photon-excited state were investigated using three distinct input frequencies and non-triangular phase matching geometry. This allows individual time-ordered terms in $\chi^{(3)}$ to be probed selectively in a transient regime. Delaying both ω_2 and ω_3 relative to ω_1 allows observation of dephasing of the intermediate state, while delaying only ω_3 allows observation of dephasing of the two-photon-excited state. The latter case is similar to Doppler free two-photon transients observed in gases,[5] but it relies on phase matching selectivity to overcome inhomogeneous broadening.

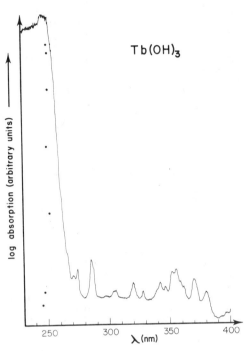

Fig. 4. Comparison of $(\omega_1+\omega_2)$ mixing resonance (points) and uv absorption spectrum of $Tb(OH)_3$ (solid line).

Experiments utilizing 5 nsec pulses have shown that the signal decays for pulse separations comparable to the laser pulse duration even at 1.3 K and 60 kG applied magnetic field. While residual inhomogeneous broadening may be causing inhomogeneous dephasing, a variety of evidence indicates that nanosecond or picosecond dephasing times are plausible in concentrated magnetic compounds such as $Tb(OH)_3$, where there is strong coupling[10] to neighboring electronic spins. If the observed width in Fig. 3 does correspond to the homogeneous limit, it gives $T_2=50$ psec.

We are aware of no other time domain coherence measurements on two-photon-excited states in rare earth compounds. The strong crystal field and interionic coupling experienced in excited configurations, however, suggest extremely rapid dephasing. Observed T_1 values are in the fast nanosecond range.[18]

Potential advantages for this method of observing coherent transients, include strong signals, spatial and spectral discrimination of signal beam from input beams, time resolution limited only by laser pulse durations, and access to new states and spectral regions with existing lasers. Experiments are now under way in dilute compounds and Eu^{3+} compounds where slower dephasing is expected and where direct comparison may be made with conventional photon echo results. Finally we point out that echo versions of these mixing transients are possible.

CONCLUSION

We have reported observations of new multiresonant four wave mixing phenomena in solids and have demonstrated the applicability of this technique to studies of highly excited states in rare earth insulators. Levels in the range of 5-7 ev arising from excited electronic configurations may be probed coherently with visible or near uv lasers due to the two photon absorption mechanism. With frequency-doubled dye lasers, much higher levels are accessible. This mechanism also allows one to study levels above the absorption edge of the material.

The multiresonant aspect of the mixing process allows it to select or label sharp spectral features in regions where overlapping bands obscure detail in conventional experiments. Crystal field components of the $4f^{7}5d$ configuration are well resolved in the mixing spectra of $Tb(OH)_3$ whereas they are merged into a strong featureless band in a conventional uv absorption experiment. We have thus provided a major new tool for analyzing spectra in this relatively unexplored region.

A mechanism for overcoming the effects of inhomogeneous broadening has also been demonstrated. As a result, novel coherent transient methods based on time-sequenced multiresonant mixing are feasible. Transient measurements were presented which indicate that both the intermediate and two-photon-excited states dephase rapidly due to strong interionic coupling.

ACKNOWLEDGEMENTS

We thank W. P. Wolf (NSF grant DMR 7918175) and S. Mroczkowski for the $Tb(OH)_3$ crystals and the NSF for support of this research at Montana State University under grant DMR 7906892.

REFERENCES

1. N. Bloembergen, in Laser Spectroscopy IV, edited by H. Walther and K. W. Rothe (Springer, Berlin, 1979), pp 340-348, and J. Opt. Soc. Am. 70, 1429 (1980).

2. M. D. Levenson and J. J. Song in Coherent Nonlinear Optics, Recent Advances (Springer, Berlin, 1980). Edited by M. S. Feld and V. S. Letokhov.

3. N. Bloembergen, H. Lotem, R. T. Lynch, Jr, Indian J. Pure Appl.

Phys. <u>16</u>, 151 (1978), S. A. J. Druet, B. Attal, T. K. Gustafson, and J. P. Taran, Phys. Rev. A<u>18</u> 1529 (1978) and J.L. Oudar and Y. R. Shen, Phys. Rev. A<u>22</u>, 1141 (1980).

4. J. R. Andrews, R. M. Hochstrasser, and H. P. Trommsdorff, Chem. Phys. <u>62</u>, 87 (1981).

5. J. E. Bjorkholm and P. F. Liao, Phys. Rev. Lett. <u>33</u>, 128 (1974).

6. R. G. Brewer and E. L. Hahn, Phys. Rev. A<u>11</u> 1641 (1975), M. M. T. Loy, Phys. Rev. Lett. <u>36</u>, 1454 (1976), and P. F. Liao, J. E. Bjorkholm, and J. P. Gordon, Phys. Lett <u>39</u>, 15 (1977). Multiphoton coherent transients have also been observed in nmr experiments.

7. S. D. Kramer and N. Bloembergen, Phys. Rev. B<u>14</u>, 4654 (1976).

8. A. Maruani and D. S. Chemla, Phys. Rev. B<u>23</u>, 841 (1981).

9. M. L. Shand, R. R. Chance, and R. Silbey, Chem. Phys. Lett. <u>64</u>, 448 (1979) and Phys. Rev. B<u>8</u>, 3540 (1980).

10. R. L. Cone and R. S. Meltzer, J. Chem. Phys <u>62</u>, 3573 (1975) and C. A. Catanese, A. T. Skjeltorp, H. E. Meissner, and W. P. Wolf Phys. Rev. B<u>8</u>, 4223 (1973).

11. Transition moments from the states labeled $^5D_4$ will clearly be different, but spin-orbit coupling and the crystal field lead to strong S, L, and J mixing which should make comparison meaningful.

12. M. D. Levenson and R. T. Lynch, in <u>Optical Properties of Highly Transparent Solids</u>, edited by S. S. Mitra and B. Bendow (Plenum, New York, 1975), pp 329-337 and references contained therein.

13. R. L. Cone, J. M. Friedman, R. A. Stepnoski, D. A. Ender, M. S. Otteson, and Paula L. Fisher, Bull. Amer. Phys. Soc. <u>26</u>, 291 (1981), Papers #DR6 and DR7.

14. R. L. Cone, J. M. Friedman, R. A. Stepnoski, D. A. Ender, M. S. Otteson, and Paula L. Fisher (Submitted to Phys. Rev. Lett. 1982).

15. D. A. Ender, M. S. Otteson, R. L. Cone, M. B. Ritter, and H. J. Guggenheim (Submitted to Optics Letters, 1982).

16. D. A. Ender, M. S. Otteson, R. L. Cone, J. M. Friedman, and H. J. Guggenheim, Bull. Amer. Phys. Soc. <u>27</u>, 402 (1982), Paper #MF14.

17. M. J. Weber, T. E. Varitimos, and B. H. Matsinger, Phys. Rev. B <u>8</u>, 47 (1973) and J. A. Caird, W. T. Carnall, and J. P. Hessler, J. Chem. Phys. <u>74</u>, 3225 (1981).

18. R. R. Jacobs, W. F. Krupke, and M. J. Weber, Appl. Phys. Lett <u>33</u> 410 (1978) and W. J. Miniscalco, J. M. Pellegrino, and W. M. Yen, J. Appl. Phys. <u>49</u>, 6109 (1978).

RESONANT FOUR-WAVE MIXING PROCESSES IN XENON

Y. M. Yiu, K. D. Bonin and T. J. McIlrath
Institute for Physical Science and Technology
University of Maryland, College Park, MD 20742

ABSTRACT

Two-photon resonantly enhanced four-wave mixing processes in xenon involving the intermediate states were utilized to generate coherent VUV radiation at several discrete wavelengths between 125.9 nm and 101.8 nm. Maximum efficiencies of the order of 10^{-4} were achieved. The use of these processes for producing tunable VUV output with Xe is given and generation of tunable VUV using two-photon resonances in other rare gases is discussed.

INTRODUCTION

Coherent vacuum ultraviolet (VUV) radiation is usually produced by nonlinear mixing in atomic vapors. Metal vapors, except for Hg, require the use of high temperature furnaces. All of them self-absorb at wavelengths below 120 nm, and rapid variations in efficiency with tuning occur due to the dense structure in their spectra.[1,2] The use of rare gases not only eliminates these problems but, in addition, they are easy to handle and are chemically stable and inert. Resonantly enhanced tripling in Xe has been previously demonstrated using the 6p intermediate states.[3] In the work reported here, xenon gas was used as the nonlinear medium to generate radiation between 102 nm and 126 nm through two-photon enhanced four-wave mixing processes using the 7p and 6p' intermediate states. With a straightforward modification, the radiation can be made tunable. Tunable VUV generation using intermediate resonances in other rare gases will also be briefly discussed.

EXPERIMENTAL

A Q-switched Nd:YAG laser providing 1.064 μm radiation at 10 Hz was doubled and used to pump a Hansch type dye laser. The Nd:YAG laser was narrowed to ≃.1 cm^{-1} by the insertion of an intracavity etalon. The dye output was doubled and mixed with the remaining 1.06 μm (IR) beam by angle-tuned KDP crystals. The UV radiation was tuned between 224 and 226 nm by mechanically tuning the grating in the dye laser cavity. The UV and IR beams were then separated by a Pellin Broca prism and were recombined downstream through a quartz flat at the UV Brewster angle. The two beams were orthogonally polarized and the full UV, (typically ≃.5 mJ), and a small portion of the IR (typically ≃.5 mJ), were used. Both beams were tightly focused into a xenon cell and the output was sent through a Xe ionization chamber before exiting through a LiF window into a .5 m vacuum monochromator. Detection at the exit slit of the monochromator was accomplished with a solar blind photomultiplier. A VUV filter centered at 123.0 nm was placed just in front of the photomultiplier to reduce scattered light. In order to circumvent the LiF window cutoff at 105 nm, radiation below the Xe ionization limit at 102.2 nm was detected by the ionization chamber. A guard ring prevented electrons produced at the focus in the Xe cell from entering the chamber.

RESULTS

Table I shows the wavelengths generated using the intermediate states listed. Absolute intensity measurements of the difference

Table I. Relative powers P_{VUV} of VUV radiation generated via three two-photon resonant states of Xe. Incident powers at both ω_1 and $\omega_2 \sim 10^5 W$.

Two-photon resonant state	Type of processes	λ_{VUV} (nm)	P_{VUV} [Note 1] (Watt)
$7p[0\frac{1}{2}]_0$	$2\omega_1-\omega_2$	125.9	64.9
	$2\omega_1+\omega_2$*	101.8	4.3
	$3\omega_1$ *	75.0	3.3
$7p[1\frac{1}{2}]_2$	$2\omega_1-\omega_2$	126.1	71.5
	$2\omega_1+\omega_2$*	102.0	7.3
	$3\omega_1$ *	75.2	5.0
$6p'[1\frac{1}{2}]_2$	$2\omega_1-\omega_2$	125.4	36.5
	$2\omega_1+\omega_2$*	101.5	7.5
	$3\omega_1$ *	74.8	7.1

[Note 1]: Power of the generated radiation is calculated based on manufacturer's specification of the efficiencies of the grating, window and VUV filter, and the gain of the solar blind photomultiplier (EMI-G24E314LF).

*The sum and harmonic generation may be accompanied by parametric processes so that the powers quoted may be due to several wavelengths produced by mixtures of the driving wave with parametrically produced waves.

signals ($2\omega_1 - \omega_2$) were obtained from photomultiplier currents at
the monochromator exit. The output energies listed in Table 1
should be taken as order of magnitude calculations as they had to
be estimated from approximate window, filter, and grating efficiencies and geometric considerations. The sum-frequency ($2\omega_1 + \omega_2$)
signals were determined by making measurements with a Xe filled
ionization chamber. Measurements were made of the total signal and
then the background in the absence of the IR beam was subtracted to
give the contribution from the sum frequency output. The beams
were unfocused in the ionization chamber and the contribution from
multiphoton ionization was small as measured at higher pressures
where a phase mismatch suppressed coherent generation. All calculations assumed 100% efficiencies for the Xe ionization chamber
with no self-absorption before the chamber so that the tabulated
values represent lower limits. It should be mentioned that the
sum-frequency signal went away when the IR beam was slightly misa-
ligned. Figures 1 and 2 exhibit the pressure dependence of the VUV
radiation and demonstrate phase-matching of the generated beams.

From Fig. 3, the intermediate resonance enhancement is clearly
visible. The enhancement exceeded a factor of 20 and its spectral
width followed that of the dye laser which was $\simeq 0.8$ cm^{-1}. For ease
of tuning, the dye laser intracavity etalon was not used.

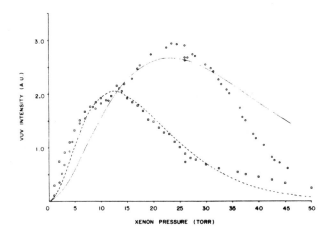

Fig. 1. Pressure dependence of VUV generation: Experimental points
0 and ☐ for $\omega_3 = 2\omega_1 - \omega_2$ due to the two-photon resonant states
$7p[0\frac{1}{2}]_0$ and $7p[1\frac{1}{2}]_2$, respectively. ___ and ---- are theoretical
curves for TEM$_{00}$ mode in functional form $A_1 \pi^2 P^2 e^{-A_2 P}$.

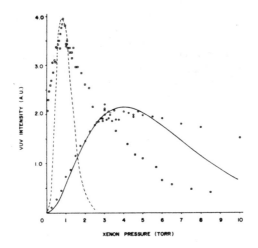

Fig. 2. Pressure dependence of VUV output: O and □ are experimental points for $\omega_3 = 2\omega_1 - \omega_2$ and $\omega_4 = 2\omega_1 + \omega_2$ due to the two-photon resonant states $6p[1\frac{1}{2}]_2$ and $7p[1\frac{1}{2}]_2$, respectively. ___ and ---- are theoretical curves for TEM_{00} mode in functional forms of $A_1\pi^2P^2 e^{-A_2P}$ and $A_1\pi^2P^4 e^{-A_2P}$, respectively.

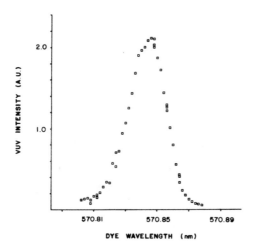

Fig. 3. Dependence of difference generation on dye laser wavelength $(2\pi c/\omega_D)$: Experimental data O for $\omega_3 = 2\omega_1 - \omega_2$ where $\omega_1 = 2\omega_D + \omega_{IR}$.

The dependence of output powers on the input powers at ω_1 and ω_2 are shown in Figs. 4 and 5. An upper limit of ≈ 1 mJ for the energy in the IR was used in order to avoid damage to the mono-chromator grating. Close to quadratic power dependence of the VUV on the UV at low powers is apparent and saturation seems to begin at input energies of $\approx .9$ mJ with the $7p[0\frac{1}{2}]_0$ intermediate resonance. The output power dependence on IR power has a slope of 1.4 as shown in Fig. 5. It is clear from these figures that the conversion efficiency, defined as the VUV output relative to either the total input energy or the IR input energy, would be increased by the use of a higher UV intensity.

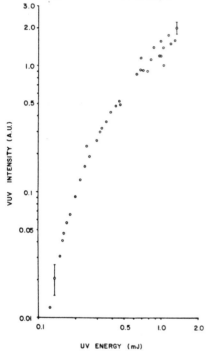

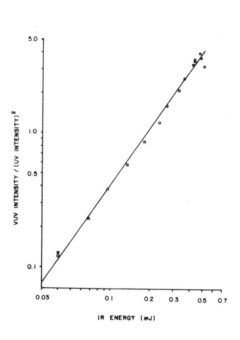

Fig. 4. UV power dependence of difference generation for $7p[0\frac{1}{2}]_0$ intermediate state: Experimental data O.

Fig. 5. IR power dependence of difference generation for $7p[0\frac{1}{2}]_0$ intermediate state: Experimental data ☐ and the linear least square fit ——.

DISCUSSION

For TEM_{00} beams tightly focused in a homogeneous medium with good beam overlap, the output power for difference-frequency gener-ation can be expressed by $A_1 P^2 \exp(-A_2 P)$, where A_1 and A_2 are positive constants and P is the pressure.[4] A glance at Figs. 1 and 2 show that the experimental points follow such a functional dependence except at the higher pressures. The deviation at higher pressures could result from the presence of other modes in the beam or the effect of Xe_2 molecules at high pressure.

The sum-generation signal should follow a functional form of $A_1P^4exp(-A_2P)$ under tight focusing. The experimental agreement in this case is poor. This may be due to several factors including the effect of self-absorption at high pressures, the inefficiency of ionization absorption at low pressures, or the existence of more than one generated wavelength. Several of these factors can be overcome in future experiments. The last two problems can be solved by replacing the LiF window with a capillary array and then differentially pumping the monochromator.[5] The ionization chamber can then be placed at the exit slit of the monochromator which would enable accurate detection of wavelengths below 105 nm. Complete ionization can be assured by simply making the pressure in the ionization chamber as high as necessary. There is an added advantage in that numbers closer to absolute intensities can be obtained. The disadvantage of differential pumping in the present setup is the relative expense of xenon compared to other rare gases.

The current experiment produces several VUV wavelengths simultaneously. This can be controlled by using circularly polarized beams. If the UV is circularly polarized then harmonic generation is suppressed. If the IR has the same helicity as the UV then only difference generation occurs while if the helicities are opposite then only sum generation occurs. The use of circularly polarized beams would eliminate the use of J=0 intermediate states.

The most useful VUV coherent source is, of course, one that is tunable. This can be acheived with the present system by the addition of another dye laser to provide the tunable ω_2 frequency for the four-wave mixing processes. Because ω_2 does not involve any intermediate resonances there should be little change of efficiency over a wide tuning range of ω_2. The dye laser can be used directly as the tunable ω_2 frequency in the visible or it can be used to generate tunable IR radiation by stimulated Raman scattering in hydrogen.[6] A typical scheme for generating tunable VUV near 121.5 nm utilizing xenon is given in Fig. 6. Ability to tune ω_2 between .5 µ and 1.5 µ, for example, would yield tunable output of 121.7-145.3 nm by difference-frequency mixing and 919-104.7 nm by sum-frequency mixing.

Two-photon resonant enhancement in argon and krypton can be reached by sum-frequency mixing in potassium pentaborate (KB5).[7] Stimulated anti-Stokes Raman scattering has been used to generate UV and VUV radiation at shorter wavelengths than achievable with crystals.[8] These anti-Stokes generated wavelengths can pump two-photon resonances in argon at 186.8 nm and 189.3 nm. Argon is transparent to 79 nm and is much less expensive than Xe or Kr so that differential pumping through a capillary array would provide a practical source of radiation below 100 nm.

484

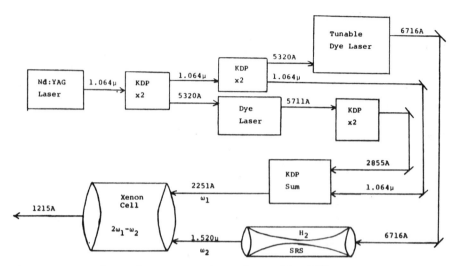

Fig. 6.   A scheme for generation of tunable VUV near 121.5 nm

CONCLUSIONS

These experiments have shown that an intense, high repetition rate source of coherent VUV radiation can be generated using intermediate resonances in xenon. Extension to intermediate resonances in other rare gases is feasible and will enable construction of a tunable source of coherent VUV from 150.0 nm to well below 100.0 nm. Work on these and other developments is continuing in our laboratory. This work was supported by NSF Grant (CPE 7918 387A01) and by the Dissertation Research Committee of the Graduate School at the University of Maryland.

REFERENCES

1. F. S. Tomkins and R. Mahon, Opt. Lett. 6, 179 (1981); J. Bokor, R. R. Freeman, R. L. Panock, and J. C. White, Opt. Lett. 6, 182 (1981).

2. R. T. Hodgson, P. P. Sorokin, and J. J. Wynne, Phys. Rev. Lett. 32, 343 (1974); H. Puell, K. Spanner, W. Falkerstein, W. Kaiser, and C. R. Vidal, Phys. Rev. A 14, 2240 (1976).

3. H. Egger, R. T. Hawkins, J. Bokor, H. Pummer, M. Rothschild and C. K. Rhodes, Opt. Lett. 5, 282 (1980).

4. G. C. Bjorklund, IEEE J. Quantum Electron. QE-11, 287 (1979); Y. M. Yiu, T. J. McIlrath, and R. Mahon, Phys. Rev. A 20, 2470 (1979).

5. T. B. Lucatorto, T. J. McIlrath, and J. R. Roberts, Appl. Opt. 18, 2505 (1979).

6. W. Hartig and W. Schmidt, Appl. Phys. 18, 235 (1979).

7. K. Kato, Appl. Phys. Lett. 30, 583 (1977); R. E. Stickel, Jr., and F. B. Dunning, Appl. Opt. 17, 981 (1978).

8. J. A. Paisner and R. S. Hargrove, Electro Optics/Laser 79 Conference, October 23-25 (1979), Anaheim, CA; V. Wilke and W. Schmidt, Appl. Phys. 18, 177 (1979).

SUM FREQUENCY GENERATION OF CW 194 nm RADIATION IN POTASSIUM PENTABORATE

H. Hemmati, J. C. Bergquist, and Wayne M. Itano
Frequency and Time Standards Group
Time and Frequency Division
National Bureau of Standards
Boulder, CO 80303

ABSTRACT

Narrowband, tunable cw radiation in the 194 nm region has been produced by sum frequency mixing in a potassium pentaborate (KB5) crystal. The input wavelengths required for 90° phase-matched sum frequency mixing (SFM) are approximately 257 nm and 792 nm. The tunable 792 nm radiation was obtained from a cw dye laser. The 257 nm radiation was obtained by frequency doubling the output of a cw argon ion laser in an ammonium dihydrogen phosphate (ADP) crystal. It is estimated that several microwatts of 194 nm radiation in a bandwidth of less than 10 MHz can be produced when all operating conditions are optimized.

INTRODUCTION

Proposals have been made for microwave and optical frequency standards based on transitions of mercury ions stored in Penning traps.[1] These standards have the potential of achieving absolute accuracies of 1 part in 10^{15} or greater, better than any standards now in existence. These proposals require a narrowband cw source of radiation tunable around the first resonance line of the ion (194.23 nm), for radiation-pressure cooling[2,3] and optical pumping.[1,4] (The wavelengths referred to in this paper are all vacuum wavelengths.) For optimum cooling, the frequency bandwidth and stability of the source must be less than the natural linewidth of the resonance line of the ion (about 70 MHz). The minimum cw power required is about 1 μW.

Our method for producing 194 nm radiation is to sum frequency mix the 257 nm second harmonic, generated in an ADP crystal, of the output of a cw 514.7 nm argon ion laser with the output of a tunable cw dye laser in the 792 nm region in a KB5 crystal (see Fig. 1). Previously, Stickel and Dunning,[5] using pulsed dye lasers, have generated coherent radiation tunable between 185 nm and 217 nm by SFM in KB5. With cw lasers, the frequency stability and bandwidth requirements can easily be satisfied. Meeting the minimum power requirement is feasible, provided that resonant cavities are used to enhance the efficiencies of the frequency doubling and mixing stages.[6]

ISSN:0094-243X/82/900485-06$3.00   1982 American Institute of Physics

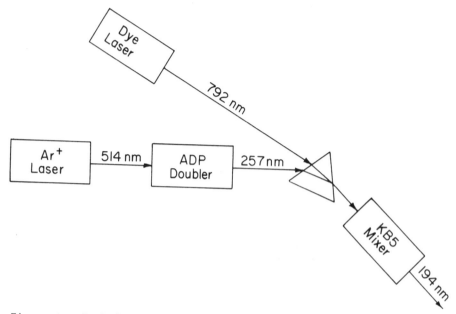

Figure 1. Technique for cw generation of 194 nm.

ADP DOUBLING STAGE

The ring cavity frequency doubler is shown in Fig. 2. The cavity is servo controlled to resonate at the fundamental wavelength (514.7 nm) in order to enhance the second harmonic generation (SHG) in the ADP crystal. A similar method has recently been used by Brieger et. al.[7] to frequency double the output of a dye laser in an ammonium dihydrogen arsenate (ADA) crystal. In our experiments, single mode operation of the argon ion laser was obtained by inserting an etalon into the cavity.

For 90° phase-matched SHG, which gives the highest efficiency, the ADP crystal must be cooled to about -11°C. Dry nitrogen was flowed across the crystal faces to prevent condensation of water. A sealed cell with Brewster windows, filled with argon or helium, was also tried, with good results. The ring cavity was servo controlled by using the polarization method of Hansch and Couillaud[8] to detect a deviation from the resonance condition. A dichroic beamsplitter was used to extract the 257 nm from the cavity. Further details of the cavity design will be given in a later publication.[9]

As much as 80 mW of steady cw second harmonic output has been observed. The circulating power inside the ring cavity was 12.5 times higher than the 2 W power of the argon ion laser. This ratio was determined by measuring the 514.7 nm power leaking through one of the cavity mirrors when the input coupling mirror was in place and when it was removed. The crystal used was 5.5 cm long. This output is considerably below that which would be expected from extrapolation of the results obtained at low input powers.

This is presumably due to absorption of the fundamental and second harmonic in the crystal, which causes a temperature gradient and disturbs the phase matching. This effect has been discussed by Okada and Ieiri.[10]

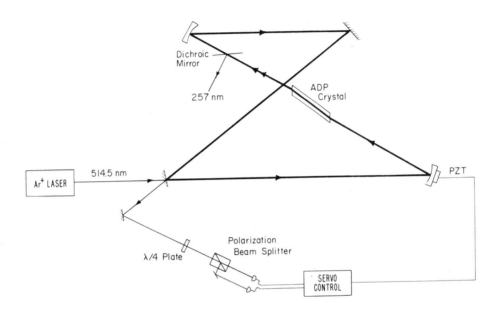

Figure 2. ADP ring cavity frequency doubler.

KB5 MIXING STAGE

The ring cavity sum frequency mixer is shown in Fig. 3. The cavity is servo controlled to resonate at the wavelength of the dye laser (about 792 nm) by the same method that was used for the doubling cavity. The KB5 crystal is 3 cm long. The 257 nm radiation is injected into the ring with a dichroic mirror and the 194 nm radiation is extracted with another dichroic mirror. The ratio of the circulating 792 power inside the cavity to the output of the dye laser has been measured to be about seven. The single-mode ring dye laser is pumped by a krypton ion laser and uses LD700 dye.

The experiments are still in progress, and so far the only SFM results have been obtained using a 1 cm KB5 crystal and without the mixing cavity. The 194 nm output was separated with a fused silica prism and detected with a photomultiplier tube. The absolute intensity was not known very accurately, but was estimated to be approximately 1 nW. The input powers, measured at the entrance window of the crystal housing, were approximately 10 mW at 257 nm and approximately 100 mW at 792 nm. For these experiments, the dye laser was multimode, with a bandwidth of about 1 cm^{-1}.

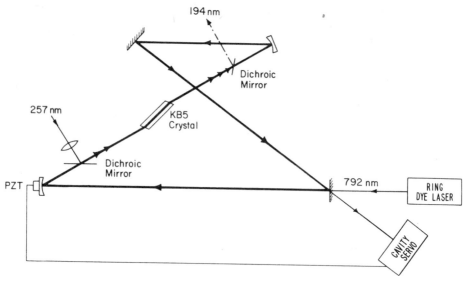

Figure 3.   KB5 ring cavity frequency mixer.

Figure 4 shows the propagation angle for the input beams in the a-b plane, relative to the b axis, versus the wavelength for which phase-matched SFM occurs. The crosses represent our data; the solid curve is calculated from previously published indices of refraction.[11]

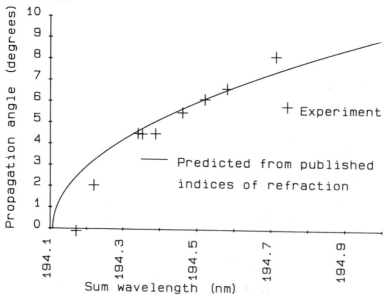

Figure 4.   Angle tuning curve for KB5 SFM.

Figure 5 shows the wavelength for which 90° phase-matched SFM
occurs, as a function of temperature. The greatest conversion
efficiency occurs when the 90° condition is met (propagation along
the b axis). The crosses represent our experimental data; the line
is a linear least-squares fit. For the mercury ion resonance line,
the required temperature is about 34°C.

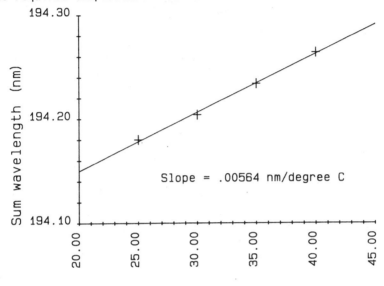

Slope = .00564 nm/degree C

Temperature (degrees C)

Figure 5. Temperature tuning curve for 90° phase-matched SFM in
KB5.

The output power can be increased by increasing the input
powers, by using the ring mixing cavity and a longer crystal, and by
optimally focusing and overlapping the 257 nm and 792 nm beams. We
estimate that it should then be possible to obtain several
microwatts of 194 nm radiation in a bandwidth of less than 10 MHz.

If the argon ion laser were replaced by a dye laser, or other ion
laser lines were used, cw radiation could be generated at shorter
wavelengths, down to 185 nm or below. We note that the only other
technique that has been demonstrated for the generation of coherent
cw radiation below 200 nm is four-wave mixing in strontium vapor, at
approximately 170 nm, by Freeman et al.[12] The lowest-wavelength cw
radiation that has previously been produced by SFM in KB5 is
approximately 211 nm.[13]

ACKNOWLEDGMENTS

We wish to acknowledge useful discussions with Drs. D. J.

Wineland and F. L. Walls and financial support from the Air Force Office of Scientific Research and the Office of Naval Research.

REFERENCES

1. D. J. Wineland, W. M. Itano, J. C. Bergquist, and F. L. Walls, in Proceedings of the 35th Annual Symposium on Frequency Control (U.S. Army Electronics Command, Fort Monmouth, N. J., 1981), p. 602 .

2. D. J. Wineland, R. E. Drullinger, and F. L. Walls, Phys. Rev. Lett. 40, 1639 (1978).

3. W. Neuhauser, M. Hohenstatt, P. Toschek, and H. Dehmelt, Phys. Rev. Lett. 41, 233 (1978).

4. D. J. Wineland, J. C. Bergquist, W. M. Itano, and R. E. Drullinger, Opt. Lett. 5, 245 (1980).

5. R. E. Stickel, Jr., and F. B. Dunning, Appl. Opt. 17, 981 (1978).

6. A. Ashkin, G. C. Boyd, and J. M. Dziedzic, IEEE J. Quantum Electron. QE-2, 109 (1966).

7. M. Brieger, H. Busener, A. Hese, F. V. Moers, and A. Renn, Opt. Commun. 38, 423 (1981).

8. T. W. Hansch and B. Couillaud, Opt. Commun. 35, 441 (1980).

9. J. C. Bergquist, H. Hemmati, and W. M. Itano, to be published.

10. M. Okada and S. Ieiri, IEEE J. Quantum Electron. QE-7, 469 (1971).

11. W. R. Cook, Jr., and L. M. Hubby, Jr., J.Opt. Soc. Am. 66, 72 (1976); F. B. Dunning and R. E. Stickel, Jr., Appl. Opt. 15, 3131 (1976).

12. R. R. Freeman, G. C. Bjorklund, N. P. Economou, P. F. Liao, and J. E. Bjorkholm, Appl. Phys. Lett. 33, 739 (1978).

13. R. E. Stickel, Jr., S. Blit, G. F. Hildebrandt, E. D. Dahl, F. B. Dunning, and F. K. Tittel, Appl. Opt. 17, 2270 (1978).

THIRD HARMONIC CONVERSION WITH A FOCUSED BEAM IN AN INFINITE MEDIUM WITH POSITIVE DISPERSION

J. Reintjes, M. Dlabal
U. S. Naval Research Laboratory, Code 6540, Washington, DC 20375

L. L. Tankersley
U. S. Navy Academy, Dept. of Physics, Annapolis, MD 21402

ABSTRACT

We demonstrate for the first time that it is possible to generate third harmonic radiation with a focused beam in an infinite medium with positive dispersion. This result is in contradiction with the usual restrictions on dispersion. Possible explanations in terms of a nonlinear refractive index are examined. Implications for generation of radiation in the XUV are discussed.

INTRODUCTION

We have studied third harmonic conversion of radiation from a XeCl laser in the vicinity of the gas-vacuum interface that is created near the entrance pinhole in a differential pumping geometry. The results show that, under certain circumstances, it is possible for harmonic conversion to be observed in a tightly focused beam in an effectively infinite medium. This result is in conflict with the restrictions on wavevector mismatch put forth by Ward and New[1] and Bjorklund.[2] We interpret our results in terms of an intensity dependent refractive index that disturbs the phase matching at large pump intensities and relaxes the restrictions that otherwise would exist. An effect of this type has been predicted by Anikin et al[3] for two photon resonant interactions, but has not been previously reported experimentally. The implications of these results for the generation of radiation at short wavelengths by frequency conversion are discussed.

LEVEL DIAGRAMS

Level diagrams for the third harmonic process in Ar and Kr and the fifth harmonic process in Ar are shown in Fig. 1. In Ar the third harmonic wavelength lies above the upper resonance line by about 2000 cm^{-1}. In this region the third harmonic process is negatively dispersive ($n_3 < n_1$) and the process can be phase optimized in a tight focus or by a mixture of Ar and a gas with positive dispersion as described by Bjorklund.[2] The fifth harmonic wavelength lies in the continuum, about 49500 cm^{-1} above the first ionization threshold. Although accurate refractive index data for argon in this wavelength range is not available, comparison with the calculated refractive index of neon in a similar region above

ISSN:0094-243X/82/900491-07$3.00   1982 American Institute of Physics

its ionization threshold indicates that the fifth harmonic process may be negatively dispersive in Ar. However the strong continuum absorption can be expected to affect the conversion process significantly. The third harmonic wavelength in Kr lies about 316 cm-1 above the 4d[1/2]o J=1 level. This is in the correct position relative to a dispersive resonance to obtain negative dispersion. However the oscillator strength of the 4p6-4d[1/2]o J=1 transition is not large enough to overcome the positive contribution of the continuum and the dispersion for the third harmonic process in Kr is positive.[4] As a result third harmonic conversion in Kr is expected to occur only if the Kr is used in a mixture with another medium with negative dispersion of if the beam is focused at the edge of a cell containing Kr.

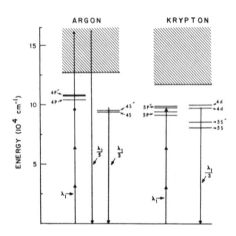

Fig. 1 Level diagrams for third and fifth harmonic conversion of XeCl laser radiation in Ar and third harmonic conversion in Kr.

EXPERIMENTAL RESULTS

In Fig. 2 we show measurements of third and fifth harmonic conversion in Ar obtained in a differential pumping geometry as a function of the position of the beam waist relative to a 250 μm diameter pinhole. As the beam waist is brought from the vacuum through the region of the pinhole and into the nonlinear medium the harmonic signal rises rapidly and then levels off at a constant value. This behavior is to be expected for conversion in a transparent negatively dispersive medium. When the beam waist is located well within the nonlinear medium the medium appears to be infinite in extent and the conversion approaches a value given by the integrals in reference 2. The drop off of the harmonic signal as the beam waist moves through the pinhole and into the vacuum is

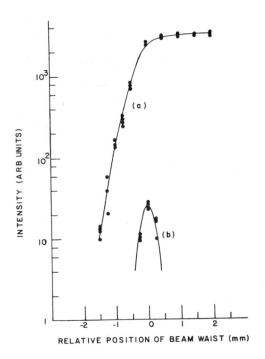

Fig. 2 Dependence of third and fifth harmonic signals in Ar as a function of position of beam waist relative to pinhole. Positive positions correspond to beam waist in nonlinear medium, negative positions to beam waist in vaccum.

indicative of a sharp gradient in the density of the gas near the pinhole. The rate of decrease of the third harmonic signal with the position of the beam waist indicates that the width of the density gradient is comparable to or smaller than the confocal parameter.

The fifth harmonic signal on the other hand exhibits a peak when the beam waist is near the pinhole. The drop off of the signal on the vacuum side is again compatible with the existence of a sharp density gradient. The decrease in signal as the beam waist moves into the nonlinear medium however, is due to absorption of the harmonic signal by the nonlinear medium. For these measurements the absorption length was 1/3 of the confocal parameter and fifth harmonic conversion could be observed only when the beam waist was located very close to the effective edge of the nonlinear medium. This behavior indicates one of the difficulties involved with frequency conversion when the wavelength of the generated radiation falls in an absorbing region of the nonlinear medium: the signal is generated only in the last absorption length. In order to have as much conversion as possible it is necessary to have the highest intensity at the edge of the medium and for the edge to be as sharp as possible. The first of these conditions can be limited by damage for media that are contained by windows while the second is difficult to satisfy for generation of short wavelengths where windows cannot be used.

Similar measurements for third harmonic conversion in Kr are shown in Fig. 3 for three different pressures. Apart from small variations at the different pressures the behavior is very similar to that observed for the third harmonic in Ar, increasing as the beam waist is moved from the vacuum into the nonlinear medium and reaching an approximate constant value when the beam waist is well within the nonlinear medium. This behavior is in contrast with that which is expected for a transparent positively dispersive medium. The calculated dependence of the harmonic signal on the position of the beam waist assuming a sharp density gradient at z=0 is shown as the dotted curve in Fig. 3. The amount of dispersion assumed for this calculation is approximately that which exists in Kr at 30 torr. The theoretical behavior for the third harmonic in Kr is similar to that observed for the fifth harmonic, namely the

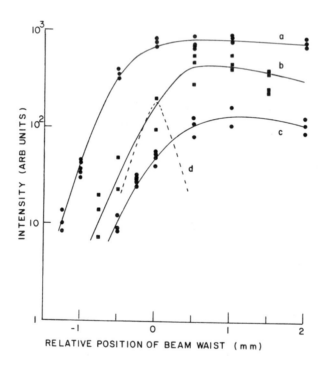

Fig. 3 Dependence of third harmonic signal in Kr as a function of position of beam waist for (a) 100 torr, (b) 30 torr, (c) 8 torr. Geometry is the same as in Fig. 2. Dotted curve is theoretical shape for a nonlinear medium with a sharp edge and positive dispersion.

conversion peaks at the edge of the nonlinear medium and falls off on either side. Now, however, the decrease in third harmonic as the beam waist moves into the nonlinear medium is a result of phase cancellations due to dispersion rather than to absorption.

We believe that the explanation of the observed behavior lies in the presence of a nonlinear refractive index that disturbs the phase mismatch in the medium. The restriction of conversion from a focused beam in an infinite medium to materials with negative dispersion arises because of the phase shift which the beam undergoes as it focuses. As a result of this phase shift the harmonic power that is generated in the front half of the focus is returned to the pump beam in the back half unless the phase shift caused by the focus is compensated by negative dispersion in the medium. In the present situation a nonlinear phase shift caused by the pump intensity can also prevent the conversion in the back half of the focus from cancelling that in the front part. Indeed, when measurements were performed at lower laser power, behavior was observed that was in qualitative agreement with the theoretical behavior in Fig. 3. In addition if a nonlinear index were important it should appear as a difference in the behavior at different pressures. Such behavior is evident in the data taken at 8 torr compared to that at 100 torr. Rather than staying constant the conversion at 8 torr starts to decrease as the beam waist moves away from the edge into the medium indicating that there is some degree of phase cancellation at this low pressure.

In order to obtain additional confirmation of the presence of a nonlinear phase shift and also to gain some insight into possible mechanisms that could be responsible for the observed effects we have examined the dependence of the harmonic intensity on the fundamental intensity. In the absence of competing effects such as a nonlinear index the harmonic power should vary as the cube of the pump power. Two types of behavior were evident in these measurements. At relatively low pump powers the harmonic signal rises as the 4.8 power of the pump signal. For pump powers above about 10MW the slope changes to about unity. Similar behavior was evident in the measurements taken at 100 torr and 30 torr.

The intensity dependent data show that another process is present in addition to the harmonic conversion process. This result supports the identification of a nonlinear refractive index indicated by the earlier measurements, although by itself it cannot show which of many possible competing process are present.

Further insight into the mechanism for this behavior can be obtained by considering the power dependence at low intensities. The fact that the third harmonic signal increases at close to the fifth power of the incident pump intensity indicates that the competing mechanism is improving the conversion process. In general this can result from either an improvement in the phase matching process or from an increase in the nonlinear susceptibility. Although the present measurements are not extensive enough to allow

unambiguous identification of the source of this behavior we can gain
some insight by noting that the break point in the power dependence
of the harmonic signal is independence of the pressure of the gas.
This result indicates that the effect responsible for this behavior
takes place within a single atom and probably changes the
susceptibility of the medium. One type of process that can be
responsible for this behavior is a Stark shift of one or more
dispersive resonance in the Kr. If both the dispersion of the medium
and the nonlinear susceptibility are determined by the detuning from
a resonance movement of the position of the resonance could move the
Kr into a position of negative dispersion and simultaneously increase
the nonlinear susceptibility. Upon continued Stark shifting the
levels can pass through the harmonic wavelength, leading to a
susceptibility that decreases with intensity, causing the high
intensity dependent behavior that was observed. Under these
assumptions the harmonic conversion would probably occur only for
part of the pulse, but such behavior cannot be resolved with the
current experimental parameters.

DISCUSSION

The results presented in this paper have some very important
implications for the generation of radiation at short wavelengths
through frequency conversion. The restriction of the sign of the
dispersion has proven in the past to correpond to a restriction on
the wavelength range that is accessible with certain nonlinear
pro-cesses in certain media. Kung[5] has reported experiments on third
harmonic generation and four wave sum and difference frequency mixing
in Xe in the range from 118.2 nm to 200 nm. In these experiments it
was demonstrated that harmonic generation and four-wave sum-frequency
mixing could be performed only for those wavelengths for which the
Xe had negative dispersion. Four-wave difference-frequency mixing
could be used, on the other hand, in spectral regions in which the Xe
had positive dispersion. Such restrictions on the generation of
radiation in a given wavelength range can usually be avoided by
choosing a medium that has the proper dispersion for the frequency
conversion process under consideration, or by using a frequency
conversion process that is compatible with dispersion that is present
in the available media.

Generation of radiation at wavelengths below 100 nm has been
proposed through processes of the form $\omega_{XUV} = q\omega$, $\omega_{XUV} = n\omega_{uv} \pm \omega_{dye}$
where ω_{XUV} is the frequency in the XUV, ω_{uv} is the frequency of a
powerful pulsed laser in the near uv and ω_{dye} is the frequency of a
tunable dye laser.[6] If the harmonic order number q and the four-wave
mixing index n go from 3 to 7 and from 2 to 6 respectively, radiation
down to about 300 nm can be generated with various of the rare gas
halide lasers. All of these processes except the four-

wave difference-frequency process $\omega_{XUV} = 2\omega_{uv} - \omega_{dye}$ require a medium with negative dispersion in order to allow conversion in an infinite medium. In the XUV however there is not a large selection of suitable media to allow an arbitrary choice of dispersion at all wavelengths. In the rare gases for example, regions of negative dispersion exist at wavelengths above certain dispersive resonances,[4] but the composite spectrum for negative dispersion in regions in which the medium is transparent leaves large regions below 100 nm open. Alternatively one can use one of the rare gases above its ionization threshold. However this usually limits the pressure of gas that can be used and puts restrictions on the type of geometry that can be used. An example of such a restriction was given in Fig. 2 in which it was demonstrated that the pump beam had to be focused very close to a gas-vacuum interface to obtain significant conversion. In addition, the effectiveness of the conversion now depends on the details of the interface.

The new results presented here indicate that it is possible to obtain significant conversion in media with positive dispersion at high pump intensity. This is just the region of pump intensities that would ordinarily be used to obtain the largest output power. This can be done with transparent media without regard to precise focusing. As a result added versatility and convenience are provided for harmonic conversion and sum frequency mixing processes in spectral regions in which negatively dispersive media are not available.

REFERENCES

1. J. F. Ward and G. H. C. New, Phys. Rev. 185, 57 (1969).
2. G. C. Bjorklund, IEEE J. Quant. Elect. QE-11, 287 (1975).
3. V. I. Anikin, V. D. Gora, K. N. Drabovich and A. N. Dubovik, Sov. J. Quant. Elect. 6, 174 (1976).
4. R. Mahon, T. J. McIlrath, V. P. Meyerscough and D. W. Koopman, IEEE J. Quant. Elect. QE-15, 444 (1979).
5. A. H. Kung, Appl. Phys. Lett. 25, 653 (1974).
6. J. Reintjes, Appl. Optics, 19, 3889 (1980).

AIP Conference Proceedings

		L.C. Number	ISBN
No.1	Feedback and Dynamic Control of Plasmas	70-141596	0-88318-100-2
No.2	Particles and Fields - 1971 (Rochester)	71-184662	0-88318-101-0
No.3	Thermal Expansion - 1971 (Corning)	72-76970	0-88318-102-9
No.4	Superconductivity in d-and f-Band Metals (Rochester, 1971)	74-18879	0-88318-103-7
No.5	Magnetism and Magnetic Materials - 1971 (2 parts) (Chicago)	59-2468	0-88318-104-5
No.6	Particle Physics (Irvine, 1971)	72-81239	0-88318-105-3
No.7	Exploring the History of Nuclear Physics	72-81883	0-88318-106-1
No.8	Experimental Meson Spectroscopy - 1972	72-88226	0-88318-107-X
No.9	Cyclotrons - 1972 (Vancouver)	72-92798	0-88318-108-8
No.10	Magnetism and Magnetic Materials - 1972	72-623469	0-88318-109-6
No.11	Transport Phenomena - 1973 (Brown University Conference)	73-80682	0-88318-110-X
No.12	Experiments on High Energy Particle Collisions - 1973 (Vanderbilt Conference)	73-81705	0-88318-111-8
No.13	π-π Scattering - 1973 (Tallahassee Conference)	73-81704	0-88318-112-6
No.14	Particles and Fields - 1973 (APS/DPF Berkeley)	73-91923	0-88318-113-4
No.15	High Energy Collisions - 1973 (Stony Brook)	73-92324	0-88318-114-2
No.16	Causality and Physical Theories (Wayne State University, 1973)	73-93420	0-88318-115-0
No.17	Thermal Expansion - 1973 (lake of the Ozarks)	73-94415	0-88318-116-9
No.18	Magnetism and Magnetic Materials - 1973 (2 parts) (Boston)	59-2468	0-88318-117-7
No.19	Physics and the Energy Problem - 1974 (APS Chicago)	73-94416	0-88318-118-5
No.20	Tetrahedrally Bonded Amorphous Semiconductors (Yorktown Heights, 1974)	74-80145	0-88318-119-3
No.21	Experimental Meson Spectroscopy - 1974 (Boston)	74-82628	0-88318-120-7
No.22	Neutrinos - 1974 (Philadelphia)	74-82413	0-88318-121-5
No.23	Particles and Fields - 1974 (APS/DPF Williamsburg)	74-27575	0-88318-122-3
No.24	Magnetism and Magnetic Materials - 1974 (20th Annual Conference, San Francisco)	75-2647	0-88318-123-1
No.25	Efficient Use of Energy (The APS Studies on the Technical Aspects of the More Efficient Use of Energy)	75-18227	0-88318-124-X